Actinobacteria - Diversity, Applications and Medical Aspects

Edited by Wael N. Hozzein

Published in London, United Kingdom

IntechOpen

Supporting open minds since 2005

Actinobacteria - Diversity, Applications and Medical Aspects
http://dx.doi.org/10.5772/intechopen.95202
Edited by Wael N. Hozzein

Contributors

Frans Maruma, Rhulani Edward Ngwenya, Fiona M. Stainsby, Halina Vaughan, Janki Hodar, Adetayo Adesanya, Victor Adesanya, Onix Cantres-Fonseca, Vanessa Fonseca-Ferrer, Christian Castillo-Latorre, Vanessa Vando-Rivera, Francisco Del Olmo, Savarimuthu Ignacimuthu, Pathalam Ganesan, Aasif Majeed Bhat, Qazi Parvaiz Hassan, Aehtesham Hussain, Sunita Bundale, Aashlesha Pathak, Maria Elena Flores, Toshiko Takahashi, Jonathan Alanis, Polonia Hernández, Sangeeta D. Gohel, Vaishali R. Majithiya, Avery August, Jessica Elmore, Shalini Swami, Priyanshu Walia, Saloni Jain, Ishita Gupta, Arka Pratim Chakraborty, Sumi Paul, Erika T. Quintana, Luis A. Maldonado, Luis Contreras-Castro, Amanda Alejo-Viderique, Martha E. Esteva-García, Claudia J. Hernández-Guerrero, Juan C. Cancino-Diaz, Luis A. Ladino, Juan Esteban Martínez-Gómez, Noemi Matias-Ferrer, Carlos Sánchez

Notice
Statements and opinions expressed in the chapters are these of the individual contributors and not necessarily those of the editors or publisher. No responsibility is accepted for the accuracy of information contained in the published chapters. The publisher assumes no responsibility for any damage or injury to persons or property arising out of the use of any materials, instructions, methods or ideas contained in the book.

First published in London, United Kingdom, 2022 by IntechOpen
IntechOpen is the global imprint of INTECHOPEN LIMITED, registered in England and Wales, registration number: 11086078, 5 Princes Gate Court, London, SW7 2QJ, United Kingdom
Printed in Croatia

British Library Cataloguing-in-Publication Data
A catalogue record for this book is available from the British Library

Additional hard and PDF copies can be obtained from orders@intechopen.com

Actinobacteria - Diversity, Applications and Medical Aspects
Edited by Wael N. Hozzein
p. cm.
Print ISBN 978-1-80355-096-1
Online ISBN 978-1-80355-097-8
eBook (PDF) ISBN 978-1-80355-098-5

We are IntechOpen,
the world's leading publisher of
Open Access books
Built by scientists, for scientists

6,100+
Open access books available

149,000+
International authors and editors

185M+
Downloads

156
Countries delivered to

Our authors are among the

Top 1%
most cited scientists

12.2%
Contributors from top 500 universities

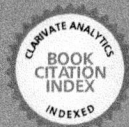

Interested in publishing with us?
Contact book.department@intechopen.com

Numbers displayed above are based on latest data collected.
For more information visit www.intechopen.com

Meet the editor

Wael N. Hozzein is a professor of microbiology at the Faculty of Science, Beni-Suef University, Egypt. He received his Ph.D. from Cairo University, Egypt, then worked as a visiting scientist at Newcastle University, UK, Michigan State University, USA, and most recently as Bioproducts Research Chair at King Saud University, Saudi Arabia. He has vast experience in bacterial taxonomy, microbial biodiversity, and biotechnological applications of bacteria. Prof. Hozzein has authored more than 180 publications and he is a guest editor, editorial board member and reviewer for several international journals. Recently he was included in the Stanford University list of the top 2 percent of the world's most-cited scientists. He has been the principal investigator for several funded grants and has received several awards, including the State Encouragement Prize in Biological Sciences in 2015. Prof. Hozzein has been involved in many academic activities and educational reform projects and initiatives. He has recently been appointed an adviser to Nahda University's President for Development, Research and Quality.

Contents

Preface

Actinobacteria represent one of the most diverse phyla in the bacterial domain, which is widely distributed in terrestrial and aquatic environments. This interesting group of bacteria is well known for their unique characteristics and biotechnological applications, especially their ability to produce a vast range of bioactive molecules, including antibiotics and other industrially important secondary metabolites, making them an important area of study. This book aims to provide in-depth insights into the diversity of actinobacteria, their different medical aspects, and their biotechnological applications.

This book is for students starting their research work on actinobacteria, for researchers working with actinobacteria who need to know recent advances in the field, and for teachers teaching topics related to actinobacteria.

The book is divided into four sections and contains 13 chapters. The first section looks at the diversity of actinobacteria, with three chapters on the selective isolation of actinobacteria from different Mexican ecosystems, the diversity of actinobacteria associated with marine invertebrates, and endophytic actinobacteria. This section covers the different methods and techniques applied for selective isolation, molecular identification, diversity evaluation, and potential applications of actinobacteria. It is particularly important for students and researchers to be aware of the different techniques they are going to be using. The three chapters also discuss the diversity of actinobacteria in special ecosystems of interest as a potential source for the discovery of novel taxa and novel bioactive compounds.

The second section concerns bioactive metabolites from actinobacteria, with four chapters on antimicrobials from extreme actinobacteria, gene multiplicity in antibiotic-producing Streptomyces, metabolites from actinobacteria for mosquito control, and anti-quorum-sensing compounds from rare actinobacteria. As the chapter titles indicate, this section covers several aspects of the application of actinobacteria in the production of diverse metabolites. The four chapters provide a detailed account of the isolation, characterization, structure elucidation, and biological activities of the metabolites produced by actinobacteria.

The chapters in the third section, which is entitled "Applications of Actinobacteria", examine applications of actinobacteria in agriculture, biosurfactants from mycolic acid-containing actinobacteria, and biodegradation of bisphenol A. The three chapters review recent findings and discuss future prospects for these important applications.

Finally, the fourth section concerns actinobacteria and diseases, an especially important topic that is rarely covered, especially by physicians. Two of the chapters in this section are about actinomycosis and the third discusses the immune response during induced farmer's lung disease. Diagnostic methods, clinical features, and the immune response for two diseases of concern caused by some pathogenic genera of actinobacteria, actinomycosis and hypersensitivity pneumonitis are discussed.

This book is a compilation of scientific articles that should be of great interest to those interested in actinobacteria research. Students and researchers with an interest in general microbiology will also find the book an interesting read. The book offers a unique resource for all graduate students and researchers in the fields of diversity, applications, and the medical aspects of actinobacteria.

Finally, I would like to thank all the authors who have shared their excellent research work results, knowledge, and ideas. I would like also to warmly thank the author service manager Ms. Romina Rovan for her support, commitment, and attention to detail which made this book possible. My gratitude is also extended to members of the staff of IntechOpen, especially Ms. Lucija Tomicic-Dromgool, for their support.

Wael N. Hozzein
Professor,
Faculty of Science,
Botany and Microbiology Department,
Beni-Suef University,
Beni-Suef, Egypt

Section 1

Diversity of Actinobacteria

Chapter 1

On the Selective Isolation of Actinobacteria from Different Mexican Ecosystems

Erika T. Quintana, Luis A. Maldonado,
Luis Contreras-Castro, Amanda Alejo-Viderique,
Martha E. Esteva García, Claudia J. Hernández-Guerrero,
Juan C. Cancino-Díaz, Carlos Sánchez, Luis A. Ladino,
Juan Esteban Martínez-Gómez and Noemí Matías-Ferrer

Abstract

Actinobacteria isolated from less studied sites on our planet represent a huge opportunity for the discovery of novel microorganisms that may produce unique compounds with biological activity. The class actinobacteria encompasses 80% of the microbes that produce the antibacterial compounds used in medicine today. However, the resistance acquired/showed by pathogenic microorganisms opens the opportunity to explore Mexican ecosystems as a source of novel actinobacteria. Air samples have shown to be an excellent site of study, marine ecosystems which include sediments and marine organisms are important sources of novel actinobacteria and soil samples are still a promising source to isolate this microbial group. The isolation of novel actinobacteria is a dynamic strategy that depends on the expertise, patience, and talent of the techniques applied and needs to be fully explored to untap the unknown actinobacterial diversity with potential in biology.

Keywords: actinobacteria, air samples, discovery, marine resources, soil samples

1. Introduction

Megadiverse countries constitute exceptional areas on Earth where most of the planetary biodiversity is present. The complexity of these areas is huge, but in most of the cases, two major points are key: (1) the geographical location, and (2) the abiotic and biotic elements present. Mexico is one of the top five megadiverse countries in the world and its macrodiversity and endemism are well represented by amphibians, mammals, plants, and reptiles [1]. However, the knowledge of microscopic organisms such as archaea, bacteria, protozoa, microscopic algae, and microscopic fungi, that inhabit aquatic, atmospheric, marine, and soil ecosystems is neither poorly known, studied nor understood.

The vision of this chapter is to contribute to the knowledge, research, and study of microscopic life in different Mexican ecosystems, as they are often ignored or poorly mentioned in federal texts or even in biotic inventories. Our examples are some members of the class actinobacteria [2], and we aim to demonstrate why it is so important to study these bacteria in such detail to fully explore and untap the unknown actinobacterial diversity with potential in biology. Using a dynamic isolation strategy on air, soil, and marine sediments and sponges collected from yet unexplored sites of the Mexican territory, we have been able to cultivate novel actinobacteria. Our findings showed that expertise, patience, and talent of the techniques applied are keys in the hunt for new potential microbes.

The isolation of microorganisms, including actinobacteria, is not new but a dynamic strategy that is continuously changing, and the developed to date is a powerful tool. For more than two centuries, researchers from Japan, the UK, and USA have shown that beneficial microorganisms isolated from the soil are important to Biology. In recent years the isolation of the first genus of actinobacteria from the marine origin [3] and novel marine species [4] have shown the importance of exploring the marine environment. Extreme or unexplored sites have also shown the isolation of actinobacteria including putative novel actinobacteria [5].

Our research studying actinobacteria started in 1999 [6, 7], but until 2009 we properly started the exploration of the Mexican (marine) ecosystems [8] as an independent group. We followed bioprospecting, diversity, and systematic approach but designing a selective isolation strategy was the first step for a complete full project or protocol [9].

Actinobacteria is a complex group of bacteria, they present forms such as rods or bacilli, many differentiate in vegetative mycelium, aerial hyphae, and chain of spores, and in a few genera fragmentation of the hyphae is present. In general, the Gram reaction is positive and the content of guanine plus cytosine is above 69%mol. The morphological characteristics within the class showed how complex this group is. Actinobacteria are considered saprophytes or beneficial microbes, but a small number of species have been shown to be either pathogenic [10] or opportunistic [11]. This microbial group has been isolated or cultivated using classical methods from almost every sample taken on Earth and they are always detected when using molecular methods to study this group in a given environmental sample.

Actinobacteria also have the innate ability to produce secondary metabolites with biological activity, to date, this class encompasses 80% of the microbes that produce the antibacterial compounds used in medicine. Complete Genome Sequencing of some genera of actinobacteria such as *Streptomyces* [12] and *Salinispora* [3, 4] have shown the biotechnological potential that these organisms contain and maybe explored and exploited for human wellbeing.

The more we study and discover actinobacteria the more important they become in pass, present, and future assignments. Microorganisms and microbial biomass, including actinobacteria, represent the major resource for biotechnology and biological areas. We should continue exploring their role in nature in order to understand their biology, ecology, and bioprospecting potential [13–16].

2. Selective isolation of actinobacteria from different Mexican ecosystems

2.1 The atmosphere as a source of novel actinobacteria

The Earth's atmosphere is divided into six specific layers with completely different characteristics: (1) Troposphere, (2) Stratosphere, (3) Mesosphere, (4) Thermosphere, (5) Ionosphere, and (6) Exosphere. It has been established that the atmosphere plays an important role to transport microorganisms, place to place, continent to continent. The latter has been established using scientific tools in the last 200 years and in the last 15 years, NASA has monitored mineral dust particles from the Sahara desert with a robust precision using spaceborne satellites. These Saharan dust plumes contain microorganisms and enter mainland Mexico by the Yucatan Peninsula [17].

The atmosphere is a hostile environment for microorganisms though there are a significant number of them in the troposphere, with air as their main dispersion pathway. The abundance, diversity, survival, and transport of microorganisms, as passive drivers, and how they get stressed severely by the conditions presented in the atmosphere have fully been reported [18]. Most of the microorganisms in the atmosphere are present as spores, while others have adapted to resist desiccation or high/low temperatures [19]. Recent reports have also shown that some microorganisms (i.e., by using specific proteins) can act as ice nucleating particles [20] and that they may play an important role in cloud formation [21]. In general, bacteria (including actinobacteria) present in the atmosphere are attached to suspended particles [17], and their concentration change notably during the dry or wet seasons of each year [22].

2.1.1 Isolation of a streptomycete from air samples of Merida-Yucatan

As part of the African Dust and Biomass Burning Over Yucatan (ADABBOY) Project [23] in the city of Merida (N 21°02′75.4′′ W 89°65′44.8′′) a selective isolation strategy was carried out in order to cultivate/recovered putative actinobacteria in May 2017. Air samples were impacted using a Quick Take 30 Sample Pump® and a BioStage® SKC (**Figure 1**) in Petri dishes prepared with a slightly modified Glucose Yeast Malt extract agar (GYM medium; Appendix A; Medium 65: DSMZ; www.dsmz.de) supplemented with Rifampicin (5 μg/mL; Sigma-Aldrich, USA) and Nystatin (50 μg/mL; MICOSTATIN® Bristol Myers Squibb, Mexico).

Plates were incubated in two different laboratories and conditions. The first laboratory was in the city of Merida at the Universidad Autónoma de Yucatán, using an aerobic incubator set at 25°C and the plates were incubated for 24 hours. For the second procedure, the plates were transported to a laboratory in Mexico City where the incubation time continued aerobically at 30°C (IncuMax IC-320, Amerex USA) for 8 weeks with eye observation each week. One microorganism with the production of aerial hyphae, a gray mass of spores, and a very deep purple diffusible pigment (**Figure 2**) was selected from the isolation plates for further studies.

Figure 1.
The device used for the air particles.

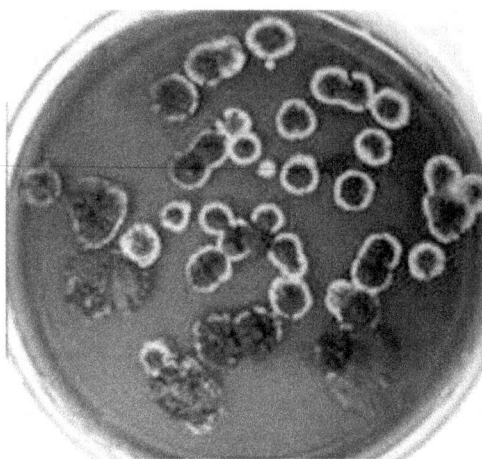

Figure 2.
Morphology and purple diffusible pigment of an airborne streptomycete.

The selected isolate was coded C6-CCA-May-1. After a purification process using GYM medium and two different techniques (cross streak and serial dilutions), bacterial biomass and spores of the strain were ultra-preserved in 20% glycerol. Morphological characterization was carried out using a GYM medium (**Figure 3**) and a Gram staining procedure (**Figure 4**) was carried out following well-known universal protocols.

Molecular identification of strain C6-CCA-May-1 was carried out following protocols previously published [8, 24]. First, the DNA of strain C6-CCA-May-1 was extracted and used as a template for PCR amplification using the 16S rRNA gene (Appendix B). The sequence of the 16S rRNA gene PCR product confirmed that strain C6-CCA-May-1 belongs to the genus *Streptomyces*. According to the EZbiocloud phylogenetic approach *Streptomyces* sp. C6-CCA-May-1 was related to *Streptomyces viridiviolaceus* (NBRC 133559[T]), *Streptomyces werraensis* (NBRC 13404[T]), *S. asenjonii* (KNN35.1b[T]), *Streptomyces minutiscleroticus* (NBRC 13000[T]) and *S. levis* (NBRC 15423[T]) (**Table 1**).

A Bayesian phylogenetic tree was constructed in order to establish the taxonomic position of *Streptomyces* sp. C6-CCA-May-1 shows that *Streptomyces* sp.

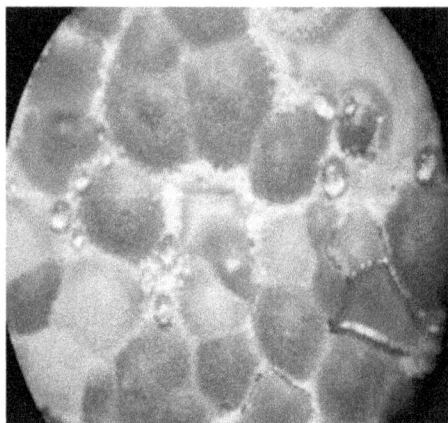

Figure 3.
Aerial hyphae and spore mass of isolate C6-CCA-May-1.

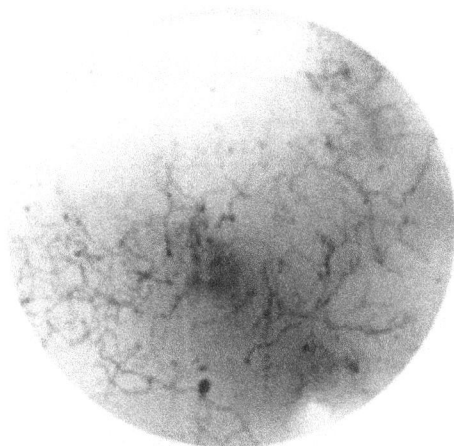

Figure 4.
Gram staining of the airborne streptomycete.

Hit taxon name	Hit strain name	Accesion	Similarity	Hit taxonomy	Completeness (%)
Streptomyces viridiviolaceus	NBRC 133559^T	AB18350	99.51	Bacteria; Actinobacteria;	99.6
Streptomyces werraensis	NBRC 13404^T	AB18381	98.88	Actinobacteria_c; Streptomycetales;	99.9
Streptomyces asenjonii	KNN 35.1b^T	LT621750	98.77	Streptomycetaceae;	95.5
Streptomyces minutiscleroticus	NBRC 13000^T	AB184249	98.66	Streptomyces	99.9
Streptomyces levis	NBRC 15423^T	AB184670	98.66		99.0

Table 1.
List of hits from the EZbiocloud 16S database.

C6-CCA-May-1 is related to *Streptomyces viridiviolaceus* (**Figure 5**). Moreover *Streptomyces* sp. C6-CCA-May-1 belongs to the *S. glaucus* subclade [25] and according with the probability (number 1) showed in the cluster formed in the phylogenetic tree, *Streptomyces* sp. C6-CCA-May-1 may well represent a novel species. The similarity value amongst *Streptomyces* sp. C6-CCA-May-1 and *S. viridiviolaceus* is 98.2%. A full comparison study based on the phenotypic, morphological microscopic characteristics and chemotaxonony amongst *Streptomyces* sp. C6-CCA-May-1 and *S. viridiviolaceus* could further clarify their accurate taxonomic position and status. Furthermore, genomic analyses are required to fully understand the putative unique biotechnological potential of this airborne streptomycete.

Streptomycetes are an ecologically important group capable of producing diverse bioactive compounds. However, their taxonomy and diversity in air samples remain unknown. For almost two centuries the genus *Streptomyces* has been considered a goldmine and the major producer of bioactive natural products (i.e. antibiotics) [13, 26]. The recent discovery of a new member of the actinomycin family of antibiotics shows the potential to explore old streptomycetes [26], but there still is an open door for the discovery of new bioactive molecules through novel species [9], recovered from "unusual" environments.

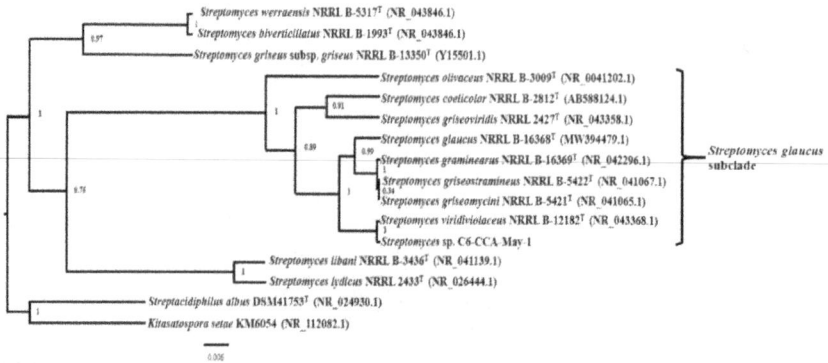

Figure 5.
Phylogenetic tree of the 16S rRNA gene of the airborne streptomycete.

2.2 Marine Mexican resources home of novel actinobacteria

Seventy percent of our planet is covered by the ocean but from one marine research project, there are 10 of terrestrial origin. Little is still known about marine biodiversity (including microorganisms) though their potential is extraordinary and needs to be fully studied and exploited. Mexico is surrounded by the Pacific Ocean, the Sea of Cortez (*aka.* Gulf of California), the Gulf of Mexico, and the Caribbean Sea and this is the main reason why the country shows, at least potentially, a high number of species richness, diversity, and endemism in the coastal areas. The study of the Mexican marine ecosystems and their marine resources is still poorly studied. In contrast, the potential of actinobacteria isolated from marine sediments collected in Mexico has been reported [8, 27–29] and showed that they produce novel and potent compounds with biological activity. Major European research marine programs have shown the importance to study and protect the marine ecosystem but in Latin America, the efforts of conservation and protection of unique marine Mexican sites are urgently needed. It has been recognized that the ancient life of planet Earth started in an aquatic system and that the immense microbial diversity present on it plays an important role in the biochemical cycles. In this subsection, two projects are presented: (a) the isolation of marine actinobacteria from sediments and (b) sponges. Marine sediments were collected from the Revillagigedo Archipelago National Park (RANP) in December 2017 and January 2018. The exploration of microbes from marine sediments of this pristine and unique place has never been studied. Species of the *Aplysina* sponge are ubiquitous inhabitants of tropical and subtropical marine locations [30]. In recent years our group described novel marine sponges of the sponge *Aplysina* (order Verongida) [31] and since 2005 the exploration of the microbiota associated with five different species is an undergoing study.

2.2.1 Actinobacteria isolated from marine Mexican resources

In the present project, a total collection of 34 marine sediments or sponges were collected at RANP during two expeditions (December 2017 and January 2018). A selective isolation strategy using 11 of the marine sediments and two different media was developed following a previously reported study [24]. In order to isolate marine obligate and nonobligate actinobacteria, 1 g of wet sediments was transferred to tubes containing 9 mL of saline solution (0.9%; NaCl; Sigma-Aldrich,

Mexico), four dilutions were prepared (10^{-1} to 10^{-4}) and 100 μL (Gilson, France) of each dilution were spread onto marine GYM medium and 1:10 marine GYM medium (Appendix A); both media supplemented with Rifampicin [15, 25 and 50 μg/mL] and Nystatin (100 μg/mL). Plates were then aerobically incubated at 30° C (IncuMax IC-320, Amerex USA) for up to 16 weeks. Starting at week eight, the isolation plates were checked by eye looking for actinobacterial colonies. Once putative colonies were noticed each was then streaked in new GYM plates without antibiotics or antifungal compounds until an axenic culture was obtained. The conditions of incubation were as mentioned above. Because of the pressure set in the isolation strategy, not many microbes were able to grow but actinobacteria were successfully cultivated.

A preliminary test to quickly select marine obligate actinobacteria was carried out using marine GYM medium (**Figure 6A** and **C**) and GYM medium (**Figure 6B** and **D**). A positive result was considered when nonmicrobial biomass was observed growing on the surface of GYM medium after 4 weeks of incubation (**Figure 6B**). The ones that presented growth only in the marine GYM medium were considered those marine obligate actinobacteria (**Figure 6A**). It should be pointed out, however, that we were also able to isolate nonobligate actinobacteria that showed the typical characteristics of members of the family Micromonosporaceae [32] (**Figure 6C** and **D**). Up to date there is only one genus that is considered halophile within the Phylum Actinobacteria, and this is *Salinispora* [3, 4]. For 15 years there were only three *Salinispora* species, namely, *S. arenicola*, *S. tropica* [3] and *S. pacifica* [33], but last year seven new species were formally described [4]. The isolation of salinisporae has never been reported from sediments taken from RANP.

The molecular identification using the 16S rRNA gene of obligate and nonobligate marine actinobacteria confirmed that they belong to the genera *Micromonospora*, *Salinispora* and *Williamsia* (**Table 2**). In addition, other bacteria such as species of the genus *Erythrobacter* were also identified (**Table 2**).

To isolate obligate and nonobligate marine actinobacteria from the sponge samples, five different species of *Aplysina* (*A. airapii*, *A. clathrata*, *A. encarnacionae*, *A. gerardogrenii*, and *A. sinuscaliforniensis*) were selected. The selective isolation strategy was based on that previously reported by [3, 24]. Ten grams of each sponge were transferred into 90 mL of saline solution that was previously added to a plastic bottle with a wide mouth (**Figure 7**). The sponge was disintegrated for 5 min using an electric mixer at maximum speed (**Figure 7**). One milliliter of each suspension

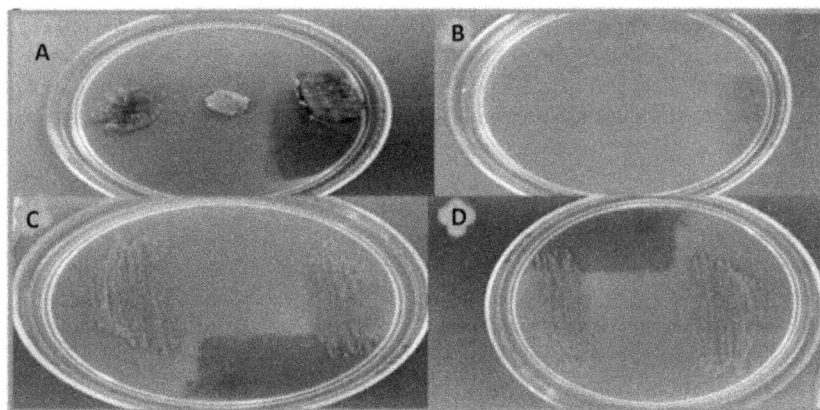

Figure 6.
Screening of obligate and nonobligate marine actinobacteria.

Code of microorganism	Identity (%)	Hit taxonomy	Accesion
C114 col. 1	99	*Micromonospora* sp.	GD145235.1
C60 b bca col. 1	100	*Salinispora* sp.	MH299440.1
C60a col. 3	100	*Salinispora arenicola*	KX394599.1
C60 col. 4	99	*Salinispora arenicola*	KX394598.1
C72 col. 2	100	*Williamsia* sp.	AG506245.1
C134 col. 4	99	*Erythrobacter litoralis*	CF133005.1

Table 2.
Taxonomic identification of some of the obligate and nonobligate marine actinobacteria from RANP.

Figure 7.
Strategy to isolate actinobacteria from six marine sponges.

was then used to prepare serial dilutions (up to 10^{-4}) in tubes containing 9 mL of saline solution. One hundred milliliters were spread in marine GYM medium and 1:10 marine GYM medium (Appendix A), both media were supplemented with Rifampicin [5 and 15 µg/mL] and Nystatin (100 µg/mL). Plates were aerobically incubated at 30°C (IncuMax IC-320, Amerex USA) for up to 16 weeks.

Starting at week eight, the isolation plates were checked by eye looking for actinobacterial colonies (**Figure 8**). Once putative colonies were selected they were streaked in new GYM plates until axenic culture were obtained. The conditions of incubation were the same as mentioned before.

The preliminary test to select marine obligate actinobacteria was carried out as mentioned previously. Obligate marine actinobacteria (**Figure 9A**) were isolated from *A. clathrata* and *A. gerardogrenii* and non-obligate marine actinobacteria from *A. gerardogrenii* and *A. encarnacionae* (**Figure 9B** and **C**). One isolate presented a red-wine color diffusible pigment (**Figure 9B**) and another was a symbiont (**Figure 9C**).

We isolated marine obligate actinobacteria that were preliminarily assigned to the genus *Salinispora*, one nonobligate marine actinobacteria with typical characteristics of members of the family Micromonosporaceae [32] and one symbiont actinobacteria. The isolation of salinisporae has never been reported from sponges collected or studied in Mexico and there are also non-reports about symbionts marine actinobacteria.

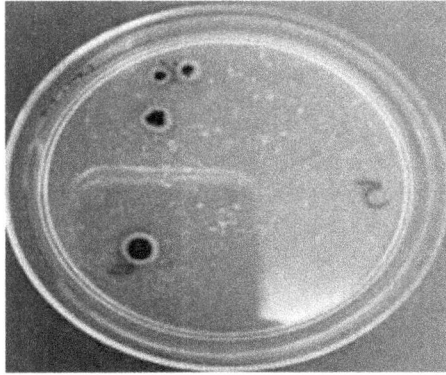

Figure 8.
Morphology of the marine obligate actinobacteria.

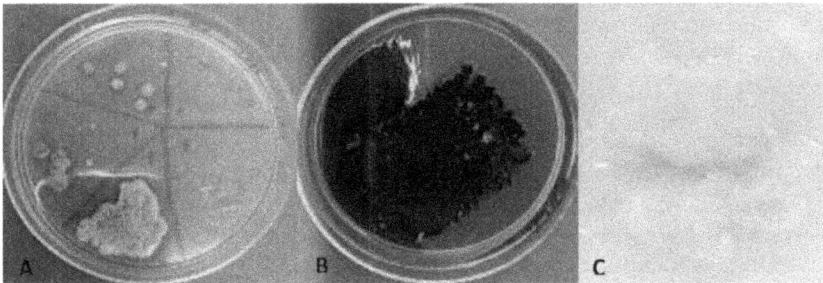

Figure 9.
Morphology of obligate and nonobligate marine actinobacteria.

The microbial communities of marine obligate and nonobligate actinobacteria associated with marine sediments remain poorly characterized [34] and we must continue searching for these gifted microorganisms [35]. Culture-dependent methods captured approximately 3% of the total count of the microbes and in some reports around 39 genera have been only detected in culture. The latter shows the importance to carry out/improve, innovative and original selective isolation techniques since these may be more effective than previously recognized.

2.3 Soil is still an extraordinary resource to isolate important actinobacteria

Soil is one of the most complex ecosystems on Earth and its amount of organic matter, mineral composition, and diversity of microorganisms will determine its ecology. There are different kinds of soils but in general, those with less anthropogenic impact will be richer in microorganisms. Mexico encompasses 26 types of soil out of the 32 recognized in the world [36, 37] and this is due to several causes, namely: (1) the complexity of the topography originated from the volcanic activity in the Cenozoic Era, (2) the wide altitudinal gradient (from 0 to 5, 600 m.a.s.l.), (3) by the five main climates present according to the Köppen classification [38], (4) the enormous diversity of landscapes present and, finally (5) the different kind of rocks that the Mexican territory enclose. It is well recognized that actinobacteria are abundant in soils and that they play an important role in the degradation and recycling of organic matter. Soil microorganisms have a remarkable ability to produce compounds with biological activity such as antibiotics, and historically this has been exemplified by streptomycin which is produced by a streptomycete named

Streptomyces griseus subsp. *griseus* recovered from soil. Since 1940 soil has been a major resource to selective isolate important actinobacteria, not only streptomycetes but also several other biotechnological genera such as *Amycolatopsis* and *Saccharopolyspora*, the producers of the glucopeptide -vancomycin-, and the macrolide -erythromycin-, respectively.

Mexico encompasses nearly 4, 000 insular regions of outstanding natural beauty, their biodiversity is remarked by high number of endemism (plants and animals) and most of these regions are federal protected. Revillagigedo Archipelago National Park (RANP) [39] encompasses four tropical volcanic Islands: (1) Socorro, (2) Clarion, (3) San Benedicto, and (4) Roca Partida. The RANP is considered as one of the best-preserved areas in the world.

2.3.1 Actinobacteria isolated from unexplored soils of Mexico

In the present project, soil samples were collected at five different sites of Socorro Island (SI) (**Figure 10**) in 2016 and 2017, respectively (**Table 3**). A selective isolation strategy using five soil samples and three different media was developed in order to isolate actinobacteria. One gram of each soil was transferred to tubes containing 9 mL of saline solution (0.9%; NaCl; Sigma-Aldrich, Mexico). Modified Pikovskaya agar (Appendix A), GYM without antibiotics or antifungal compounds, and marine GYM supplemented with Rifampicin [5 µg/mL] and Nystatin (100 µg/mL) were used as isolation media. The dilutions used for modified Pikovskaya and marine GYM were 10^{-1} to 10^{-4} and for GYM 10^{-6} to 10^{-8}. One hundred microliters (Gilson, France) of each dilution were spread in the media and then aerobically incubated at 30°C (IncuMax IC-320, Amerex USA) for up to 4 weeks.

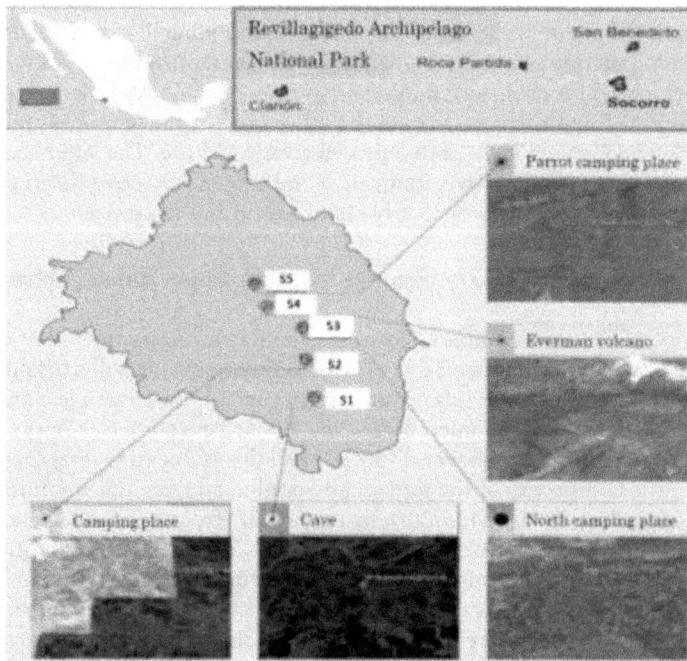

Figure 10.
Sites of sampling in Socorro Island.

Code of sampling site	Name of sampling site		Meter of above sea level (masl)	Date of sampling
S1	Camping place	South of Socorro Island	450	25 December 2016
S2	Cave		550	28
S3	North camping place	North of Socorro Island	941	30
S4	Everman volcano	South East of Everman volcano	850	3 January 2017
S5	Parrot camping place		600	

Table 3.
General information of the sampling sites.

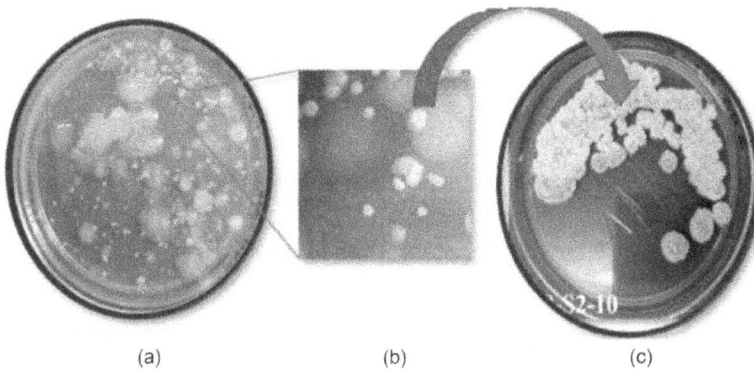

Figure 11.
Isolation plate (a), selected actinobacteria (b), and axenic culture (c).

The isolation plates were checked by eye looking for actinobacteria and selected colonies were streaked in new GYM media plates until axenic cultures were obtained (**Figure 11**). The conditions of incubation were the same as mentioned above. The cultivable actinobacteria diversity was remarkable but only 215 isolates were selected from the isolation plates. Eighty-six isolates presented morphological differentiation based on aerial hyphae and spore mass so that they were considered as streptomycetes (**Figure 12**).

Molecular identification using the 16S rRNA gene of 10 selected strains morphologically resembling streptomycetes confirmed that they indeed belong to this extensive and important genus (**Table 4**). The phylogenetic tree constructed using five of sequences of the selected strains showed that they are different amongst them and from those more related *Streptomyces* type species selected (**Figure 13**).

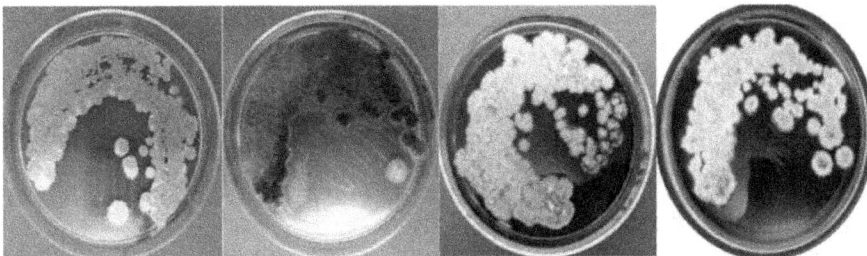

Figure 12.
Morphological diversity of actinobacteria from Socorro Island soils.

Code of microorganism	Hit taxonomy	Identity (%)
C1-S1-1	*Streptomyces charreusis* F22	99
C10-S2-14	*Streptomyces shaanxiensis* CCNWHQ	
C67-S2-1	*Streptomyces* sp. TK08046	
C43-S3-1	*Streptomyces* sp. TK08046	
C43-S3-2	*Streptomyces* sp. A124	
C58-S5-2	*Streptomyces* sp. 746G1	
C59-S5-1	*Streptomyces* sp. HP1	
C59-S5-2	*Streptomyces* sp. MJM2443	
C27-S4-3	*Streptomyces tsukubensis* 9993	
C57-S5-1	*Streptomyces* sp. A124	

Table 4.
Preliminary identification of selected streptomycetes from the different sites.

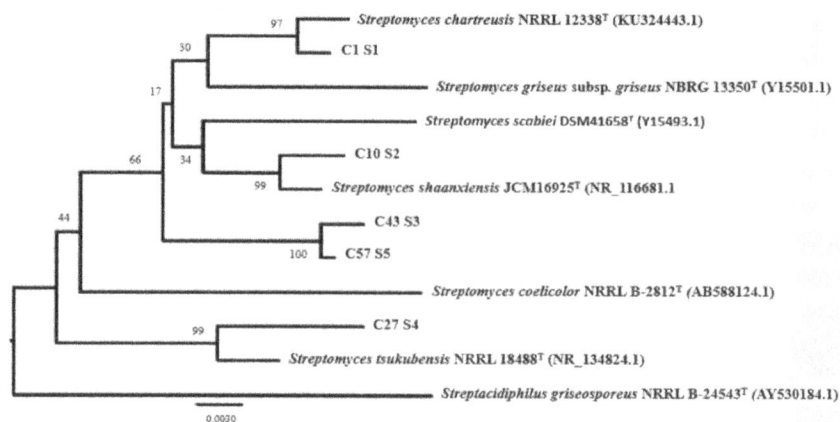

Figure 13.
Phylogenetic tree of the 16S rRNA gene selected streptomycetes isolates.

The strains from the soil samples collected at RANP may well represent novel species, but more studies to delineate its novelty are certainly needed. According to our results site 5 (S5) presented the higher percentage of actinobacteria like-strains (26%), S1 23.7%, S2 20.5%, S3 18.1% and S4 11.2%, respectively. Soil samples from unexplored sites with outstanding natural beauty represent a significant resource to isolate potential streptomycetes and use their genetic reservoir. Soil is one of the most complex, dynamic, and in constant evolution ecosystem and so it surely is its actinobacteria portion.

Soil actinobacteria, particularly *Streptomyces* is still a never-ending source for the discovery of secondary metabolites with diverse biological activities [13, 14, 26] and due to its ubiquity, survival capabilities, and metabolic versatility, it is the most studied genus within the class Actinobacteria [2, 15]. The diversity of actinobacteria isolated from SI at RANP shows the potential of our study [40] and represents the entree to explore its secondary or specialized metabolite potential [13–16, 41, 42] and diversity. Actinobacteria are amongst the most abundant organisms in soil but their potential role in this environment is still unknown. Socorro Island harbors diverse actinobacteria and according to the selected streptomycetes (**Figure 13**) they are unique and may well represent candidates of novel taxa. Improved isolation strategies are needed to recover culturable and unculturable yet soil actinobacteria to fully explore their diversity and biotechnological applications for human and environmental welfare.

3. Conclusions

More than two centuries of work to isolate actinobacteria from natural resources have ended in major discoveries, academic contributions, and important recognitions. Novel actinobacteria represent the entree of new natural compounds of significant importance. For more than 15 years our research group has developed and applied selective isolation strategies exploring distinct natural unexplored sites or less studied ecological niches in Mexico. To avoid the isolation of the "same bugs," it is needed to use different selective isolation strategies, pre-treatments of the environmental sample, supplementation of the media with a specific concentration of antibiotics and antifungal compounds while carefully selecting the target organisms.

The results presented in this chapter support the proposal that the isolation of microorganisms is not *"an old fashioned and boring work"*; it requires creativity, experience, knowledge, and patience. Moreover, to study physiology or genetic potential of a specific microorganism indeed needs to be cultivated in the laboratory.

Actinobacteria is one most of the diverse and complex groups of bacteria and produces more than 80% of the antibiotics used in medicine today. Their ability to produce novel natural compounds, such as antibiotics and novel cancer compounds is widely recognized. According to the World Health Organization (WHO) there is an urgent need to discover novel antibiotics for priority pathogenic bacteria and emerging pathogenic organisms [43] and the first step to respond and to contribute with this global initiative is to isolate novel actinobacteria. To our knowledge, our reports here are one of the few research projects in our country that are dedicated to study the ecological role and the genetic potential of novel actinobacteria from unexplored sites or less studied ecological niches in Mexico.

Acknowledgements

Our research was supported by Instituto Politécnico Nacional (IPN)—Secretaría de Investigación y Posgrado (SIP), Grants SIP 20170432, 20170434, 20170410, 20181167, 20181528, 20181803, 20196605, 20196630, 20196649, 20201026, 20201893, 20202083, 20210987, 20211209, and 20211740. L.C.-C. was supported by a Ph.D. Scholarship from Consejo Nacional de Ciencia y Tecnología (CONACyT, Mexico) no. 270230 and Beca de Estímulo Institucional de Formación de Investigadores Program (BEIFI-IPN). ETQ, CJHG and JCCD acknowledge Comisión de Operación y Fomento de Actividades Académicas del Instituto Politécnico Nacional (COFAA), Estímulo al Desempeño de los Investigadores (EDI) and Sistema Nacional de Investigadores (SNI-CONACYT) fellowships. A.A-V ackcnowledges a Mexican Postdoctoral Scholarship Program 3 and 4, Program 4, 2020–2021 and Program 1 and 2, Program 2, 2021–2022 (CONACyT, Mexico).

Conflict of interest

The authors declare no conflict of interest.

Appendix A

Glucose Yeast Malt Extract Agar -GYM medium- (DSM medium 65)	
Dextrose (Bacto™, BD)	4 g
Yeast extract (Bacto™, BD)	4 g
Malt extract (Bacto™, BD)	10 g
Calcium carbonate (SIGMA-ALDRICH)	2 g
Agar (Bacto™, BD)	12 g
Distilled water	1000 mL
pH 7.2	

Marine media was prepared by replacing distilled water with artificial seawater (Ocean™). 1:10 marine GYM medium was prepared using an aliquot of marine GYM. All the media was sterilized at 121°C, 1.5 Lb. for 15 min.

Appendix B

Reagents	Volume
10× DNA polymerase buffer [50 mM stock solution] (Bioline, USA)	5 μL
MgCl2 [50 mM] (Bioline, USA)	1.5 μL
dNTPs [10 mM stock mixture] (Bioline, USA)	1.25 μL
Primer 27f [20 μM stock solution] (Invitrogen)	0.5 μL
Primer 1525r [20 μM stock solution] (Invitrogen)	0.5 μL
DNA [100 ng/μL]	1 μL
Taq polymerase [5 U] (Bioline, USA)	1 Unit
Ultra-pure Milli-Q water	Up to 50 μL

Amplification was achieved using a Techno 512 gradient PCR machine.

Author details

Erika T. Quintana[1*], Luis A. Maldonado[2], Luis Contreras-Castro[1],
Amanda Alejo-Viderique[1,3], Martha E. Esteva García[1],
Claudia J. Hernández-Guerrero[4], Juan C. Cancino-Díaz[1], Carlos Sánchez[5],
Luis A. Ladino[6], Juan Esteban Martínez-Gómez[7] and Noemí Matías-Ferrer[7]

1 Instituto Politécnico Nacional, Escuela Nacional de Ciencias Biológicas,
México City, Mexico

2 Facultad de Química, Universidad Nacional Autónoma de México, Mexico City,
Mexico

3 Facultad de Ciencias, Universidad Nacional Autónoma de México, Mexico City,
Mexico

4 Instituto Politécnico Nacional, Centro Interdisciplinario de Ciencias Marinas,
La Paz, Baja California Sur, Mexico

5 Universidad Autónoma de Baja California Sur, La Paz, Baja California Sur, Mexico

6 Instituto de Ciencias de la Atmósfera y Cambio Climático, Universidad Nacional
Autónoma de México, Mexico City, Mexico

7 Instituto de Ecología, A.C., Xalapa, Veracruz, Mexico

*Address all correspondence to: equintanac@ipn.mx

IntechOpen

References

[1] Levin GA, Moran R. The vascular flora of Isla Socorro, Mexico. San Diego Society of Natural History Memoir. 1989;**16**:1-71

[2] Stackebrandt E, Rainey FA, Ward-Rainey NL. Proposal for a new hierarchic classification system, Actinobacteria classis nov. International Journal of Systematic Bacteriology. 1997;**47**:479-491. DOI: 10.1099/00207713-47-2-479

[3] Maldonado LA, Fenical W, Jensen PR, Kauffman CA, Mincer TJ, Ward AC, et al. Salinispora arenicola gen. nov., sp. nov. and Salinispora tropica sp. nov., obligate marine actinomycetes belonging to the family Micromonosporaceae. International Journal of Systematic and Evolutionary Microbiology. 2005;**5**:1759-1766. DOI: 10.1099/ijs.0.63625-0

[4] Román-Ponce B, Millán-Aguiñaga N, Guillen-Matus D, Chase AB, Ginigini JGM, Soapi K, et al. Six novel species of the obligate marine actinobacterium Salinispora, Salinispora cortesiana sp. nov., Salinispora fenicalii sp. nov., Salinispora goodfellowii sp. nov., Salinispora mooreana sp. nov., Salinispora oceanensis sp. nov. and Salinispora vitiensis sp. nov., and emended description of the genus Salinispora. International Journal of Systematic and Evolutionary Microbiology. 2020;**70**:4668-4682. DOI: 10.1099/ijsem.0.004330

[5] Quintana ET, Badillo RF, Maldonado LA. Characterisation of the first actinobacterial group isolated from a Mexican extremophile environment. Antonie Van Leeuwenhoek. 2013;**104**: 63-70. DOI: 10.1007/s10482-013-9926-0

[6] Maldonado L, Hookey JV, Ward AC, Goodfellow M. The Nocardia salmonicida clade, including descriptions of Nocardia cummidelens

sp. nov., Nocardia fluminea sp. nov. and Nocardia soli sp. nov. Antonie Van Leeuwenhoek. 2000;**78**:367-377. DOI: 10.1023/a:1010230632040

[7] Lu Z, Wang L, Zhang Y, Shi Y, Liu Z, Quintana ET, et al. Actinomadura catellatispora sp. nov. and Actinomadura glauciflava sp. nov., from a sewage ditch and soil in southern China. International Journal of Systematic and Evolutionary Microbiology. 2003;**53**:137-142. DOI: 10.1099/ijs.0.02243-0

[8] Maldonado LA, Fragoso-Yáñez D, Pérez-García A, Rosellón-Druker J, Quintana ET. Actinobacterial diversity from marine sediments collected in Mexico. Antonie Van Leeuwenhoek. 2009;**95**:111-120. DOI: 10.1007/s10482-008-9294-3

[9] Goodfellow M, Fiedler HP. A guide to successful bioprospecting: Informed by actinobacterial systematics. Antonie Van Leeuwenhoek. 2010;**98**:119-142. DOI: 10.1007/s10482-010-9460-2

[10] Quintana ET, Wierzbicka K, Mackiewicz P, Osman A, Fahal AH, Hamid ME, et al. Streptomyces sudanensis sp. nov., a new pathogen isolated from patients with actinomycetoma. Antonie Van Leeuwenhoek. 2008;**93**:305-313. DOI: 10.1007/s10482-007-9205-z

[11] Vázquez-Boland JA, Giguère S, Hapeshi A, MacArthur I, Anastasi E, Valero-Rello A. Rhodococcus equi: The many facets of a pathogenic actinomycete. Veterinary Microbiology. 2013;**29**:9-33. DOI: 10.1016/j.vetmic.2013.06.016

[12] Bentley SD, Chater KF, Cerdeño-Tárraga AM, Challis GL, Thomson NR, James KD, et al. Complete genome sequence of the model actinomycete Streptomyces coelicolor A3(2). Nature. 2002;**417**:141-147. DOI: 10.1038/417141a

[13] Sharma P, Thakur D. Antimicrobial biosynthetic potential and diversity of culturable soil actinobacteria from forest ecosystems of Northeast India. Scientific Reports. 2020;**10**:4104. DOI: 10.1038/s41598-020-60968-6

[14] De Simeis D, Serra S. Actinomycetes: A never-ending source of bioactive compounds—An overview on antibiotics production. Antibiotics (Basel). 2021;**5**:483. DOI: 10.3390/antibiotics10050483

[15] Farda B, Djebaili R, Vaccarelli I, Del Gallo M, Pellegrini M. Actinomycetes from caves: An overview of their diversity, biotechnological properties, and insights for their use in soil environments. Microorganisms. 2022;**1**:453. DOI: 10.3390/microorganisms10020453

[16] Xie F, Pathom-Aree W. Actinobacteria from desert: Diversity and biotechnological applications. Frontiers in Microbiology. 2021;**12**:765531. DOI: 10.3389/fmicb.2021.765531

[17] Rodriguez-Gomez C, Ramirez-Romero C, Cordoba F, Raga BG, Salinas E, Martinez E, et al. Characterization of culturable airborne microorganisms in the Yucatan peninsula. Atmospheric Environment. 2020;**223**:117-183. DOI: 10.1016/j.atmosenv.2019.117183

[18] Yoo K, Lee TK, Choi EJ, Yang J, Shukla SK, Hwang SI, et al. Molecular approaches for the detection and monitoring of microbial communities in bioaerosols: A review. Journal of Environmental Sciences (China). 2017;**51**:234-247. DOI: 10.1016/j.jes.2016.07.002

[19] Tian B, Hua Y. Carotenoid biosynthesis in extremophilic Deinococcus-Thermus bacteria. Trends in Microbiology. 2010;**18**:512-520. DOI: 10.1016/j.tim.2010.07.007

[20] Ariya PA, Kos G, Mortazavi R, Hudson ED, Kanthasamy V, Eltouny N, et al. Bio-organic materials in the atmosphere and snow: Measurement and characterization. Topics in Current Chemistry. 2014;**339**:145-199. DOI: 10.1007/128_2013_461

[21] Wilson TW, Ladino LA, Alpert PA, Breckels MN, Brooks IM, Browse J, et al. A marine biogenic source of atmospheric ice-nucleating particles. Nature. 2015;**525**:234-238. DOI: 10.1038/nature14986

[22] De la Rosa MC, Mosso MA, Ullán C. El aire: hábitat y medio de transmisión de microorganismos. Observatorio Medioambiental. 2002;**5**:375-402

[23] Raga GB, Ladino LA, Baumgardner D, Ramirez-Romero C, Córdba F, Álvarez-Ospina H, et al. ADABBOY: African dust and biomass burning over Yucatan. Bulletin of the American Meteorological Society. 2021;**102**:1543-1556. DOI: 10.1175/BAMS-D-20-0172.1

[24] Contreras-Castro L, Martinez-García S, Cancino-Díaz JC, Maldonado LA, Hernández-Guerrero CJ, Martínez-Diaz SF, et al. Marine sediment recovered salinispora sp. inhibits the growth of emerging bacterial pathogens and other multi-drug-resistant bacteria. Polish Journal of Microbiology;**69**:321-330. DOI: 10.33073/pjm-2020-035

[25] Kämpfer P. Streptomyces. In: Whitman WB, Rainey F, Kämpfer P, Trujillo M, Chun J, DeVos P, Hedlund B, Dedysh S, editors. Bergey's Manual of Systematics of Archaea and Bacteria. Vol. 4. Hoboken (USA): John Wiley & Sons, Inc; 2015. pp. 1-414. DOI: 10.1002/9781118960608.gbm00191

[26] Machushynets NV, Elsayed SS, Du C, Siegler MA, de la Cruz M, Genilloud O, et al. Discovery of actinomycin L, a new member of the

actinomycin family of antibiotics. Scientific Reports. 2022;2:2813. DOI: 10.1038/s41598-022-06736-0

[27] Jensen PR, Moore BS, Fenical W. The marine actinomycete genus Salinispora: A model organism for secondary metabolite discovery. Natural Product Reports. 2015;32:738-751. DOI: 10.1039/C4NP00167B

[28] Becerril-Espinosa A, Guerra-Rivas G, Ayala-Sánchez N, Soria-Mercado IE. Antitumor activity of actinobacteria isolated in marine sediment from Todos Santos Bay, Baja California, Mexico. Revista de Biología Marina y Oceanografía. 2012;47:317-325

[29] Parera-Valadez Y, Yam-Puc A, López-Aguiar LK, Borges-Argáez R, Figueroa-Saldivar MA, Cáceres-Farfán M, et al. Ecological strategies behind the selection of cultivable Actinomycete strains from the Yucatan peninsula for the discovery of secondary metabolites with antibiotic activity. Microbial Ecology. 2019;77: 839-851. DOI: 10.1007/s00248-019-01329-3

[30] Hentschel U, Fieseler L, Wehrl M, Gernert C, Steinert M, Hacker J, et al. Microbial diversity of marine sponges. Progress in Molecular and Subcellular Biology. 2003;37:59-88. DOI: 10.1007/978-3-642-55519-0_3

[31] Gómez P, González-Acosta B, Sánchez-ortíz C, Hoffman Z, Hernández-Guerrero CJ. Amended definitions for Aplysinidae and Aplysina (Porifera, Demospongiae, Verongiida): On three new species from a remarkable population in the Gulf of California. Zootaxa. 2018;455:322-342. DOI: 10.11646/zootaxa.4455.2.4

[32] Genilloud O. Micromonosporaceae. In: Whitman WB, Rainey F, Kämpfer P, Trujillo M, Chun J, DeVos P, Hedlund B, Dedysh S, editors. Bergey's Manual of Systematics of archaea and bacteria. Hoboken (USA): John Wiley & Sons, Inc; 2015. pp. 1-7

[33] Ahmed L, Jensen PR, Freel KC, Brown R, Jones AL, Kim BY, et al. Salinispora pacifica sp. nov., an actinomycete from marine sediments. Antonie Van Leeuwenhoek. 2013;103: 1069-1078. DOI: 10.1007/s10482-013-9886-4

[34] Demko AM, Patin NV, Jensen PR. Microbial diversity in tropical marine sediments assessed using culture-dependent and culture-independent techniques. Environmental Microbiology. 2021;11:6859-6875. DOI: 10.1111/1462-2920.15798

[35] Abdel-Mageed WM, Al-Wahaibi LH, Lehri B, Al-Saleem MSM, Goodfellow M, Kusuma AB, et al. Biotechnological and ecological potential of Micromonospora provocatoris sp. nov., a gifted strain isolated from the challenger deep of the Mariana trench. Marine Drugs. 2021;19(243). DOI: 10.3390/md19050243

[36] INEGI. Conjunto de Datos Vectorial Edafológico, Serie II, escala 1: 250 000 (Continuo Nacional). 2007. México

[37] 10 IUSS, Grupo de Trabajo WRB. Base Referencial Mundial del Recurso Suelo. Primera actualización 2007. Informes sobre Recursos Mundiales de Suelos No. 103. FAO. 2007. Roma

[38] García E. Modificaciones al Sistema climático de Köppen Adaptado Para México. México: Instituto de Geografía, UNAM; 1988

[39] Secretaría de Medio Ambiente y Recursos Naturales. Firma el Presidente Enrique Peña Nieto el Decreto del Parque Nacional Revillagigedo. 2017. Availale from: https://www.gob.mx/semarnat/prensa/firma-el-presidente-enrique-pena-nieto-el-decreto-del-parque-nacional-revillagigedo

[40] Almuhayawi MS, Mohamed MSM, Abdel-Mawgoud M, Selim S, Al Jaouni SK, AbdElgawad H. Bioactive potential of several Actinobacteria isolated from microbiologically barely Explored Desert habitat, Saudi Arabia. Biology (Basel). 2021;**10**:235. DOI: 10.3390/biology10030235

[41] Blin K, Shaw S, Steinke K, Villebro R, Ziemert N, Lee SY, et al. antiSMASH 5.0: Updates to the secondary metabolite genome mining pipeline. Nucleic Acids Research. 2019; **47**:W81-W87. DOI: 10.1093/nar/gkz310

[42] Miethke M, Pieroni M, Weber T, Brönstrup M, Hammann P, Halby L, et al. Towards the sustainable discovery and development of new antibiotics. Nature Reviews Chemistry. 2021;**5**: 726-749. DOI: 10.1038/s41570-021-00313-1

[43] WHO. Global Priority List of Antibiotic-Resistant Bacteria to Guide Research, Discovery, and Development of New Antibiotics [Internet]. Geneva (Switzerland): World Health Organization; 2017. Available from: https://www.who.int/medicines/publications/global-priority-list-antibiotic-resistant-bacteria/en

Chapter 2

Actinobacteria Associated with Marine Invertebrates: Diversity and Biological Significance

Vaishali R. Majithiya and Sangeeta D. Gohel

Abstract

The ocean harbors a wide diversity of beneficial fauna offering an enormous resource for novel compounds, and it is classified as the largest remaining reservoir of natural molecules to be evaluated for biological activity. The metabolites obtained from marine invertebrate-associated actinobacteria have different characteristics compared to terrestrial actinobacteria as marine environments are exigent and competitive. Actinobacteria produce a wide range of secondary metabolites, such as enzymes, antibiotics, antioxidative, and cytotoxic compounds. These allelochemicals not only protect the host from other surrounding pelagic microorganisms but also ensure their association with the host. The harnessing of such metabolites from marine actinobacteria assures biotechnological, agricultural, and pharmaceutical applications.

Keywords: actinobacteria, marine invertebrates, diversity, biological activity

1. Introduction

Actinobacteria are gram-positive bacteria with high G + C DNA content [1] that can live in different habitats including the marine environment. These bacteria enclose significant biotechnological potential as they produce complex biopolymers, such as polysaccharides, extracellular and intracellular enzymes, antibiotics, inhibitors, and various metabolic products [2]. Since the discovery of streptomycin from **Streptomyces griseus**, actinobacteria are recognized for antibiotic production. Approximately two-thirds of world bioactive compounds are isolated from these phyla. Therefore, actinobacteria are considered as a potential source for the development of new drugs [3]. It has been emphasized that actinobacteria from marine habitats may be valuable for the isolation of novel strains that can potentially produce secondary metabolites like enzymes, cosmetics, antibiotics, anti-parasitic, vitamins, nutritional material, and immunosuppressive agents having great economical and biotechnological importance [4]. Nowadays, research has been diverted toward marine niches to screen that actinobacteria can produce bioactive compounds with different metabolic characteristics as the frequency of novel bioactive compounds from terrestrial actinobacteria decreases with time [5]. Marine flora (mangroves, seaweed, seagrasses, and algae) and fauna (Porifera, coelenterates, ascidians, crustaceans, and mollusks) are part of highly productive ecosystems and are habitats of numerous bioactive compounds producing microorganisms. Bioactive compounds obtained from associated microorganisms are known for a broad range of biological effects, such as antimicrobial, antifungal, insecticides, antiprotozoal, antiparasitic, anti-inflammatory, and antitumor [6–8]. Sponges (phylum: Porifera) and corals (phylum: Cnidaria) are the most predominant source of marine natural products. Ascidians (phylum: Chordata) have been classified as the second predominant source of natural products [9, 10]. Seaweeds-associated bacterium produces a diverse range of biologically active compounds that specifically target fouling organisms and also show antifungal and antibacterial effects [11]. The rare actinobacterial species are explored from marine environments based on conventional and molecular approaches. The changes in salinity along with the nutrient value within the ocean make the marine environment a good source of novel actinobacteria. Due to the nutrient diversity present in the oceans, one can expect that the organisms that inhabit the marine environment would be very diverse. The different zones of the ocean contain different nutrients and minerals. Therefore, the microorganisms that inhabit various zones generally have metabolites as well as functions that are conducive to their specific zone. It indicates a high probability of microbial diversity that leads to the production of novel antibiotics and enzymes in the marine environment [12, 13]. The present investigation has emphasized invertebrate-associated actinobacteria as a source for novel natural products. The rationale behind this is because invertebrate-associated microorganisms are capable of producing various secondary metabolites and enzymes that can be used in predator defense, antifouling, inhibition of overgrowth, protection from ultraviolet radiation as well as acting as mediators in the competition for settling space [14].

2. Diversity of actinobacteria associated with marine invertebrates

Invertebrate-associated bacterial communities have a significant ability to produce bio-medically relevant micro molecules. Cultivable approaches and metagenomic analysis show that many invertebrates harbor actinobacterial species. These actinobacterial species mainly belong to genera Streptomyces, Nocardiopsis, Kocuria, Salinospora, Nocardia, Rhodococcus, Nonomuraea, Actinokinespora, and Saccharopolyspora [15–17].

2.1 Sponges

The phylum Porifera harbor dense and diverse bacterial communities. According to the literature review, about 40% of the sponge biomass was due to their associated bacteria [18]. The sponges are recognized for their potential source of bioactive metabolites. These bioactive metabolites are generally produced by their associated microbial communities which suggest that associated microorganisms might play a role in the chemical defense of their host [19]. Till today, about 60 actinobacterial genera have been isolated from marine sponges [20]. Sponge-associated actinobacteria dwell in the mesohyl matrix of sponges. They may be true halobiont or taken up from nearby water through the filtration process. About 20 actinobacterial genera belonging to genera *Kocuria, Micromonospora, Nocardia, Nocardiopsis, Saccharopolyspora, Salinispora,* and *Streptomyces* were isolated from South China sea sponges (Genera: *Haliclona, Amphimedon, Phyllospongia, Agelas, Hippospongialachne, Cinachyrsina, Arenosclera, Phakellia, Cliona*) [18, 20].

2.2 Coral reefs

The bacterial communities associated with coral have unique properties. Bacteria inhabit coral in three different parts of the coral body that includes the surface of the mucus layer, the interior of coral tissue, and the calcium carbonate skeleton. Each of them harbors unique beneficial properties. The skeletons of corals are porous which provides micro niches for a variety of bacterial communities for colonization [21]. Cyanobacteria in the skeleton of *Oculina Patagonica* provide organic compounds to the coral tissue which helps them to survive during losses of endosymbiotic algae [21, 22]. In addition, a recent study shows that bacteria isolated from the mucus of healthy *Acropora palmata* produce antibiotics inhibiting the growth of potentially pathogenic microorganisms. This shows that the diversity of bacterial species that are associated with a particular coral species is high, including many novel species. A number of mutualistic benefits have been suggested [23, 24]. The other microbial lives, such as bacteria are associated with coral halobiont for their nutritional requirement. In return, the associated bacteria protect the host by the production of secondary metabolites, such as antifungal, antibacterial, and antihelminth. The antimicrobial peptide damicornin produced by coral-associated bacteria was active against fungi and selective gram-positive bacteria but not against *Vibrios sp.* while the organic extract of *Siderastrea sidereal* coral showed anti-bacterial activity against two of the four strains of gram-positive bacterial isolates from coral surfaces [25, 26].

The actinobacterial communities associated with corals can fix nitrogen which explains their dominance in healthy corals [21, 27]. Lampert et al. reported mucus-associated bacterial diversity among which 23% were Actinobacteria [28]. Mahmoud and Kalendar reported *Brachybacterium, Brevibacterium, Cellulomonas, Dermacoccus, Devriesea, Kineococcus, Kocuria, Marmoricola, Micrococcus, Micromonospora, Ornithinimicrobium, Renibacterium,* and *Rhodococcus* actinobacterial genera that belong to three corals (*Coscinaraea columna, Platygyra daedalea,* and *Porites harrisoni*) among which *Kocuria* and *Brevibacterium* were dominant genera [22]. Kuang et al. reported *Fridmanniella* and *Propionibacterium* as major groups associated with *P. lutea* while genera *Demetria, Fodinicola, Friedmanniella, Geodermatohilus, Iamia, Modestobacter, Ornithinimicrobium, Tersicoccus,* and *Yonghaparikia* were detected for the first time from *P. lutea* through culture-independent study [23]. The novel actinobacterial species *Nocardiopsis coralli* HNM0947[T] isolated from Hainan province, PR China inhabitants of coral *Galaxeaastreata* show optimum growth at pH 7, temperature 28°C, and 3% NaCl(w/v). The strain HNM0947[T] contained 71.3% mol G + C, MK-10(H8), MK-10(H6), and MK-10(H4) as major menaquinones, iso-C16:0, anteiso-C17:0, C18:0, C18:0 10-methyl (TBSA), and anteiso-C15:0 as major fatty acids [29]. Among the cultivable actinobacterial genera *Jiangella, Micromonospora, Nocardia, Nocardiopsis, Rhodococcus, Verrucosispora, Salinispora,* and *Streptomyces* showed potential actinobacterial activity against various test pathogens consequently contributing to coral health [23].

2.3 Shrimps

Shrimps are considered one of the most famous seafood consumed worldwide. They belong to the phylum Arthropoda, subphylum Crustacea. The exoskeleton of Shrimps contains chitin, structural proteins, and mineral deposits, and its construction is an energy-demanding process. Till today, very few studies regarding actinobacterial diversity associated with Shrimps are published. The *Streptomyces californicus* isolated from Shrimp farming displayed the ability to inhibit the growth of *Vibrio sp.*, one of the disease-causing pathogens [30]. You et al. demonstrated the significance of actinobacteria in shrimp farming due to their antimicrobial, antifungal, and antioxidative ability [31]. Kumar et al. prepared extract from actinobacterial sp. and observed its in vivo effect by supplementing extract with feed to black tiger shrimp having white sport syndrome [32]. You et al. reported actinobacteria with the ability to inhibit the formation of biofilm produced by *Vibrio sp*. All these studies indicate the significance of actinobacteria in aquaculture [33].

2.4 Ascidians

Ascidians belong to phylum Chordata, subphylum Tunicate. More than 1000 bioactive metabolites have been isolated from ascidians and their associated microbial communities. The indolocarbazoles having anticancer ability were produced by *E. toealensis* associated with actinobacterial genera *Salinispora* and *Verrucosispora* [34]. Lee et al. reported actinobacterial species, including *Arthrobacter rhomb,* Brachybacterium muris, Micrococcus lylae, and *Nocardiopsissynnemataformans* from squid collected from Jumunjin, Gangwon-do, Korea [35].

3. Biological significance of actinobacteria associated with marine invertebrates

The actinobacteria are omnipresent in all environmental conditions. They produce a wide range of metabolites having biotechnological applications as described in **Table 1**. Actinobacteria is a group of organisms having the ability to produce inhibitors, **immunomodifiers, biosurfactants, antioxidative, anti-inflammatory,** antimicrobial, antifungal, and anticancer compounds along with intracellular and extracellular enzymes with unique characteristics in terms of substrate selectivity, stability in presence of salts, temperature tolerance, pH variation, etc. The significance of host-associated marine actinobacteria is also described in **Table 1**.

3.1 Antimicrobial ability

The **Micrococcus luteus** isolated from sponge shows strong inhibition against **Staphylococcus aureus, Vibrio anguillarum,** and **Candida albicans** [53]. Mayamycin and Microluside isolated from sponge-associated *Streptomyces sp*. HB202 and *Micrococcus sp*. EG45 inhibited **S. aureus** with IC50 (1.16 µg/mL) and (12.42 µg/mL) respectively [54, 55]. *Salinispora sp*. isolated from *Pseudoceratina clavate*, as well as *Micromonospora sp*. *CPI 12* and *Saccharomonospora sp*. *CPI 13* isolated from *Callyspongia diffusa* showed antibacterial activity against **S. epidermidis** [56]. Actinobacterial genera *Pseudonocardia, Streptomyces, Kocuria, Aeromicrobium, Brachybacterium*, and *Nocardiopsis* were isolated from sponges, such as *Haliclona sp., Callyspongia sp.,* and *Desmacella sp*. Among 92 isolated actinobacterial strains, 52 actinobacterial strains exhibited antibacterial activity against **E. coli, P. fluorescens, V. alginolyticus,** and **V. splendidus**. Further analysis revealed that 18% of actinobacterial strains contained NRPS gene clusters while 10% harbor PKS-KS gene and 6% have PKS-NRPS gene clusters [57]. Different actinobacterial genera of marine origin consist of nontoxic antibiotics able to inhibit the growth of *Vibrio sp*. resulting in a potential source for aqua culturing [58, 59].

3.2 Antioxidative ability

Ferric reducing antioxidant power (FRAP), nitric oxide (NO) scavenging, and DPPH radical scavenging activity were extensively used to measure antioxidant capacity. *Nocardiopsis sp*. PU3 isolated from the coral reef of the Pullivasal Island, Gulf of Mannar, and India was detected as a potential source of antioxidative agent that can be used to treat various oxidative stress-related disorders with 53.6%

Actinobacteria	Host	Location	Significance	References
Sponge				
Nocardiopsis dassonvillei MAD08	D. nigra	Southwest coast, India	Antibacterial activity and anticandidal activity	[36]
Nocardiopsis sp., Micromonospora sp., Rhodococcus sp. and Streptomyces sp.	Iotrochota sp., and Hymeniacidonperleve	Eastern Mediterranean coast, Turkey	15 strains exhibited antimicrobial activity against methicillin-resistant Staphylococcus aureus (MRSA) and vancomycin-resistant Enterococcus faecium and *three strains* showed anticandidal activity against Candida albicans	[37]
Actinoalloteichus sp. Micrococcus sp., Micromonospora sp., Nocardiopsis sp., Streptomyces sp.	Halichondriapanicea	Baltic Sea, Germany	Presence of polyketide synthases (PKS) and nonribosomal peptide synthases (NRPS)	[38]
Streptomyces poriferorum sp. nov	Geodiabarretti and Anthodichotoma	Tautra island, Trondheim fjord of Norway	Anti MRSA activity	[39]
Salinispora arenicola and *Salinispora pacifica*	Sponges	Great Barrier Reef	N-acyl homoserine lactones	[40]
]Actinobacterial strains	Callyspongia sp., Callyspongiaaerizusa, Carteriospongia contorta, Chelanoplysilla sp., and Diacarnushismarckensis	Panggang Island, Taman Nasional KepulauanSeribu, Indonesia	Plant growth promoting bioactivity: Indolacetic acid production, HCN production, phosphate solubilization, and antimicrobial activity against Xanthomonas oryzae and Pyricularia oryzae	[41]
Williamsiaaurantiacus sp. nov	Glodiacorticostylifera	Praia Guaeca, Sao Paulo, Brazil	Antimicrobial activity against S. aureus and Colletotrichum gloeosporioides	[42]
Streptomyces olivaceus	Dysideaavara	Larak Island, Persian Gulf	Olivomycin A: cytotoxic activity against SW480, HepG2, and MCF7 cell line	[43]
Corals				
S. arenicola	P. lobata and P. panamensis	Tropical central Pacific	Promotes plant growth in salt stress conditions	[44]
Micromonospora marina	Soft coral	Indian Ocean coast, Mozambique	Thiocoraline: anticancer activity against LoVo and SW620 human colon cancer cell lines	[45]

Actinobacteria	Host	Location	Significance	References
Mycobacterium sp	Coral	Kurusadai Island, Gulf of Mannar	Antioxidant Activity: Scavenging of Hydrogen peroxide, Nitric Oxide Radical	[46]
Actinobacterial sp.	*Sarcophyton glaucum*	Red sea	Antibacterial activity against S. aureus, *K. pneumonia*, P. aeruginosa, V. fluvialis, *antifungal activity against A. niger, Penicillium sp.,* and *C. albicans*	[47]
Myceligeneranscantabricum sp. nov	Fam. Caryophillidae	Aviles Canyon, Cantabrian Sea, Asturias, Spain	Antimicrobial activity against *E. coli,* Micrococcus luteus, and Saccharomyces cerevisiae	[48]
Micromonosporaaurantiaca, Nocardiopsisdassonvillei subsp. Albirubida, Nocardiopsissymnemataformans, Streptomyces rutgersensis, Streptomyces viridodiastaticus	Scleractinian corals	Luhuitou fringing reef	Presence of NRPS, PKS type I and II biosynthetic gene cluster	[49]
Ascidians				
Salinisporasp. and *Verrucosispora sp.*	*E. toealensis*	Micronesian Islands, Chuuk	Antibiotic: Indolocarbazoles and Staurosporines	[34]
Mollusks				
Streptomyces sp. CP32	*Conus pulicarius*	Mactan Island, Philippines	Derivatives of Pulicatins with neurological activity	[50]
Nocardiopsis albaCRt67	*C. rolani,*	Mactan Island, Philippines	Nocapyrones with neurological activity	[51]
Streptomyces sp.	*Crassostrea sikamea*	Bahia de La Paz, Baja California Sur, Mexico	—	[52]

Table 1.
Significance of actinobacteria associated with marine invertebrates.

DPPH, 74.2% hydrogen peroxide, and 56% nitric oxide radical scavenging activity at 100 μg/mL concentration [60]. Recently, 54.50% ABTS free radical scavenging activity with 100 μg/mL was reported from coral reef-associated *Saccharopolyspora sp.* IMA1 [16]. While *Streptomyces sp.* NMF6 associated with marine sponge *Diacarnus ardoukobae* possessed significant phosphomolybdenum, ferric-reducing power, and DPPH free radical scavenging activity [61]. Ser et al. reported *Streptomyces malaysiense sp.* having 27.24% DPPH radical scavenging activity at 2 mg/mL of 0.016% DPPH solution [62].

3.3 Cytotoxic activity

Cancer treating drugs have elevated toxic effects with undesirable side effects. Therefore, the reach of new and less harmful drugs has a high demand. Progress had been made recently to reach antitumor compounds from marine actinobacteria [63]. Violapyrone H and I were isolated from *Streptomyces sp.* associated with starfish. Violapyrone exhibited cytotoxicity against 10 human cancer cell lines with a GI_{50} value from 1.10 M to 26.12 M. This is the first report of violapyrones tested for their cytotoxicity potential [64]. *Streptomyces sp.* isolated from *Acanthaster planci* collected from the Federated States of Micronesia produced violapyrone H, Iα pyrones derivatives that displayed cytotoxicity against 10 human cancer cell lines with GI_{50} value from 1.10 μM to 26.12 μM [63]. *Donaxtrunculus anatinus*-associated Actinobacterial genera *Nocardioides*, *Kitasatosporia*, and *Streptomyces* showed strong antitumor activity against human carcinoma of the liver (HEPG2), cervix (HELA), and breast (MCF7) cell line with IC_{50}: 3.89, 9.4, and 10 μg/ml, respectively [65]. Steffimycin produced by *Streptomyces* sp. 0630 exhibited cytotoxic effect against a panel of human cancer cells MCF-7, HepG-2, and A2780 with IC_{50} values of 5.05, 5.57, and 1.91 μM, respectively [66]. *Streptomyces sp.* TMKS8 associated with sea slug exhibited cytotoxicity against murine leukemia P388 cells with IC_{50} 9.8 μM [67, 68].

3.4 Enzymes

The significance of enzymes in food, textile, detergent, and pharmaceutical has been known for a long. The harnessing of actinobacterial-derived enzymes comprises of cost-effective and eco-friendly nature due to its mild fermentation condition, such as the use of agricultural waste as a source of nutrients, temperature, pH, agitation, and less production time [69]. Actinobacteria associated with marine hosts produce various enzymes including amylase, protease, keratinase, lipase, L asparaginase, Xylanase, chitinase, cellulase, and dextranase that embrace industrial significance [70].

Proteases hydrolyze protein molecules to peptides and eventually to free amino acids. Protease plays a significant role in the metabolic cycles of all living forms. There are several types of proteases including serine, carboxy serine, cystine, metallo, carboxy metallo, and aspartic proteases [71]. The application of protease includes animal fodder preparation, silk degumming, detergent formulation (stain remover), dehairing and dewooling (leather industry), and silver recovery from X-ray film [72–74]. Actinobacterial genera *Microbacterium* isolated from stony coral *Pocillopora sp.* and *Faviia sp.* produced proteases [75]. Similarly, *Micrococcus sp.* and *Brevibacterium sp.* isolated from *Faviia sp.* produced proteases [75, 76]. Whereas 12.6% and 10.9% of bacteria associated with marine sponge **D. granulosa** and *S. fibulata*, respectively produced proteases [77].

Amylase is one of the most demanding enzymes used mainly for scarification of starch, pulp processing, bread dough making, winery, and detergent industry whereas solvent tolerant amylase was used mainly for bioremediation and improvement of detergent [78, 79]. Meena et al. isolated 10 actinobacterial genera associated with *Phallusia nigra* ascidian. Among them, *Kinecoccus mangrovi, Kocuria polaris, Salinospora sp.,* and *Nocardiopsis exalbidus* were amylase positive with 12.29, 8.85, 6.61, and 5.13 U/mL activity, respectively [80]. According to the earlier report, bacteria associated with sponges exhibited a higher percentage of amylase production followed by phosphatase and protease while the least urease-positive isolates were obtained from *Dysidea granulose* and *Sigmadocia fibulata* sponge [81].

Chitinase involves in hydrolyses of chitin polymer by cleaving β 1–4 linkages. Chitin is a polymer present in the cell wall of fungi, shells of marine invertebrates, and few insects. The use of chitinase involves the development of pesticides, management of marine wastes, biofuel production, and food and pharma industries [82]. Many actinobacterial genera, including *Streptomyces, Micromonospora, Nocardiopsis,* and *Nocardia* have been reported [83]. Recombinant chitinase from **S. griseus** showed enhanced hydrolysis of α and β chitin with a higher rate of activity using shrimp shells as substrate [84]. SaChiB chitinase isolated and cloned from *S. alfalfa* ACCC400021 showed maximal activity at 45°C temperature with pH 8 while SaChiB exhibited antifungal activity so considered as a biocontrol agent in agriculture [85].

The pharmaceutical demand for L-asparaginase is high due to its anti-carcinogenic ability. L-asparaginase inhibits the growth of cancerous cells by cleaving L-asparagine into ammonia and aspartic acid. Mainly L-asparaginase is produced by fungi while few actinobacterial species are also reported [86]. *Streptomyces noursei* MTCC 10469 associated with marine sponge *Callyspongia diffusa* produced 102 kDa of L-asparaginase that displayed optimum activity in pH 8 at 50°C [87]. Also, **S. fradiae** NEAE-82-derived L-asparaginase showed anti-proliferative activity on cancer cells (HepG2, Hep2, and Caco2) [88]. Currently, *Aspergillus* and *Bacillus*-derived L-asparaginase are available commercially while actinobacterial-derived L-asparaginase is under investigation [86].

Actinobacteria are a distinctive group of prokaryotes having similarities to both fungi and bacteria. The actinobacterial-derived cellulase has unique features in terms of adaptation to extreme environmental parameters and degradation of plant-based biomass. Cellulase is a carbohydrate degrading enzyme that hydrolyzes cellulose into mono- and oligosaccharides [89]. The *Kineococcus mangrove, Kocuria Polaris, Nocarciopsis exalbidus,* and *Salinispora sp.* belonging to the phylum Actinobacteria were isolated from marine ascidian *Phallusia nigra* exhibited cellulase production [80]. Most terrestrial microbial-derived cellulase showed an inhibitory effect in presence of glucose whereas marine halophilic *Streptomyces*-derived cellulase retained more than 60% activity in presence of 0.5 M glucose which makes the enzyme feasible to conduct high biomass saccharification [90]. Also, *Actinoalloteichus cyanogriseus* strain MHA15 isolated from marine habitat showed higher cellulase activity 14.378 U/mL in carboxy methyl cellulose medium suggesting ideal bacteria for cellulose bioconversion [89].

3.5 Helometabolites

Halometabolites produced by marine organisms play a significant role in host defense mechanisms by quorum sensing and production of toxins, growth hormones, or antibiotics. Helometabolites, such as chloride, bromide, and iodine are omnipresent in marine environments whereas fluoride is present in Earth's crust [91]. *Micromonospora echinospora*-derived Calicheamicin is a group of enediyne metabolites with iodine and has remarkable anticancer activity [92]. Whereas, halogenated glycopeptide Vancomycin produced by *S. lavendulae* showed antibacterial activity against various test pathogens [93] followed by the production of indimicine A-E chlorinated bisindolie alkaloid from deep-sea *Streptomyces sp.* showed strong antimicrobial activity along with cytotoxic activity against MCF-7 cell line with IC$_{50}$ 10 μM [94]. FU et al. reported chlorinated streptochlorides derived from coral-associated *Streptomyces sp.* with antimicrobial activity against proteobacteria and cytotoxicity against breast cancer cell line MCF-7 [95]. The research of halo metabolites through genome mining resulted in 26 FADH$_2$dependanthalogenase positive actinobacterial strains associated with mangroves [96]. The antifungal kutzneride involves in dichlorination at C6 and C7 position of tryptophan has been isolated from *Kutzneria sp.* whereas pyrroindomycin isolated from *S. rugosporus* involves tryptophan dichlorination at position 5 [97]. Further, three novel halogenase gene clusters were identified from sponge-associated actinobacteria that suggest the importance of actinobacteria as a remarkable source for harnessing halometabolites [98].

3.6 Enzymes inhibitors

Enzyme inhibitors have a pivotal position in agriculture to protect crops from predators. Allosamidin derived from *Streptomyces sp.* can inhibit chitinase and shows potent insecticidal activity against *Bombyx mori* and Silkworm [99]. Several α-glucosidase and α-amylase inhibitor has been reported from marine habitat, for instance, *Streptomyces* sp.PW638-derived Acarviostatin 103 inhibit α-amylase with IC50 value 12.23 μg/mL and α-glucosidase with IC$_{50}$ 1.25 μg/mL [100]. Whereas amino-oligosaccharide α-glucosidase inhibitors produced by *Streptomyces sp.* CKD-711 showed potent inhibitory activity against **Comamonas terrigena** [101]. Protease inhibitor has valuable antiviral activity against Zika, Dengue, hepatitis C virus, and many more. Kamarudheen et al. reported protease inhibitor from marine sponge *Callyspongiasp*-associated **S. griseoincarnatus** HK12 having antiviral activity against Chikungunya [102]. Leupeptin, Pepstatin, Antipain, Phosphoramidon, Talopeptin, and Diketopiperazine are well-known actinobacteria-derived protease inhibitors having remarkable pharmaceutical significance [103, 104].

4. Conclusion

The marine actinobacterial provides vast scope for therapeutically active macro-molecules, such as antibiotics and halometabolites with the addition of industrially significant enzymes, such as amylase, protease, asparaginase, xylanase, cellulase, chitinase, and lipase. The presence of actinobacteria in marine habitats plays a pivotal role in their associated organisms by providing protection against harmful moieties. Harnessing host-specific actinobacterial diversity from the marine ecosystem will result in novel species with the ability to produce new and diverse secondary metabolites which will be beneficial for detergent, food, medicine, agriculture, cosmetics, paper, and pulp industries. Furthermore, the exploitation of bioactive compounds from marine microorganisms will fulfill the current demand for drug-resistant microorganisms as many more marine niches are still unexplored.

Acknowledgements

Ms. Vaishali R. Majithiya acknowledges SHODH (ScHeme Of Developing High quality research- MYSY), Gujarat, India for a research fellowship. The authors are grateful to DST-SERB, New Delhi, India for financial support (Sanction order No. ECR/2016/000928 Dated: 20.03.2017). The authors are also thankful to Saurashtra University, Gujarat, India for infrastructural facilities.

Conflict of interest

The authors declare no conflict of interest.

Acronyms and abbreviations

FRAP	Ferric reducing antioxidant power
NO	Nitric oxide
VCPO	Vanadium dependant chloroperoxidase
MRSA	Methicillin-resistant Staphylococcus aureus
PKS	Polyketide synthases
NRPS	Nonribosomal peptide synthases
IC_{50}	Half maximal inhibitory concentration

Author details

Vaishali R. Majithiya and Sangeeta D. Gohel*
UGC-CAS Department of Biosciences, Saurashtra University, Rajkot, Gujarat, India

*Address all correspondence to: sangeetagohel@gmail.com

IntechOpen

References

[1] Goodfellow M. Phylum XXVI. *Actinobacteria* phyl. nov. In: Goodfellow M, Kampfer P, Trujillo ME, Suzuki K, Ludwig W, Whitman WB, editors. Bergey's Manual of Systematic Bacteriology. second ed. Vol. 5. New York: Springer; 2012. pp. 33-34

[2] Hauhan JV, Gohel SD. Molecular diversity and pharmaceutical applications of free-living and Rhizospheric marine Actinobacteria. In: Marine Niche: Applications in Pharmaceutical Sciences. Singapore: Springer; 2020. pp. 111-131

[3] Sivakumar K, Sahu MK, Thangaradjou T, Kannan L. Research on marine actinobacteria in India. Indian Journal of Microbiology. 2007;**47**(3):186-196. DOI: 10.1007/s12088-007-0039-1

[4] Gohel SD, Singh SP. Molecular phylogeny and diversity of the salt-tolerant alkaliphilic actinobacteria inhabiting coastal Gujarat. India. Geomicrobiology Journal. 2018;**35**(9): 775-789. DOI: 10.1080/01490451. 2018.1471107

[5] Quinn GA, Banat AM, Abdelhameed AM, Banat IM. Streptomyces from traditional medicine: Sources of new innovations in antibiotic discovery. Journal of Medical Microbiology. 2020;**69**(8):1040. DOI: 10.1099/jmm.0.001232

[6] Egan S, James S, Holmström C, Kjelleberg S. Inhibition of algal spore germination by the marine bacterium Pseudoalteromonastunicata. FEMS Microbiology Ecology. 2001;**35**(1): 67-73. DOI: 10.1111/j.1574-6941.2001. tb00789.x

[7] Egan S, Thomas T, Kjelleberg S. Unlocking the diversity and biotechnological potential of marine surface associated microbial communities. Current Opinion in Microbiology. 2008;**11**(3):219-225. DOI: 10.1016/j.mib.2008.04.001

[8] Penesyan A, Tebben J, Lee M, Thomas T, Kjelleberg S, Harder T, et al. Identification of the antibacterial compound produced by the marine epiphytic bacterium Pseudovibrio sp. D323 and related sponge-associated bacteria. Marine Drugs. 2011;**9**(8):1391-1402. DOI: 10.3390/md9081391

[9] Gohel SD, Sharma AK, Thakrar FJ, Singh SP. Endophytic actinobacteria and their interactions with plant host systems. In: Understanding Host-microbiome Interactions-an Omics Approach. Singapore: Springer; 2017. pp. 247-246

[10] Shady NH, Tawfike AF, Yahia R, Fouad MA, Brachmann AO, Piel J, et al. Cytotoxic activity of actinomycetes Nocardia sp. and Nocardiopsi s sp. associated with marine sponge Amphimedon sp. Natural Product Research. 2021;**19**:1-6. DOI: 10.1080/14786419.2021.1931865

[11] Hindu SV, Chandrasekaran N, Mukherjee A, Thomas J. A review on the impact of seaweed polysaccharide on the growth of probiotic bacteria and its application in aquaculture. Aquaculture International. 2019;**27**(1):227-238. DOI: 10.1007/s10499-018-0318-3

[12] De Voogd NJ, Cleary DF. An analysis of sponge diversity and distribution at three taxonomic levels in the Thousand Islands/Jakarta Bay reef complex, West-Java, Indonesia. Marine Ecology. 2008;**29**(2):205-215. DOI: 10.1111/j.1439-0485.2008.00238.x

[13] Baker PW, Kennedy J, Morrissey J, O'Gara F, Dobson AD, Marchesi JR. Endoglucanase activities and growth of marine-derived fungi isolated from the sponge Haliclonasimulans. Journal of Applied Microbiology. 2010;**108**(5): 1668-1675. DOI: 10.1111/j.1365-2672. 2009.04563.x

[14] Ziegler M, Seneca FO, Yum LK, Palumbi SR, Voolstra CR. Bacterial community dynamics are linked to patterns of coral heat tolerance. Nature Communications. 2017;**8**(1):1-8. DOI: 10.1038/ncomms14213

[15] Hifnawy M, Hassan HM, Mohammed R, Fouda M, Sayed AM, Hamed A, et al. Induction of Antibacterial Metabolites by Co-Cultivation of Two Red-Sea-Sponge-Associated Actinomycetes Micromonospora sp. UR56 and Actinokinespora sp. EG49. Marine Drugs. 2020;**18**(5):243

[16] Krishnamoorthy M, Dharmaraj D, Rajendran K, Karuppiah K, Balasubramanian M, Ethiraj K. Pharmacological activities of coral reef associated actinomycetes, Saccharopolyspora sp. IMA1. Biocatalysis and Agricultural Biotechnology. 2020;**28**:101748. DOI: 10.1016/j.bcab.2020.101748

[17] Ramesh C, Tulasi BR, Raju M, Thakur N, Dufossé L. Marine natural products from tunicates and their associated microbes. Marine Drugs. 2021;**19**(6):308. DOI: 10.3390/md19060308

[18] Campana S, Busch K, Hentschel U, Muyzer G, de Goeij JM. DNA-stable isotope probing (DNA-SIP) identifies marine sponge-associated bacteria actively utilizing dissolved organic matter (DOM). Environmental Microbiology. 2021;**2022**. DOI: 10.1111/1462-2920.15642

[19] Ravikumar S, Krishnakumar S, Inbaneson SJ, Gnanadesigan M. Antagonistic activity of marine actinomycetes from Arabian Sea coast. Archives of Applied Science Research. 2010;**2**(6):273-280

[20] Li Z, Sun W, Zhang F, He L, Loganathan K. Actinomycetes from the South China Sea sponges: Isolation, diversity and potential for aromatic polyketides discovery. Frontiers in Microbiology. 2015;**6**:1048

[21] Sun W, Liu C, Zhang F, Zhao M, Li Z. Comparative genomics provides insights into the marine adaptation in sponge-derived Kocuriaflava S43. Frontiers in Microbiology. 2018;**9**:1257. DOI: 10.3389/fmicb.2015.01048

[22] Mahmoud HM, Kalendar AA. Coral-associated Actinobacteria: Diversity, abundance, and biotechnological potentials. Frontiers in Microbiology. 2016;**7**:204. DOI: 10.3389/fmicb.2016.00204

[23] Kuang W, Li J, Zhang S, Long L. Diversity and distribution of Actinobacteria associated with reef coral Porites lutea. Frontiers in Microbiology. 2015;**6**:1094. DOI: 10.3389/fmicb.2015.01094

[24] Nithyanand P, Manju S, Karutha PS. Phylogenetic characterization of culturable actinomycetes associated with the mucus of the coral Acropora digitifera from Gulf of Mannar. FEMS Microbiology Letters. 2011;**314**(2):112-118. DOI: 10.1111/j.1574-6968.2010.02149.x

[25] Tremblay P, Maguer JF, Grover R, Ferrier-Pagès C. Trophic dynamics of scleractinian corals: Stable isotope evidence. The Journal of experimental biology. 2015;**218**(8):1223-1234. DOI: 10.1242/jeb.115303

[26] Little AF, Van Oppen MJ, Willis BL. Flexibility in algal endosymbioses shapes growth in reef corals. Science. 2004;**304**(5676):1492. DOI: 10.1126/science.1095733

[27] Lesser MP, Morrow KM, Pankey SM, Noonan SH. Diazotroph diversity and nitrogen fixation in the coral Stylophorapistillata from the Great Barrier Reef. The ISME Journal. 2018;**12**(3):813-824. DOI: 10.1038/s41396-017-0008-6

[28] Lampert Y, Kelman D, Dubinsky Z, Nitzan Y, Hill RT. Diversity of culturable bacteria in the mucus of the Red Sea coral Fungiascutaria. FEMS Microbiology Ecology. 2006;**58**(1):99-108. DOI: 10.1111/j.1574-6941.2006.00136.x

[29] Li F, Xie Q, Zhou S, Kong F, Xu Y, Ma Q, et al. Nocardiopsiscoralli sp. nov. a novel actinobacterium isolated from the coral Galaxeaastreata. International Journal of Systematic and Evolutionary Microbiology. 2021;**71**(6):004817

[30] Gozari M, Mortazavi MS, Bahador N, Rabbaniha M. Isolation and screening of antibacterial and enzyme producing marine actinobacteria to approach probiotics against some pathogenic vibrios in shrimp Litopenaeusvannamei. Iranian Journal of Fisheries Sciences. 2016;**15**(2):630-644

[31] You JL, Cao LX, Liu GF, Zhou SN, Tan HM, Lin YC. Isolation and characterization of actinomycetes antagonistic to pathogenic Vibrio spp. from nearshore marine sediments. World Journal of Microbiology and Biotechnology. 2005;**21**(5):679-682. DOI: 10.1007/s11274-004-3851-3

[32] Kumar SS, Philip R, Achuthankutty CT. Antiviral property of marine actinomycetes against white spot syndrome virus in penaeid shrimps. Current Science. 2006;**25**:807-811. DOI: stable/24093913

[33] You J, Xue X, Cao L, Lu X, Wang J, Zhang L, et al. Inhibition of Vibrio biofilm formation by a marine actinomycete strain A66. Applied Microbiology and Biotechnology. 2007;**76**(5):1137-1144. DOI: 10.1007/s00253-007-1074-x

[34] Steinert G, Taylor MW, Schupp PJ. Diversity of actinobacteria associated with the marine ascidian Eudistomatoealensis. Marine Biotechnology. 2015;**17**(4):377-385. DOI: 10.1007/s10126-015-9622-3

[35] Lee NY, Cha CJ, Im WT, Kim SB, Seong CN, Bae JW, et al. A report of 42 unrecorded actinobacterial species in Korea. Journal of Species Research. 2018;**7**(1):36-49

[36] Selvin J, Shanmughapriya S, Gandhimathi R, Kiran GS, Ravji TR, Natarajaseenivasan K, et al. Optimization and production of novel antimicrobial agents from sponge associated marine actinomycetes Nocardiopsisdassonvillei MAD08. Applied Microbiology and Biotechnology. 2009;**83**(3):435-445. DOI: 10.1007/s00253-009-1878-y

[37] Jiang S, Li X, Zhang L, Sun W, Dai S, Xie L, et al. Culturable actinobacteria isolated from marine sponge Iotrochota sp. Marine Biology. 2008;**153**(5):945-952. DOI: 10.1007/s00227-007-0866-y

[38] Ekiz G, Hames EE, Demir V, Can F, Yokes MB, Uzel A, et al. Cultivable sponge-associated actinobacteria from coastal area of Eastern Mediterranean Sea. Advances in Microbiology. 2014;**2014**

[39] Schneemann I, Nagel K, Kajahn I, Labes A, Wiese J, Imhoff JF. Comprehensive investigation of marine Actinobacteria associated with the sponge Halichondriapanicea. Applied and Environmental Microbiology. 2010;**76**(11):3702-3714. DOI: 10.1128/AEM.00780-10

[40] Sandoval-Powers M, Králová S, Nguyen GS, Fawwal DV, Degnes K, Lewin AS, et al. Streptomyces poriferorum sp. nov., a novel marine sponge-derived Actinobacteria species expressing anti-MRSA activity. Systematic and Applied Microbiology. 2021;**44**(5):126244

[41] Bose U, Ortori CA, Sarmad S, Barrett DA, Hewavitharana AK, Hodson MP, et al. Production of N-acyl homoserine lactones by the sponge-associated marine actinobacteria

Salinisporaarenicola and
Salinisporapacifica. FEMS Microbiology
Letters. 2017;1:364. DOI: 10.1093/
femsle/fnx002

[42] Retnowati D, Solihin DD,
Ghulamahdi M, Lestari Y. New
information on the potency ofthe
potency of sponge-associated
actinobacteria as producer of plant
growth-promoting bioactive
compounds. Malaysian Applied Biology.
2018;**47**(6):127-135

[43] De Menezes CB, Afonso RS, de
Souza WR, Parma MM, de Melo IS,
Fugita FL, et al. Williamsiaaurantiacus
sp. nov. a novel actinobacterium
producer of antimicrobial compounds
isolated from the marine sponge.
Archives of Microbiology.
2019;**201**(5):691-698. DOI: 10.1007/
s00203-019-01633-z

[44] Ocampo-Alvarez H, Meza-
Canales ID, Mateos-Salmón C,
Rios-Jara E, Rodríguez-Zaragoza FA,
Robles-Murguía C, et al. Diving into
reef ecosystems for land-agriculture
solutions: Coral microbiota can alleviate
salt stress during germination and
photosynthesis in terrestrial plants.
Frontiers in Plant Science.
2020;**25**(11):648. DOI: 10.3389/
fpls.2020.00648

[45] Erba E, Bergamaschi D, Ronzoni S,
Faretta M, Taverna S, Bonfanti M, et al.
Mode of action of thiocoraline, a natural
marine compound with anti-tumour
activity. British Journal of Cancer.
1999;**80**(7):971-980. DOI: 10.1038/
sj.bjc.6690451

[46] Poongodi SU, Karuppiah VA,
Sivakumar KA, Kannan LA. Marine
actinobacteria of the coral reef
environment of the gulf of Mannar
biosphere reserve, India: A search for
antioxidant property. International
Journal of Pharmacy and
Pharmaceutical Sciences.
2012;**4**:316-321

[47] ElAhwany AM, Ghozlan HA,
ElSharif HA, Sabry SA. Phylogenetic
diversity and antimicrobial activity of
marine bacteria associated with the soft
coral Sarcophyton glaucum. Journal of
Basic Microbiology. 2015;**55**(1):2-10.
DOI: 10.1002/jobm.201300195

[48] Sarmiento Vizcaíno A,
González Iglesias V, Fernández
Braña AJ, Molina Ramírez A, Acuña
Fernández JL, García Díaz LA, et al.
Myceligeneranscantabricum sp. nov., a
barotolerant actinobacterium isolated
from a deep cold-water coral.
International Journal of Systematic and
Evolutionary Microbiology. 2015;**65**(4)

[49] Li J, Dong JD, Yang J, Luo XM,
Zhang S. Detection of polyketide
synthase and nonribosomal peptide
synthetase biosynthetic genes from
antimicrobial coral-associated
actinomycetes. Antonie Van
Leeuwenhoek. 2014;**106**(4):623-635.
DOI: 10.1007/s10482-014-0233-1

[50] Lin Z, Antemano RR, Hughen RW,
Tianero MD, Peraud O, Haygood MG,
et al. Pulicatins A– E, neuroactive
thiazoline metabolites from cone
snail-associated bacteria. Journal of
Natural Products. 2010;**73**:1922-1926.
DOI: 10.1021/np100588c

[51] Lin Z, Torres JP, Ammon MA,
Marett L, Teichert RW, Reilly CA, et al.
A bacterial source for mollusk pyrone
polyketides. Chemistry & Biology.
2013;**20**(1):73-81. DOI: 10.1016/j.
chembiol.2012.10.019

[52] García Bernal M, Trabal
Fernández N, Saucedo Lastra PE, Medina
Marrero R, Mazón-Suástegui JM.
Streptomyces effect on the bacterial
microbiota associated to Crassostrea
sikamea oyster. Journal of Applied
Microbiology. 2017;**122**(3):601-614.
DOI: 10.1111/jam.13382

[53] Indraningrat AA, Smidt H,
Sipkema D. Bioprospecting

sponge-associated microbes for antimicrobial compounds. Marine Drugs. 2016;**14**(5):87. DOI: 10.3390/md14050087

[54] Schneemann I, Kajahn I, Ohlendorf B, Zinecker H, Erhard A, Nagel K, et al. Mayamycin, a cytotoxic polyketide from a Streptomyces strain isolated from the marine sponge Halichondriapanicea. Journal of Natural Products. 2010;**73**(7):1309-1312. DOI: 10.1021/np100135b

[55] Eltamany EE, Abdelmohsen UR, Ibrahim AK, Hassanean HA, Hentschel U, Ahmed SA. New antibacterial xanthone from the marine sponge-derived Micrococcus sp. EG45. Bioorganic & Medicinal Chemistry Letters. 2014;**24**(21):4939-4942

[56] Gandhimathi R, Arunkumar M, Selvin J, Thangavelu T, Sivaramakrishnan S, Kiran GS, et al. Antimicrobial potential of sponge associated marine actinomycetes. Journal de MycologieMedicale. 2008;**18**(1):16-22. DOI: 10.1016/j.mycmed.2007.11.001

[57] Liu T, Wu S, Zhang R, Wang D, Chen J, Zhao J. Diversity and antimicrobial potential of Actinobacteria isolated from diverse marine sponges along the Beibu Gulf of the South China Sea. FEMS Microbiology Ecology. 2019;**95**(7):fiz089

[58] Jagannathan SV, Manemann EM, Rowe SE, Callender MC, Soto W. Marine actinomycetes, new sources of biotechnological products. Marine Drugs. 2021;**19**(7):365. DOI: 10.3390/md19070365

[59] Mahapatra GP, Raman S, Nayak S, Gouda S, Das G, Patra JK. Metagenomics approaches in discovery and development of new bioactive compounds from marine actinomycetes. Current Microbiology. 2020;**77**(4):645-656. DOI: 10.1007/s00284-019-01698-5

[60] Poongodi S, Karuppiah V, Sivakumar K, Kannan L. Antioxidant activity of *Nocardiopsis sp.*, a marine actinobacterium, isolated from the Gulf of Mannar Biosphere Reserve, India. National Academy Science Letters. 2014;**37**(1):65-70. DOI: 10.1007/s40009-013-0193-4

[61] Fahmy NM, Abdel-Tawab AM. Isolation and characterization of marine sponge–associated Streptomyces sp. NMF6 strain producing secondary metabolite (s) possessing antimicrobial, antioxidant, anticancer, and antiviral activities. Journal, Genetic Engineering & Biotechnology. 2021;**19**(1):1-4. DOI: 10.1186/s43141-021-00203-5

[62] Webb HE, Brichta-Harhay DM, Brashears MM, Nightingale KK, Arthur TM, Bosilevac JM, et al. Salmonella in peripheral lymph nodes of healthy cattle at slaughter. Frontiers in Microbiology. 2017;**9**(8):2214. DOI: 10.3389/fmicb.2017.02214

[63] Hassan SS, Anjum K, Abbas SQ, Akhter N, Shagufta BI, Shah SA, et al. Emerging biopharmaceuticals from marine actinobacteria. Environmental Toxicology and Pharmacology. 2017;**49**:34-47. DOI: 10.1016/j.etap.2016.11.015

[64] Shin HJ, Lee HS, Lee JS, Shin J, Lee MA, Lee HS, et al. Violapyrones H and I, new cytotoxic compounds isolated from Streptomyces sp. associated with the marine starfish Acanthasterplanci. Marine Drugs. 2014;**12**(6):3283-3291. DOI: 10.3390/md12063283

[65] El-Shatoury SA, El-Shenawy NS, Abd El-Salam IM. Antimicrobial, antitumor and in vivo cytotoxicity of actinomycetes inhabiting marine shellfish. World Journal of Microbiology and Biotechnology. 2009;**25**(9):1547-1555. DOI: 10.1007/s11274-009-0040-4

[66] Liu CY, Li YL, Lu JH, Qian LL, Xu K, Wang NN, et al. A new

steffimycin-type derivative from the lichen-derived actinomycetes steptomyces sp. Journal of Molecular Structure. 2021;**1227**:129352. DOI: 10.1016/j.molstruc.2020.129352

[67] Zhang Z, Sibero MT, Kai A, Fukaya K, Urabe D, Igarashi Y. TMKS8A, an antibacterial and cytotoxic chlorinated α-lapachone, from a sea slug-derived actinomycete of the genus Streptomyces. The Journal of Antibiotics. 2021;**74**(7):464-469. DOI: 10.1038/s41429-021-00415-4

[68] Zhang X, Song C, Bai Y, Hu J, Pan H. Cytotoxic and antimicrobial activities of secondary metabolites isolated from the deep-sea-derived Actinoalloteichuscyanogriseus 12A22. 3. Biotech. 2021;**11**(6):1-8. DOI: 10.1007/s13205-021-02846-0

[69] Gohel SD, Sharma AK, Dangar KG, Thakrar FJ, Singh SP. Biology and applications of halophilic and haloalkaliphilic actinobacteria. In: Extremophiles. CRC Press; 2018. pp. 103-136

[70] Majithiya V, Gohel S. Isolation and screening of extracellular enzymes producing Actinobacteria associated with sea weed. In: Proceedings of the National Conference on Innovations in Biological Sciences (NCIBS). 2020

[71] Gohel SD, Sharma AK, Dangar KG, Thakrar FJ, Singh SP. Antimicrobial and biocatalytic potential of haloalkaliphilic actinobacteria. In: Halophiles. Cham: Springer; 2015. pp. 29-55

[72] Singh SP, Thumar JT, Gohel SD, Kikani B, Shukla R, Sharma A, et al. Actinomycetes from marine habitats and their enzymatic potential. In: Marine Enzymes for Biocatalysis. Woodhead Publishing; 2013. pp. 191-214

[73] Gohel SD, Singh SP. Thermodynamics of a Ca2+−dependent highly thermostable alkaline protease

from a haloalkliphilic actinomycete. International Journal of Biological Macromolecules. 2015;**72**:421-429. DOI: 10.1016/j.ijbiomac.2014.08.008

[74] Gohel SD, Singh SP. Cloning and expression of alkaline protease genes from two salt-tolerant alkaliphilic actinomycetes in *E. coli*. International Journal of Biological Macromolecules. 2012;**50**(3):664-671. DOI: 10.1016/j.ijbiomac.2012.01.039

[75] Bourne DG, Munn CB. Diversity of bacteria associated with the coral *Pocilloporadamicornis* from the Great Barrier Reef. Environmental Microbiology. 2005;7(8):1162-1174. DOI: 10.1111/j.1462-2920.2005.00793.x

[76] Feby A, Nair S. Sponge-associated bacteria of Lakshadweep coral reefs, India: Resource for extracellular hydrolytic enzymes. Advances in Bioscience and Biotechnology. 2010;**1**:330-337. DOI: 10.4236/abb.2010.14043

[77] Su H, Xiao Z, Yu K, Huang Q, Wang G, Wang Y, et al. Diversity of cultivable protease-producing bacteria and their extracellular proteases associated to scleractinian corals. PeerJ. 2020;**8**:e9055

[78] Kikani BA, Singh SP. Amylases from thermophilic bacteria: Structure and function relationship. Critical Reviews in Biotechnology. 2021;**20**:1-7. DOI: 10.1080/07388551.2021.1940089

[79] Strahler J, Skoluda N, Kappert MB, Nater UM. Simultaneous measurement of salivary cortisol and alpha-amylase: Application and recommendations. Neuroscience & Biobehavioral Reviews. 2017;**83**:657-677. DOI: 10.1016/j.neubiorev.2017.08.015

[80] Meena B, Anburajan L, Nitharsan K, Vinithkumar NV, Dharani G. Existence in cellulose shelters: Industrial and pharmaceutical leads of symbiotic actinobacteria from ascidian Phallusia

nigra, Andaman Islands. World Journal of Microbiology and Biotechnology. 2021;**37**(7):1-22. DOI: 10.1007/s11274-021-03090-7

[81] Feby A, Nair S. Sponge-associated bacteria of Lakshadweep coral reefs, India. Enzymes. 2010;**ct1**:330-337. DOI: 10.4236/abb.2010.14043

[82] Ober C, Chupp GL. The chitinase and chitinase-like proteins: A review of genetic and functional studies in asthma and immune-mediated diseases. Current Opinion in Allergy and Clinical Immunology. 2009;**9**(5):401. DOI: 10.1097%2FACI.0b013e3283306533

[83] Zhang W, Liu Y, Ma J, Yan Q, Jiang Z, Yang S. Biochemical characterization of a bifunctional chitinase/lysozyme from Streptomyces sampsonii suitable for N-acetyl chitobiose production. Biotechnology Letters. 2020;**1**:42

[84] Nakagawa YS, Kudo M, Onodera R, Ang LZ, Watanabe T, Totani K, et al. Analysis of four chitin-active lytic polysaccharide monooxygenases from Streptomyces griseus reveals functional variation. Journal of Agricultural and Food Chemistry. 2020;**68**(47):13641-13650. DOI: 10.1021/acs.jafc.0c05319

[85] Lv C, Gu T, Ma R, Yao W, Huang Y, Gu J, et al. Biochemical characterization of a GH19 chitinase from Streptomyces alfalfae and its applications in crystalline chitin conversion and biocontrol. International Journal of Biological Macromolecules. 2021;**167**:193-201. DOI: 10.1016/j.ijbiomac.2020.11.178

[86] Chand S, Mahajan RV, Prasad JP, Sahoo DK, Mihooliya KN, Dhar MS, et al. A comprehensive review on microbial l-asparaginase: Bioprocessing, characterization, and industrial applications. Biotechnology and Applied Biochemistry. 2020;**67**(4):619-647. DOI: 10.1002/bab.1888

[87] Dharmaraj S. Study of L-asparaginase production by Streptomyces noursei MTCC 10469, isolated from marine sponge Callyspongiadiffusa. Iranian Journal of Biotechnology. 2011;**9**(2):102-108

[88] Meena B, Anburajan L, Sathish T, Raghavan RV, Dharani G, Vinithkumar NV, et al. L-asparaginase from Streptomyces griseus NIOT-VKMA29: Optimization of process variables using factorial designs and molecular characterization of l-asparaginase gene. Scientific Reports. 2015;**5**(1):1-2. DOI: 10.1038/srep12404

[89] Rajagopal G, Kannan S. Systematic characterization of potential cellulolytic marine actinobacteria Actinoalloteichus sp. MHA15. Biotechnology Reports. 2017;**13**:30-36. DOI: 10.1016/j.btre.2016.12.003

[90] Bhattacharya AS, Bhattacharya A, Pletschke BI. Synergism of fungal and bacterial cellulases and hemicellulases: A novel perspective for enhanced bio-ethanol production. Biotechnology Letters. 2015;**37**(6):1117-1129. DOI: 10.1007/s10529-015-1779-3

[91] Kasanah N, Triyanto T. Bioactivities of halometabolites from marine actinobacteria. Biomolecules. 2019;**9**(6):225. DOI: 10.3390/biom9060225

[92] Menon BR, Richmond D, Menon N. Halogenases for biosynthetic pathway engineering: Toward new routes to naturals and non-naturals. Catalysis Reviews. 2020;**16**:1-59. DOI: 10.1080/01614940.2020.1823788

[93] Barka EA, Vatsa P, Sanchez L, Gaveau-Vaillant N, Jacquard C, Klenk HP, et al. Taxonomy, physiology, and natural products of Actinobacteria. Microbiology and Molecular Biology Reviews. 2016;**80**:1-43. DOI: 10.1128/MMBR.00019-15

[94] Zhang W, Ma L, Li S, Liu Z, Chen Y, Zhang H, et al. Indimicins A–E, bisindole alkaloids from the deep-sea-derived Streptomyces sp. SCSIO 03032. Journal of Natural Products. 2014;**77**(8):1887-1892. DOI: 10.1021/np500362p

[95] Fu P, Kong F, Wang Y, Wang Y, Liu P, Zuo G, et al. Antibiotic metabolites from the coral-associated actinomycete Streptomyces sp. OUCMDZ-1703. Chinese Journal of Chemistry. 2013;**31**(1):100-104. DOI: 10.1002/cjoc.201201062

[96] Li XG, Tang XM, Xiao J, Ma GH, Xu L, Xie SJ, et al. Harnessing the potential of halogenated natural product biosynthesis by mangrove-derived actinomycetes. Marine Drugs. 2013;**11**:3875-3890. DOI: 10.3390/md11103875

[97] Ismail M, Frese M, Patschkowski T, Ortseifen V, Niehaus K, Sewald N. Flavin-dependent Halogenases from Xanthomonas campestris pv. Campestris B100 prefer bromination over chlorination. Advanced Synthesis and Catalysis. 2019;**361**(11):2475-2486. DOI: 10.1002/adsc.201801591

[98] Bayer K, Scheuermayer M, Fieseler L, Hentschel U. Genomic mining for novel FADH 2-dependent halogenases in marine sponge-associated microbial consortia. Marine Biotechnology. 2013;**15**(1):63-72. DOI: 10.1007/s10126-012-9455-2

[99] Sakuda S, Isogai A, Matsumoto S, Suzuki A, Koseki K. The structure of allosamidin, a novel insect chitinase inhibitor, produced by Streptomyces sp. Tetrahedron Letters. 1986;**27**(22):2475-2478. DOI: 10.1016/S0040-4039(00)84560-8

[100] Meng P, Xie C, Geng P, Qi X, Zheng F, Bai F. Inhibitory effect of components from Streptomyces species on α-glucosidase and α-amilase of different origin. Applied Biochemistry and Microbiology. 2013;**49**(2):160-168. DOI: 10.1134/S0003683813020099

[101] Xu JL, Liu ZF, Zhang XW, Liu HL, Wang Y. Microbial oligosaccharides with biomedical applications. Marine Drugs. 2021;**19**(6):350. DOI: 10.3390/md19060350

[102] Siddharth S, Vittal RR. Evaluation of antimicrobial, enzyme inhibitory, antioxidant and cytotoxic activities of partially purified volatile metabolites of marine Streptomyces sp. S2A. Microorganisms. 2018;**6**(3):72

[103] Kamarudheen N, Khaparde A, Gopal DS, Rao KB. Unraveling a natural protease inhibitor from marine Streptomyces griseoincarnatus HK12 active against chikungunya virus. Microbiological Research. 2021;**31**:126858. DOI: 10.1016/j.micres.2021.126858

[104] Kamarudheen N, Rao B. An overview of protease inhibitors from Actinobacteria. Research Journal of Biotechnology. 2018;**13**:1

Chapter 3

Studies on Endophytic Actinobacteria as Plant Growth Promoters and Biocontrol Agents

Sumi Paul and Arka Pratim Chakraborty

Abstract

The exploration of microbial resources is necessary for plant growth promotion, biological control, and reducing the agrochemicals and fertilizers for sustainable agriculture. Bacteria and fungi are distributed in the biosphere including the rhizosphere and help the host plants by alleviating biotic and abiotic stress through different mechanisms and can be used as bioinoculants for biocontrol and plant growth promotion. Actinobacteria are among the most abundant groups of soil microorganisms. They have been studied for their function in the biological control of plant pathogens, interactions with plants, and plant growth promotion. *Streptomyces* is the largest genus of actinobacteria. *Streptomyces* acts as both plant growth promoter and also as plant disease suppressor by various mechanisms like an increase in the supply of nutrients such as phosphorus, iron, production of IAA, and siderophore production. Endophytic actinobacteria help in plant growth-promoting through multiple ways by producing plant hormones; controlling fungal disease through antibiosis and competition. This review briefly summarizes the effects of actinobacteria on biocontrol, plant growth promotion, and association with plants as endophytes.

Keywords: actinobacteria, endophytic in nature, growth promoters, biocontrol agents, disease suppression

1. Introduction

Agricultural activity is hampered by various plant diseases and non-living factors i.e., temperature, drought, salinity, etc. [1]. To prevent plant diseases several pesticides are used in the present day. The reason behind environmental pollution and the loss of soil fertility in crop fields is due to excessive use of chemical products in agriculture [2]. In recent years, due to environmental pollution, the use of chemical pesticides and fertilizers has been canceled in several countries. But nowadays many workers have given attention on the utilization of microbial antagonists to reduce the unrestricted use of chemical products which are applied to prevent plant disease. According to Vurukonda et al. [3] in place of chemical pesticides plant growth-promoting microbes are approved as a safe substitute in the agricultural field. Several microorganisms are known to act as a plant growth promoter and they have the capability to suppress plant disease [4–7]. Among microbes, actinobacteria are known to produce secondary metabolites, antimicrobial compounds, and plant growth-promoting regulators to improve agricultural developments [8–10].

Actinobacteria are gram-positive bacteria. Various plant pathogens are controlled by different types of antibiotics which are generally obtained from actinomycetes. Extensive use of chemical products in agriculture imparts deleterious effect on the environment and on the health of human too. Microbial pesticides act as a better and safer alternative way of chemical pesticides. For the growth of plants, the production of biological pesticides from actinobacteria is considered to be a more economical and safe method. The formation of biological pesticides is more useful in function compared to chemical pesticides. These harmful chemicals can be replaced by biological products of actinomycetes. The workers have found another way to obtain large vigor in vegetables with safety by applying the group of actinobacteria to avoid chemical fertilizers [11]. These biopesticides maintain the quality of crops as well as productivity of crops without any harmful effect on plants. In nature, actinobacteria are mostly distributed group of microorganisms. Almost 80% of drugs in the world are known to come from species of actinobacteria like *Streptomycetes* and *Micromonospora* [12]. According to Qin et al. [13] the rate of discovery of naturally antibiotics derived from the actinobacteria is increasing continuously. Actinobacteria have been reported to be an important producer of secondary metabolites [14] and these metabolites are utilized for different biological activities, such as antibacterial, antifungal, and insecticidal activities. According to Jog et al. [15], Actinobacteria also produce phytohormones. Various secondary metabolites, cell wall degrading enzymes, and antibiotics are produced from different species of actinobacteria like *Streptomyces*.

The antagonistic activity of *Streptomyces* is due to the production of the antifungal compound, antibacterial compound, and extra cellular enzymes facilitate [16]. These genera have been found to show a great potential to improve the future of agriculture [17]. The actinobacteria taxa are diverse as composed of streptomycetes and non-streptomycetes, the latter being uncommon, and classified as rare taxa.

An endophyte is a bacterial (including actinomycetes) or fungal microorganism, which spends the whole or part of its life cycle inside the healthy tissues of the host plant by colonizing inter- or intracellularly, typically without causing any harm to the host plant [18, 19]. Thus, an endophyte is an organism, which lives inside a plant [20]. Host plant becomes benefited from entophytic actinobacteria, which can inhibit the other harmful microbes and helps the host plants by increasing nutrient uptake like iron, phosphorus, etc. [21]. Endophytes make a colony in the internal tissue of the plants and are able to accelerate physiological plant responses [22, 23]. Endophytic actinobacteria in plants can produce different types of metabolites which are used for different applications, such as plant growth promoters [24, 25], biocontrol agents [26, 27], antimicrobials [28–30].

Endophytic actinobacteria help the host plants by means of growth promotion, stress tolerance, and reduction in disease symptoms [31]. From the tissues of the medicinal plants, actinobacteria are being consistently discovered [32–35]. For pharmaceutical industries and agricultural applications, endophytic actinobacteria could be a potential source of novel antimicrobial compounds [36]. In developing sustainable systems of crop production, endophytic bacteria–plant interactions have an important role [37].

2. Distribution of actinobacteria

According to Oskay et al. [38], actinobacteria are globally distributed soil-inhabiting microorganisms. Basilio et al. [39] reported that lots of actinobacteria including *Micromonospora*, *Streptomyces* were obtained from soil, and also many workers have identified *Nocardia*, *Actinoplanes*, *Streptosporangium*, *Streptomyces* as members of actinobacteria [40]. Pandey et al. [12] reported about the isolation of strains of actinobacteria from Lucknow. Actinobacteria were isolated from the soil samples of various regions around Jaipur, Sikar of Rajasthan [41]. According to Ababutain et al. [42], in Saudi Arabia, the existence of actinobacteria was observed from soil samples of different places and from sea sediment of Caspian [43]. Srinivasan et al. [44] reported that actinobacteria are a heterogeneous and widely distributed group of bacteria in nature. They grow as hyphae like fungi accountable for the characteristically "earthy odor of freshly turned healthy soil" [45].

In several habitats the actinobacteria are found to survive in nature [46]. They are originally soil inhabitants [47]. But they have been reported in various range of an ecosystem, like from deep-sea [48], in terrestrial soil as well as in extreme environments. Takami et al. [49] reported actinomycetrs from greatest depth Mariana Trench. According to Williams et al. [50], actinobacteria can be found in a wide range of soils. *Microbispora, Nocardia, Microtetraspora, Amycolaptosis, Actinomadura, and Saccharothrix* are thermo-tolerant (up to 50°C) actinobacteria, reported by Takahashi et al. [51].

2.1 Endophytic actinobacteria

The word endophyte means "in the plant" (endon Gr. = within, phyton = plant). In 1866, de Bary had given the term endophyte. According to his definition, "Endophytes are the microorganisms, which reside inside the plant tissues and are significantly different from those found on the plant surface". Microorganisms that live within the host place either intra or intercellularly, known as endophytes [52] without causing any harmful effect on their host, and have proven to be the richest source of bioactive natural products. By secreting phytohormones entophytes help the plants in nutrition improvement and enhancement the growth of plants by protecting them against phytopathogens [53]. According to Petrini et al. [54], all organisms inhabiting plant organs can colonize internal plant tissues without causing harm to the host at some time in their life. According to Singh and Dubey [55], several microorganisms like bacteria, fungi, as well as actinobacteria form symbiotic associations within the host plant cell.

Normally the endophytes without subjecting the plant to any disadvantage complete their life cycle within the host plants. When groups of actinobacteria reside within living plant cells cooperatively that is called endophytic actinobacteria, such as nitrogen-fixing endophytes *Frankia*. It is reported that endophytic actinobacteria help to promote the growth of host plants and can reduce disease symptoms. Endophytes are ubiquitous in nature and they produce phytohormones and other growth-promoting factors to enhance the growth of the host plants. In return, the host plant helps the endophytes with nutrients and shelter. The endophytic actinobacteria form one of the interesting groups of microorganisms that is associated with a wide range of plant species. Recently the scientific community have shown interest in research on endophytic actinobacteria due to produce novel and host-origin natural compounds, and various other benefits like growth enhancement and herbivore resistance.

Endophytic actinobacteria may be of two types "obligate" and "facultative".The growth of obligate endophytes depend on the host plant. Facultative endophytes can exist outside the host plant [22]. Endophytic actinobacteria have been isolated from different plant parts, such as roots [35, 56], stem [57], leaves [58], and fruits [59] . Endophytic actinobacteria in plants are found to produce different types of metabolites that can be used for different applications, such as antimicrobials [28], plant growth promoters [25], and biocontrol agents [27]. According to Passari et al. [60], the presence of PKS/NRPS gene clusters in endophytic actinobacteria is responsible for secondary metabolite biosynthesis. Endophytic actinobacteria are reported to produce several plant growth promotion compounds such as auxins, cytokinins, and gibberellins or producing siderophore to improve nutrient uptake [61, 62]. Coombs and Franco [63] reported that different strains of actinobacteria including *Microbispora, Nocardia, Streptomyces* were recognized from the tissues of vigorous wheat plants. *Streptomyces aureofaciens* is one of the endophytic actinobacteria which were obtained from the root of *Zingiber Officinale* and that endophyte was found to inhibit the growth of *Candida albicans* [64]. Endophytic actinobacteria form a symbiotic relationship by the formation of the colony within plant cells to get nutrition, shelter from host plants and in return, they produce several secondary compounds which is used by the plant for its growth and productivity [65]. These metabolites prevent the growth of other harmful pathogens in host plants. According to Loria et al. [66], *Streptomyces* species can produce active secondary metabolites like antibiotics. According to Nalini and Prakash [67], Masand et al. [41], endophytic actinobacteria are diversely distributed in the ecosystem. In China, Qin et al. [68] studied different strains of endophytic actinobacteria that were recognized from several medicinal and crop plants. *Streptomyces* spp. and non-*Streptomyces* spp. are the two types of endophytic actinobacteria. Yandigeri et al. [69] reported in the plants of arid regions drought tolerant endophytic actinobacteria like *Streptomyces olivaceus* DE10, *Streptomyces geysiriensis* DE27 and *Streptomyces coelicolor* DE07. Information on the diversity of endophytic actinobacteria and their organ-specificity is significant for helping in the screening of beneficial strains and also for understanding their ecological roles. It is reported that most of the endophytic actinobacteria are generally available in roots than in other plant parts [70, 71]. The density of endophytic actinobacteria in wheat roots was demonstrated by Conn and Franco [72].

Endophytes are reported in plants that are growing in tropical and temperate forests with the hosts ranging from herbaceous plants in various habitats such as extreme arctic, alpine, and xeric environments. Many studies have reported that endophytic actinobacteria are found in different types of plant tissue such as seeds and ovules, fruits, stems, roots, root nodes, leaves, flowers, tubers, buds, xylem, rachis, and bark [60, 73].

3. Plant growth promoting (PGP) activities byendophytic actinobacteria

Roots are the most favorite part of the plants to be colonized by the microbes. Such interaction between the plants and the microbes may result in an endosymbiotic relationship between them. In many cases, the endophytic microbes play a significant role in the protection of plants against pathogenic agents [74, 75]. Studies have been performed with endophytes by inoculating the host plant with endophytes [76] for evaluation of the colonization pattern of vegetative tissues and the effect of endophytes on the host plant. This technique comprehends plant biology and microbial ecology [74].

Actinobacteria are found as symbionts or parasites within plants. According to Hallmann et al. [77], endophytic actinobacteria usually originated from epiphytic actinobacteria colonizing soil, and through any wound or opening on the plant surface, they might have got the opportunities to enter the plant tissues and become endophytes. Individual bacterial cells are not able to penetrate intact epidermal cells as they do not posses mycelium like fungi while actinobacteria colonize on the external part and grow on plant surface by forming branching hyphae and penetrate through natural or by mechanical openings injury [78]. Petrini et al. [54] suggested that the endophytes produce enzymes that are able to degrade

most substrates present on the surfaces or in the cell wall of the host. According
to Gohain et al. [79], colonization of endophytic actinobacteria is influenced by
different climatic conditions and the rate of colonization is high in summer than in
winter. The genera *Microbispora, Micromonospora, Saccharopolyspora, Micrococcus,
Amycolatopsis, Microbacterium, and Nocardia* were isolated only in summer; however, the genus *Streptomyces* was often isolated in both the seasons. By producing
plant hormones, fixing nitrogen and by preventing the growth of phytopathogens,
endophytes help to increase plant growth. Antibiotics are produced by endophytes
with the help of an induced resistance system [80, 81]. Endophytic actinobacteria
can offer the opportunity for further research aimed at understanding the correlation between the metabolism of plants and their endophytes.

Stimulation of plant growth by endophytic actinobacteria are of two types,
direct and indirect. In the first mechanism, phytohormones such as IAA, cytokinins
are produced along with solubilization of minerals like iron, and phosphorus by the
production of siderophores for enhancing plant nutrition and 1-aminocyclopro-
pane-1-carboxylic acid (ACC) deaminase [82]. Indirectly endophytic actinobacteria
help the plants as a biocontrol agent. They can destroy the harmful phytopathogen
by stimulating the resistance system of the plant. Besides it, they can also produce
extracellular enzymes which can lysis the cell wall of dangerous fungus [83].
Different unique secondary metabolites have been produced from endophytes that
are associated with medicinal plants and these secondary metabolites can be applied
in pharmaceutical, agricultural and other industries. According to Cattelan et al.
[84], endophytic actinobacteria can increase plant growth promotion by several
way. They are able to form phytohormones, they can fix nitrogen also and they can

Plant growth-promoting attributes	Endophytic Actinobacteria	Isolated from	References
IAA, siderophore	*Streptomyces* sp. CMU PA 101	*Carcuma mangga*	Khamna et al. [10]
IAA, hydrxymate and catechol type siderophore, protease	*Streptomyces* sp. S4202, *Nonomuraea* sp. S3304, *Actinomadura* sp.S4215	*Aquilaria crassna*	Nimnoi et al. [62]
Solubilization of phosphate	*Streptomyces* sp. Nhcr0816	*Triticum aestivum*	El-Tarabily et al. [87]
Production of IAA and ACC deaminase	*Actinoplanes campanulatus, Streptomyces spirilis*	*Cucumis sativus*	El-Tarabily et al. [87]
Production of chitinase, phosphatase activity, and siderophore	*Streptpmyces* sp. AB131–1,LBRO2	Isolates of microbiology laboratory, Bogal Agricultural University	Hastuti et al. [88]
Siderophore production	*Streptomyces* sp. GMKU3100	*Oriza sativa* L.cv. KDML 105	Rungin et al. [89]
Production of IAA	*Streptomyces* sp. PT2	Plants of Algerian Sahara	Goudjal et al. [90]
IAA production	*Streptomyces* sp. *Nocardia* sp.	*Citrus reticulata*	Shutsrirung et al. [25]
Solubilization of phosphate, production of siderophores	*Streptomyces* sp. BPSAC34	Medicinal plants	Passari et al. [60]
Phosphate solubilization, siderophore production	*Streptomyces* sp. UKCW/B, *Nocardia* sp. TP1BA1B	*Pseudowintera colorata*	Purushotham et al. [91]

Table 1.
Plant growth promotion by endophytic actinobacteria.

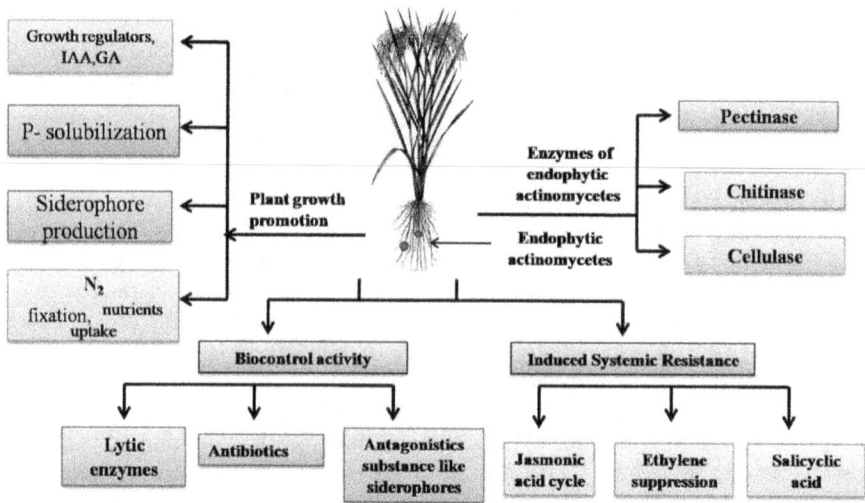

Figure 1.
Plant-endophytic actinomycetes interactions favoring plant growth and biocontrol of phytopathogens [92].

prevent the growth of phytopathogen by their antagonism activity and they help in the solubilization of phosphate also.

Numerous actinobacterial species such as endophytes with plants have been reported to have various plant growth-promoting (PGP) properties [85] s. They have also been found to show antagonistic properties against many root-borne and disease-causing plant pathogens [86]. Plant growth-promoting actinobacteria (PGPA) have been reported to be mostly endophytic (**Table 1**). Plant growth-promoting attributes have been presented in **Figure 1**.

3.1 Production of plant growth hormone - indole acetic acid (IAA) by endophytic actinobacteria

In leguminous plants and in cereals, endophytic actinobacteria function as a plant growth promoter; as a result, they have the capacity to influence plant growth and can increase the ability of nutrition absorption by plants [85].

According to Khamna et al. [93], Palaniyandi et al. [94], indole acetic acid (IAA) is a highly reported growth regulator which is produced by endophytic actinobacteria. The naturally-occurring auxin, indole-3- acetic acid (IAA) is produced by plants through different tryptophan-dependent IAA production pathways and also by bacteria and fungi [95]. The type of pathway that bacterium uses to produce IAA within plants can determine the nature of the resulting plant-microbe interactions [22]. The primary form of auxin is indole-3-acetic acid (IAA) which have an important contribution to control the different cellular process of plants. IAA helps in elongation, cell division. To form the root hair and to make short root length IAA performs very important functions. IAA helps to increase the nutrient absorption ability of the plant. Some strains of endophytic actinobacteria were reported to produce IAA to enhance the growth of cucumber plants [4, 5]. Passari et al. [96] reported various strains of actinobacteria including *Micromonospora, Streptomyces, Microbacterium, Pseudonocardia* which can produce plant growth phytohormone IAA. According to Madhurama et al. [97], actinobacteria *Streptomyces* sp. has the capability to produce IAA in the high range. In another study by Khamna et al. [93], it is reported that in many medicinal plants IAA was produced by *Streptomyces* sp.

It is reported that to improve plant growth, genus *Streptomyces,* such as *Streptomyces olivaceoviridis, S. rimosus, S. rochei, and Streptomyces spp.* have the ability to produce IAA from tomato rhizosphere [98, 99]. According to Verma et al. [100], *Streptomyces* strains of endophytic actinobacteria were obtained from *Azadirachta indica* and in tomato plants, they were found to increase the plants' growth. *Streptomyces* strain En-1 had been studied to produce IAA and to stimulate the growth of *Arabidopsis* plantlets [101]. Many endophytic, as well as rhizo-spheric actinobacteria, possess the ability to produce IAA, cytokinins, and GA3 [102]. Nimnoi et al. [62] reported that endophytic actinobacteria from eaglewood (*Aquilaria crassna*) had shown a trait of plant growth promotion by the production of indole-3-acetic acid (IAA) and ammonia. IAA and siderophore-producing actinobacteria that colonize the root in the rhizosphere are studied to promote root elongation and plant growth [103]. Several endophytic actinobacteria including *Streptomyces viridis, S. rimosus, S.olivaceoviridis, S. atrovirens, and S. rochei* have been exhibited to improve germination as well as root and shoot elongation [104].

3.2 Phosphate solubilization

Phosphorus is an important component's that is involved in a wide range of cellular processes by developing plant organs and increasing cell enlargement in plants [105].

Phosphorus (P) content is generally very low in soil and it is available in the form of insoluble metallic complexes. For that reason, plants can absorb from soil, a little amount of phosphorus for their growth [106]. Endophytic actinobacteria support the plants to get phosphorus in soluble form through acidification and mineralization of insoluble soil phosphorus to increase the growth of plants [107, 108]. According to Jog et al. [109], endophytic actinobacteria *Streptomyces* sp. obtained from *Triticum aestivum* was found to soluble the phosphate to promote the plant growth.

Various genera of actinobacteria such as *Streptomyces, Rhodococcus, Arthrobacter, Micromonospora* were reported to have P-solubilization potential under *in vivo* as well as *in vitro* [109]. Under P-deficient soils, *Streptomyces griseus, Micromonospora auranti-aceae* has been reported to help in the P-solubilization of wheat crop [109]. According to Hamdali et al. [110], actinobacteria such as *S. griseus, Micromonospora aurantiaca* were found to soluble rock phosphate to stimulate the plant growth of wheat. In a recent study, it is reported that endophytic actinobacteria *Nocardia* sp. TP1BA1B and *Streptomyces* sp. UKCW/B isolated from the native medicinal plant *Pseudowintera colorata* (Horopito) were found to solubilize phosphate in New Zealand [91].

3.3 Production of siderophore and enhanced iron availability by endophytic actinobacteria

Siderophores are iron-chelating secondary metabolites produced by various microorganisms in order to scavenge iron from their surrounding environment to make this essential element available to the cell. Due to the high affinity for ferric iron, siderophores are secreted out to form soluble ferric complexes that can be taken up by the organisms. According to Bothwell [111], iron plays an important role in the physiological processes of plants. It is available in the soil as insoluble Fe^{3+} form and plants need soluble Fe2$^+$ form to uptake from soil [112]. Actinobacteria can converts iron from Fe^{3+} to Fe^{2+} form and it can increase the bioavailability of iron in the plant rhizosphere by the production of siderophores and help the plant uptake of iron.

The mechanism of siderophore was reported by endophytic actinobacteria to stimulate plant growth [100]. *Streptomyces acidiscabies* E13 is an excellent example of siderophore producer that promotes the growth of *Vigna unguiculata* under abiotic stress conditions [113]. Several recent studies demonstrated the production of plant growth-promoting compounds such as siderophores *in vitro* by endophytic actinobacteria [114, 115]. Khamna et al. [93] studied to produce a high amount of siderophore by *Streptomyces* CMU-SK 126 that was isolated from *Curcuma mangga* in rhizospheric soil.

3.4 ACC deaminase producing strains of endophytic actinobacteria

The enzyme ACC deaminase can cleave the plant ethylene precursor ACC, and thereby lower the level of ethylene in a developing or stressed plant [116]. Under unfavorable conditions, plant growth becomes reduced and, in that condition, bacterial ACC deaminase performs an important function to increase the plant growth [117]. Nascimento et al. [118] reported that actinobacteria including *Mycobacterium, Streptomyces, Rhodococcus* were found to contain ACC deaminase producing genes [118].

By the study, it was proved that when ACC deaminase producing endophytic *Streptomyces* sp. GMKU 336 was inoculated into Thai jasmine rice Khao Dok Mali 105 cultivar (*Oryza sativa* L. cv. KDML105), *Streptomyces* sp. GMKU 336 significantly increased plant growth and decreased ethylene under salt stress (150 mM NaCl) conditions. This work demonstrates that ACC deaminase produces *Streptomyces* sp. GMKU 336 enhances the growth of rice and increases salt tolerance by reduction of ethylene by the action of ACC deaminase [119].

4. Endophytic actinobacteria as biocontrol agents

According to Lee et al. [120], endophytic actinobacteria such as *Microbispora rosea, Streptomyces olivochromogenes* prevented the growth of phytopathogen of clubroot of Chinese cabbage effectively. Coombs et al. [121] examined the endophytic actinobacteria as a biocontrol agent against *Gaeumannomyces graminis var. tritici* of wheat. The endophytic actinobacteria can control *Pythium aphanidermatum* in cucumber which was described by El-Tarabily et al. [4, 5]. Cao et al. [122] reported that endophytic actinobacteria such as *Streptomyces spiralis*, *Micromonospora chalcea* were isolated from cucumber root. They were found to promote plant growth by decreasing plant disease like damping off and crown rot. They were identified as biocontrol agents due to the formation of enzymes that can destroy the cell wall of fungal phytopathogen. These endophytic actinobacteria significantly reduced the incidence of damping-off, crown, and root-rot of cucumber roots. Phytopathogenic fungus *Sclerotinia sclerotiorum* causes stem rot which is a very harmful disease for economically important crops like soybean and sunflower worldwide [123]. *Streptomyces* sp. NEAU-S7GS2 was obtained from the root cells of *Glycine max*. In a study, it was observed that the mycelial growth and germination of *S. sclerotiorum* (99.1%) were inhibited by *Streptomyces* sp. NEAU-S7GS2 [124]. Shimizu et al. [81] first reported the powerful activity of endophytic actinobacteria biocontrol agent to decrease the foliar disease. The strain MBR-5 identified as *Streptomyces galbus*, among ten actinobacterial strains, isolated from field-grown *Rhododendron* plants showed significant antagonistic activities against *Phytophthora cinnamomi* and *Pestalotiopsis sydowiana*. According to Cao et al. [122], the growth of plant pathogens was prevented by endophytic actinobacteria to save the host plant from the attack of harmful microbes. Strain CEN26, an endophyte was isolated from *Centella asiaticato* and the strain was found to inhibit the germination of conidia and morphological development of the fungal pathogen *Alternaria brassicicola* [73]. It was studied that most of the endophytic actinobacteria were seen to protect the hosts from diseases by inhibiting plant pathogens [94]. In pot experiments, it was observed that the extract of *Streptomyces* sp. MR14 cells significantly suppressed *Fusarium moniliforme* [125].

Maggini et al. [126] also discussed the relationship between actinobacteria and their host plants to protect the host from the disease that is caused by the phytopathogen.

According to Wan et al. [127], leaf blight disease of rice was suppressed by *Streptomyces platensis*. In a study, the inhibition activities against various phytopathogens such as *Neonectria ditissima* ICMP 14417, *Ilyonectria liriodendri* WPA1C,

Endophytic actinobacteria	Host plant	Pathogen	References
Streptomyces sp. S30	*Solanum lycopersicum*	*Rhizoctonia solani*	Cao et al. [122]
Streptomyces halstedii	*Capsicum*	*Phytophthora capsica*	Liang et al. [128]
Microbispora sp. A004 and A011	*Brasica rapa*	*Plasmophora brassicae*	Lee et al. [120]
Streptomyces sp. KH-614	*Oryza sativa*	*Pyricularia oryzae*	Ningthoujam et al. [129]
Streptomyces sp. S30 *Streptomyces* sp. R18(6)	*Lycopersicon esculentum*	*Rhizoctonia solani*	De Olivera et al. [114]
Streptomyces spiralis *Microsmonopora chalcea*	*Cucumis* sp.	*Pythium aphanidermatum*	El-Tarabily et al. [87]
Streptomyces sp.	*Cicer arietinum*	*Fusarium oxysporum* f. sp. *ciceri*	Gopalakrishnan et al. [130]
Streptomyces sp. AzR-051, AzR – 049	*Azadirachta indica* A. Juss	*Alternaria alternata*	Verma et al. [100]
Streptomyces sp.	*Capsicum frutescens*	*Alternaria brassicae, Colletotrichum gloeosporioides*	Srividya et al. [16]
Streptomyces sp.	Soybean	*Xanthomonas campestris* pv. *glycines*	Mingma et al. [131]
Streptomyces indiaensis KJ872546	*Capsium*	*Fusarium oxysporum*	Jalaluldeen et al. [132]
Actinobacteria strains OUA3, OUA5, OUA18, and OUA40	*Capsicum annuum*	*Colletotrichum capsici* and *Fusarium oxysporum*	Ashokvardhan et al. [133]
Streptomyces felleus YJ1	*Brrasica napus*	*Sclerotinia sclerotiorum*	Cheng et al. [134]
Streptomyces cyaneus ZEA171	*Lactuca sativa*	*Sclerotinia sclerotiorum* FW361	Kunova et al. [135]
Streptomyces diastaticus, Streptomyces fradiae, Streptomyces collinus	Medicinal plants	*Sclerotium rolfsii, Rhizoctonia solani, Fusarium oxysporum*	Singh and Gaur [136]
Streptomyces sp. DBT204	*S. lycopersicum*	*Fusarium proliferatum*	Passari et al. [137]
Streptomyces humidus	*Brassica oleracea*	*Alternaria brassicicola*	Hassan et al. [138]
Saccharothrix algeriensis NRRL B-24137	*S. lycopersicum*	*Fusarium oxysporum*	Merrouche et al. [139]
Streptomyces sp.PRY2RB2	*Pseudowintera colorata*	*Neofusicoccum luteum* ICMP 16678	Purushotham et al. [91]

Table 2.
Endophytic actinobacteria as biocontrol agents.

Neofusicoccum luteum ICMP 16678 were shown by *Streptomyces* sp. PRY2RB2 [91]. Endophytic actinobacteria as biocontrol agents have been enlisted in **Table 2**. Prominent antagonistic potential against *Rhizoctonia solani* was found by *Streptomyces avidinii* vh32, *S. toxybicini* vh22, and *S. tricolor* vh85 which also induced the accumulation of phenolic compounds in tomato [140]. From neem (*A. indica*), endophytic actinobacteria were isolated by Verma et al. [70]. The most common genera were *Streptomyces, Streptosporangium, Microbispora, Streptoverticillium, Sacchromonospora,* and *Nocardia*, which showed antagonistic activities against root pathogens *Pythium* and *Phytophthora* sp.

The growth of the fungal pathogen *Alternaria alternata* was inhibited by endophytic actinobacteria isolated from the medicinal plant *Ferula sinkiangensis* [141]. 72 strains endophytic actinobacteria isolated from the medicinal plant *Rhynchotoechum ellipticum*, were found to inhibit the growth of *Fusarium proliferatum, F. oxysporum.* Different strains of *streptomyces* sp. such as *S. olivaceus, Streptomyces* sp. BPSAC121, *Streptomyces* sp. BPSAC101 showed antifungal activities. Antifungal antibiotics, fluconazole, ketoconazol and miconazole are produced from *S. olivaceus* and *Streptomyces* sp. BPSA 121 [96]. Endophytic *Streptomyces* sp. showed antifungal activity against *Geotrichum candidum, F. oxysporum, Alternaria* sp. [142]. According to Passari et al. [137], the growth of various phytopathogens including *Fusarium Oxysporum, Fusarium graminearum, Rhizoctonia solani, Colletotrichum capsici* were inhibited by endophytic actinobacteria such as *Nocardiopsis* sp., *Streptomyces* sp. DBT204, *Streptomyces* sp. DBT 207 by the formation of cell wall degrading enzymes and HCN. Some species of *Streptomyces* exhibit biological control activity by stimulating the plant resistance system or by the formation of secondary metabolites like antibiotics, particularly against phytopathogenic fungi such as *Fusarium oxysporum, Pythium ultimum, Phytophthora* sp. [82]. Biocontrol potentials of endophytic actinobacteria against different phytopathogens have been presented schematically in **Figure 1**.

4.1 Induction of resistance in the host by endophytic actinobacteria

Conn et al. [143] reported that by inducing system acquired resistance (SAR) and jasmonic acid (JA) or ethylene (ET) pathways, the endophytic actinobacteria were able to induce resistance against *Erwinia carotovora* and *Fusarium oxysporum* respectively. Conn et al. [143] reviewed that the growth of the pathogen *Botrytis cinerea* was inhibited by endophytic actinobacteria *Streptomyces* sp. GB4–2 by stimulating the SAR pathway. *Streptomyces* has been found to induce host plant resistance on various crops such as vegetables, forages, and eucalyptus [144]; oak [145]. Actinobacteria can act as an antagonist against pathogens due to the production of lytic enzymes that are capable of destroying fungal cell wall. Many researchers have reported the enzyme activity of actinobacteria which can prevent the growth of fungus by destroying the cell wall with their extracellular enzymes like cellulase, chitinase, amylase, etc. [146]. Taechowisan et al. [147] reported the production of chitinase from endophytic *Streptomyces aureofaciens* CMUAC130. Srividya et al. [16] discussed the enzyme activity (chitinase, glucanase) of *Streptomyces* sp. to suppress the growth of fungal phytopathogen.

Endophytic actinobacterium- *Streptomyces* sp. showed hyperparasitic activity. Compant et al. [61] reported the antimicrobial activity of strain NRRL 30562 to prevent the growth of fungal pathogens such as *Fusarium oxysporum, Pythium ultimum* by producing an antibiotic munumbicins. The strain was obtained from *Kennedia nigriscansin vitro.*

The extracellular enzymes- β-1,3-glucosidase, cellulase, and protease; produced by endophytic actinobacteria cause the lysis of hyphae to inhibit the growth of phytopathogens [148]. Hydrolytic enzymes degrade fungal cell wall, cell membrane, and extracellular virulence factors to control plant diseases [149]. According to Yandigeri et al. [69], actinobacteria produced chitinases to inhibit the growth of fungal pathogens. The extracellular antifungal metabolites especially chitinase and β-1,3 glucanase; produced by actinobacteria inhibited the growth of fungi through hyphal swelling, lysis of cell walls in *Fusarium oxysporum,* and *Sclerotium rolfsii* [150].

Endophytes are found to produce secondary metabolites, which are active at low concentrations against other microorganisms [151]. A large number of antimicrobial compounds belonging to the classes like alkaloids, peptides, steroids, terpenoids, phenols, quinines, and flavonoids were found to produce from endophytic actino-bacteria [152]. Endophytic actinobacteria were found to show antimicrobial activity against phytopathogenic fungi [153]. Another *Streptomyces* NRRL 30562, isolated from the snake vine possessed activity against many pathogenic fungi [154].

5. Conclusion and future prospects

Actinobacteria can enhance plant growth by producing growth regulators and other compounds and it is well known as a biocontrol agent for the production of antibiotics. Other properties like the production of cell wall degrading enzymes and induced systemic resistance can inhibit the growth of new plant pathogens. This review has been focused on the importance of endophytic actinobacteria as they are widely regarded as an excellent source for plant growth promotion and biocontrol agents by various mechanisms like increasing the supply of nutrients, and produc-tion of IAA, cytokinin, controlling fungal diseases through antibiosis and competi-tion. The excessive use of agrochemical is harmful for the environment. The use of biocontrol agents for the management of plant disease is very important. It is very important to review and highlight the previous achievements in endophytic research in order to draw the attention of the research community towards this emerging field. As endophytic actinobacteria help to increase plant growth, so the utilization of actinobacteria can be developed as another way for suitable organic and environmentally helpful agricultural crop production.

Author details

Sumi Paul and Arka Pratim Chakraborty*
Department of Botany, Raiganj University, Raiganj, India

*Address all correspondence to: arka.botanyrgu@gmail.com

IntechOpen

References

[1] Atkinson NJ, Urwin PE. The interaction of plant biotic and abiotic stresses: From genes to the field. Journal of Experimental Botany. 2012;**63**(10):3523-3543

[2] Djebaili R, Pellegrini M, Smati M, Del Gallo M, Kitouni M. Actinomycete strains isolated from saline soils: Plant-growth-promoting traits and inoculation effects on *Solanum lycopersicum*. Sustainability. 2020;**12**(11):4617

[3] Vurukonda SSKP, Giovanardi D, Stefani E. Plant growth promoting and biocontrol activity of *Streptomyces* spp. as endophytes. International Journal of Molecular Sciences. 2018;**19**(4):952

[4] El-Tarabily KA, Nassar AH, Hardy G, Sivasithamparam K. Plant growth promotion and biological control of *Pythium aphanidermatum*, a pathogen of cucumber, by endophytic actinomycetes. Journal of Applied Microbiology. 2009b;**107**:672-681

[5] El-Tarabily KA, Nassar AH, Hardy GES, J. and Sivasithamparam, K. Plant growth promotion and biological control of *Pythium aphanidermatum*, a pathogen of cucumber, by endophytic actinomycetes. Journal of Applied Microbiology. 2009a;**106**(1):13-26

[6] Nimaichand S, Tamrihao K, Yang LL, Zhu WY, Zhang YG, Li L, et al. *Streptomyces hundungensis* sp. nov., a novel actinomycete with antifungal activity and plant growth promoting traits. The Journal of Antibiotics. 2013;**66**(4):205-209

[7] Sadeghi A, Karimi E, Dahaji PA, Javid MG, Dalvand Y, Askari H. Plant growth promoting activity of an auxin and siderophore producing isolate of *Streptomyces* under saline soil conditions. World Journal of Microbiology and Biotechnology. 2012;**28**(4):1503-1509

[8] Barea JM, Pozo MJ, Azcon R, Azcon-Aguilar C. Microbial co-operation in the rhizosphere. Journal of Experimental Botany. 2005;**56**(417):1761-1778

[9] Franco-Correa M, Quintana A, Duque C, Suarez C, Rodríguez MX, Barea JM. Evaluation of actinomycete strains for key traits related with plant growth promotion and mycorrhiza helping activities. Applied Soil Ecology. 2010;**45**(3):209-217

[10] Khamna S, Yokota A, Lumyong S. Actinobacteriaisolated from medicinal plant rhizosphere soils: Diversity and screening of antifungal compounds, indole-3-acetic acid and siderophore production. World Journal of Microbiology and Biotechnology. 2009;**25**(4):649-655

[11] Chaurasia A, Meena BR, Tripathi AN, Pandey KK, Rai AB, Singh B. Actinomycetes: An unexplored microorganisms for plant growth promotion and biocontrol in vegetable crops. World Journal of Microbiology and Biotechnology. 2018;**34**(9):1-16

[12] Pandey A, Ali I, Butola KS, Chatterji T, Singh V. Isolation and characterization of actinobacteria from soil and evaluation of antibacterial activities of actinobacteria against pathogens. International Journal of Applied Biology and Pharmaceutical Technology. 2011;**2**(4):384-392

[13] Qin S, Li J, Chen H, Guozhen Z, Zhu WY, Jiang CL, et al. Isolation, diversity and antimicrobial activity of rare actinobacteria from medicinal plants of tropical rain forests in Xishuangbanna, China. Applied and Environmental Microbiology. 2009;**75**:6176-6186

[14] Tyc O, Song C, Dickschat JS, Vos M, Garbeva P. The ecological role of volatile

and soluble secondary metabolites produced by soil bacteria. Trends in Microbiology. 2017;**25**(4):280-292

[15] Jog R, Nareshkumar G, Rajkumar S. Enhancing soil health and plant growth promotion by actinomycetes. In: Plant Growth Promoting Actinobacteria book. Springer; 2016. pp. 33-45

[16] Srividya S, Thapa A, Bhat D, Golmei K, Dey N. *Streptomyces* sp. 9p as effective biocontrol against chilli soil borne fungal phytopathogens. European Journal of Experimental Biology. 2012;**2**(1):163-173

[17] Olanrewaju OS, Babalola OO. Streptomyces: Implications and interactions in plant growth promotion. Applied Microbiology and Biotechnology. 2019;**103**(3):1179-1188

[18] Sturz AV, Christie BR, Nowak J. Bacterial endophytes: Potential role in developing sustainable systems of crop production. Critical Reviews in Plant Sciences. 2000;**19**:1-30

[19] Wilson D. Endophyte: The evolution of a term, and clarification of its use and definition. Oikos. 1995;**73**(2):274

[20] Wilson D. Fungal endophytes: Out of sight but should not Be out of mind. Oikos. 1993;**68**(2):379

[21] Singh MJ, Sedhuraman P. Biosurfactant, polythene, plastic, and diesel biodegradation activity of endophytic *Nocardiopsis* sp. mrinalini9 isolated from *Hibiscus rosasinensis* leaves. Bioresources and Bioprocessing. 2015;**2**:1-7

[22] Hardoim PR, van Overbeek LS, van Elsas JD. Properties of bacterial endophytes and their proposed role in plant growth. Trends in Microbiology. 2008;**16**:463-471

[23] Van Wees SC, Van der Ent S, Pieterse CM. Plant immune responses triggered by beneficial microbes. Current Opinion in Plant Biology. 2008;**11**:443-448

[24] Borah A, Thakur D. Phylogenetic and functional characterization of culturable endophytic actinobacteria associated with *camellia* spp. for growth promotion in commercial tea cultivars. Frontiers in Microbiology. 2020;**11**(318):318. DOI: 10.3389/fmicb.2020.00318

[25] Shutsrirung A, Chromkaew Y, Pathom-Aree W, Choonluchanon S, Boonkerd N. Diversity of endophytic actinobacteria in mandarin grown in northern Thailand, their phytohormone production potential and plant growth promoting activity. Soil Science and Plant Nutrition. 2013;**59**(3):322-330

[26] Alblooshi AA, Purayil GP, Saeed EE, Ramadan GA, Tariq S, Altaee AS, et al. Biocontrol potential of endophytic actinobacteria against *fusarium solani*, the causal agent of sudden decline syndrome on date palm in the UAE. Journal of Fungi. 2022;**8**:8. DOI: 10.3390/jof8010008

[27] Li X, Huang P, Wang Q, Xiao L, Liu M, Bolla K, et al. Staurosporine from the endophytic *Streptomyces* sp. strain CNS-42 acts as a potential biocontrol agent and growth elicitor in cucumber. Antonie Van Leeuwenhoek. 2014;**106**(3):515-525

[28] Ding WJ, Zhang SQ, Wang JH, Lin YX, Liang QX, Zhao WJ, et al. A new di-O-prenylated flavone from an actinomycete *Streptomyces* sp. MA-12. Journal of Asian Natural Product Research. 2013;**15**:209-214

[29] Igarashi Y, Ogura H, Furihata K, Oku N, Indananda C, Thamchaipenet A. Maklamicin, an antibacterial polyketide from an endophytic *Micromonospora* sp. Journal of Natural Products. 2011;**74**(4):670-674

[30] Yan LL, Han NN, Zhang YQ, Yu LY, Chen J, Wei YZ, et al. Antimycin A18 produced by an endophytic *Streptomyces albidoflavus* isolated from a mangrove plant. Journal of Antibiotics. 2010;**63**:259-261

[31] Dudeja SS, Giri R, Saini R, Suneja-Madan P, Kothe E. Interaction of endophytic microbes with legumes. Journal of Basic Microbiology. 2012;**52**(3):248-260

[32] Chen HH, Zhao GZ, Park DJ, Zhang YQ, Xu LH, Lee JC, et al. *Micrococcus endophyticus* sp. nov., isolated from surface-sterilized *Aquilaria sinensis* roots. International Journal of Systematic and Evolutionary Microbiology. 2009;**59**(5):1070-1075

[33] Qin S, Zhu WY, Jiang JH, Klenk HP, Li J, Zhao GZ, et al. *Pseudonocardia tropica* sp. nov., an endophytic actinomycete isolated from the stem of *Maytenus austroyunnanensis*. International Journal of Systematic and Evolutionary Microbiology. 2010;**60**(11):2524-2528

[34] Rachniyom H, Matsumoto A, Indananda C, Duangmal K, Takahashi Y, Thamchaipenet A. *Actinomadurasyzygii* sp. nov., an endophytic actinomycete isolated from roots of a jambolan plum tree (*Syzygiumcumini* L. Skeels). International Journal of Systematic and Evolutionary Microbiology. 2015;**65**:1946-1949

[35] Shen Y, Zhang Y, Liu C, Wang X, Zhao J, Jia F, et al. *Micromonosporazeae* sp. nov., a novel endophytic actinomycete isolated from corn root (*Zea mays* L.). The Journal of Antibiotics. 2014;**67**(11):739-743

[36] Castillo UF, Browne L, Strobel G, Hess WM, Ezra S, Pacheco G, et al. Biologically active endophytic *Streptomycetes* from *Nothofagus* spp. and other plants in Patagonia. Microbial Ecology. 2007;**53**(1):12-19

[37] Rosenblueth M, Martínez-Romero E. Bacterial endophytes and their interactions with hosts. Molecular Plant-Microbe Interactions. 2006;**19**:827-837

[38] Oskay M, Usame T, A. and Azeri, C. Antibacterial activity of some actinobacteriaisolated from farming soils of Turkey. African Journal Biotechnology. 2004;**3**(9):441-446

[39] Basilio A, Gonzalez I, Vicente MF, Gorrochategui J, Cabello A, Gonzalez A, et al. Patterns of antimicrobial activities from soil actinobacteriaisolated under different conditions of pH and salinity. Journal of Applied Microbiology. 2003;**95**(4):814-823

[40] Wang Y, Zhang ZS, Ruan JS, Wang YM, Ali SM. Investigation of actinomycete diversity in the tropical rainforests of Singapore. Journal of Industrial Microbiology and Biotechnology. 1999;**23**(3):178-187

[41] Masand M, Jose PA, Menghani E, Jebakumar SRD. Continuing hunt for endophytic actinobacteriaas a source of novel biologically active metabolites. World Journal of Microbiology and Biotechnology. 2015;**31**(12):1863-1875

[42] Ababutain IM, Aziz ZKA, Al-Meshhen NA. Lincomycin antibiotic biosysthesis produced by *Streptomyces* sp. isolated from Saudi Arabia soil: Taxonomical, antimicrobial and insecticidal studies on the producing organism. Canadian. Journal of Pure and Applied Sciences. 2012;**6**(1):1739

[43] Mohseni M, Norouzi H, Hamedi J, Roohi A. Screening of antibacterial producing Actinobacteria from sediments of the Caspian Sea. International Journal of Molecular and Cellular Medicine. 2013;**2**(2):64-71

[44] Srinivasan MC, Laxman RS, Deshpande MV. Physiology and nutritional aspects of actinomycetes: An

overview. World Journal of Microbiology and Biotechnology. 1991;7:171-184

[45] Sprusansky O, Stirrett K, Skinner D, Denoya C, Westpheling J. The bkd R gene of *Streptomyces coelicolor* is required for morphogenesis and antibiotic production and encodes a transcriptional regulator of a branched-chain amino acid dehydrogenase complex. Journal of Bacteriology. 2005;**187**(2):664-671

[46] George M, Anjumol A, George G, Mohamed Hatha AA. Distribution and bioactive potential of soil actinobacteria from different ecological habitats. African Journal of Microbiology Research. 2012;**6**:2265-2271

[47] Kuster E. The actinomycetes. In: Burges A, Raw F, editors. Soil Biology. London: Academic Press. pp. 111-124

[48] Walker JD, Colwell RR. Factors affecting enumeration and isolation of actinobacteriafrom Chesapeake Bay and southeastern Atlantic Ocean sediments. Marine Biology. 1975;**30**(3):193-201

[49] Takami H, Inoue A, Fuji F, Horikoshi K. Microbial flora in the deepest sea mud of the Mariana trench. FEMS Microbiology Letters. 1997;**152**(2):279-285

[50] Williams ST, Goodfellow M, Alderson G. Genus *Streptomyces*. Waksman and Henrici 1943, 399AL. Vol. 2. New York: Springer Verlag; 1989. pp. 2028-2090

[51] Takahashi Y, Matsumoto A, Seino A, Iwai Y, Omura S. Rare ActinobacteriaIsolated from desert soils. Actinomycetologica. 1996;**10**(2):91-97

[52] Leuchtmann A. Systematics, distribution, and host specificity of grass endophytes. Natural Toxins. 1993;**1**(3):150-162

[53] Shen FT, Yen JH, Liao CS, Chen WC, Chao YT. Screening of rice endophytic biofertilizers with fungicide tolerance and plant growth-promoting characteristics. Sustainability. 2019;**11**:1133

[54] Petrini O, Sieber TN, Toti L, Viret O. Ecology, metabolite production, and substrate utilization in endophytic fungi. Natural Toxins. 1992;**1**(3):185-196

[55] Singh R, Dubey AK. Endophytic actinobacteria as emerging source for therapeutic compounds. Indo Global Journal of Pharmaceutical Science. 2015;**5**:106-116

[56] Indananda C, Thamchaipenet A, Matsumoto A, Inahashi Y, Duangmal K, Takahashi Y. *Actinoallomurus oryzae* sp. nov., an endophytic actinomycete isolated from roots of a Thai jasmine rice plant. International Journal of Systematic and Evolutionary Microbiology. 2011;**61**(4):737-741

[57] Gu Q, Zheng W, Huang Y. *Glycomyces sambucus* sp. nov., an endophytic actinomycete isolated from the stem of *Sambucus adnata* wall. International Journal of Systematic and Evolutionary Microbiology. 2007;**57**(9):1995-1998

[58] Kafur A, Khan AB. Influence of cultural parameters on antimicrobial activity of endophytic *Streptomyces* sp. Cr 12 from Catharanthus roseus leaves. International Journal of Drug Delivery. 2011;**3**:425-431

[59] Du H, Su J, Yu L, Zhang Y. Isolation and physiological characteristics of endophytic actinobacteria from medicinal plants. Wei Sheng Wu Xue Bao. 2013;**53**:15-23

[60] Passari AK, Mishra VK, Saikia R, Gupta VK, Singh BP. Isolation, abundance and phylogenetic affiliation of endophytic actinobacteria associated with medicinal plants and screening for

their *in vitro* antimicrobial biosynthetic potential. Frontiers in Microbiology. 2015;**6**:273

[61] Compant S, Duffy B, Nowak J, Clement C, Barka EA. Use of plant growth promoting bacteria for biocontrol of plant diseases: Principles, mechanisms of action and future prospects. Applied and Environmental Microbiology. 2005;**71**(89):4951-4959

[62] Nimnoi P, Pongsilp N, Lumyong S. Endophytic actinobacteriaisolated from *Aquilaria crassna Pierre ex Lec* and screening of plant growth promoters production. World Journal of Microbiology and Biotechnology. 2010;**26**(2):193-203

[63] Coombs JT, Franco CMM. Isolation and identification of Actinobacteria from surface-sterilized wheat roots. Applied and Environmental Microbiology. 2003;**69**(9):5603-5608

[64] Taechowisan T, Lu C, Shen Y, Lumyong S. Secondary metabolites from endophytic *Streptomyces aureofaciens* CMUAc130 and their antifungal activity. Microbiology. 2005;**151**:691-1695

[65] Tan RX, Zou WX. Endophytes: A rich source of functional metabolites (1987 to 2000). Natural Product Reports. 2001;**18**(4):448-459

[66] Loria R, Bukhalid RA, Fry BA, King RR. Plant pathogenicity in the genus *streptomyces*. Plant Disease. 1997;**81**(8):836-846

[67] Nalini MS, Prakash HS. Diversity and bioprospecting of actinomycete endophytes from the medicinal plants. Annals of Microbiology Letters in Applied Microbiology. 2017;**64**(4):261-270

[68] Qin S, Xing K, Jiang JH, Xu LH, Li WJ. Biodiversity, bioactive natural products and biotechnological potential of plant-associated endophytic actinobacteria. Applied Microbiology and Biotechnology. 2011;**89**(3):457-473

[69] Yandigeri MS, Malviya N, Solanki MK, Shrivastava P, Sivakumar G. Chitinolytic *Streptomyces vinaceusdrappus* S5MW2 isolated from Chilika lake, India enhances plant growth and bio-control efficacy through chitin supplementation against *Rhizoctonia solani*. World Journal of Microbiology and Biotechnology. 2015;**31**:1217-1225

[70] Verma VC, Gond SK, Kumar A, Mishra A, Kharwar RN, Gange AC. Endophytic Actinobacteria from *Azadirachta indica* A. Juss.: Isolation, diversity, and anti-microbial activity. Microbial Ecology. 2009;**57**(4):749-756

[71] Zin NM, Loi CS, Sarmin NM, Rosli AN. Cultivation-dependent characterization of endophytic Actinomycetes. Research Journal of Microbiology. 2010;**5**(8):717-724

[72] Conn VM, Franco CMM. Analysis of the endophytic Actinobacterial population in the roots of wheat (*Triticum aestivum* L.) by terminal restriction fragment length polymorphism and sequencing of 16S rRNA clones. Applied and Environmental Microbiology. 2004;**70**(3):1787-1794

[73] Phuakjaiphaeo C, Kunasakdakul K. Isolation and screening for inhibitory activity of *Alternaria brassicicola* endophytic actinobacteria from *Centella asiatica* (L.) urban. Journal of Agricultural Technology. 2015;**11**:903-912

[74] Bacilio-Jimenez M, Aguilar-Flores S, del Valle MV, Perez A, Zepeda A, Zenteno E. Endophytic bacteria in rice seeds inhibit early colonization of roots by *Azospirillum brasilense*. Soil Biology and Biochemistry. 2001;**33**(2):167-172

[75] Khush, G.S. (1992). Bennett, J. Nodulation and Nitrogen Fixation in Rice. Potential and Prospects IRRI Press, Manila.

[76] Poon ES, Huang TC, Kuo TT. Possible mechanisms of symptom inhibition of bacterial blight of rice by an endophytic bacterium isolated from rice. Botanical Bulletin of Academia Sinica. 1977;**18**:61-70

[77] Hallmann J, Quadt-Hallmann A, Mahaffee WF, Kloepper JW. Bacterial endophytes in agricultural crops. Canadian Journal of Microbiology. 1997;**43**(10):895-914

[78] Huang J. Ultrastructure of bacterial penetration in plants. Annual Review of Phytopathology. 1986;**24**(1):141-157

[79] Gohain A, Gogoi A, Debnath R, Yadav A, Singh BP, Gupta VK, et al. Antimicrobial biosynthetic potential and genetic diversity of endophytic actinobacteriaassociated with medicinal plants. FEMS Microbiology Letters. 2015;**362**(19):158

[80] Hasegawa S, Meguro A, Shimizu M, Nishimura T, Kunoh H. Endophytic Actinobacteria and their interactions with host plants. Actinomycetologica. 2006;**20**(2):72-81

[81] Shimizu M, Furumai T, Igarashi Y, Onaka H, Nishimura T, Yoshida R, et al. Association of induced disease resistance of *rhododendron* seedlings with inoculation of *Streptomyces* sp. R-5 and treatment with actinomycin D and amphotericin B to the tissue-culture medium. The Journal of Antibiotics. 2001;**54**(6):501-505

[82] Tamreihao K, Ningthoujam DS, Nimaichand S, Singh ES, Reena P, Singh SH, et al. Biocontrol and plant growth promoting activities of *Streptomyces corchorusii* strain UCR3-16 and preparation of powder formulation for application as biofertilizer agents for

rice plant. Microbiological Research. 2016;**192**:260-270

[83] Podile AR, Kishore GK. Plant growth-promoting rhizobacteria. In: Gnanamanickam SS, editor. Plant-Associated Bacteria. Netherlands: Springer; 2006. pp. 195-230

[84] Cattelan AJ, Hartel PG, Fuhrmann JJ. Screening for plant growth–promoting Rhizobacteria to promote early soybean growth. Soil Science Society of America Journal. 1999;**63**(6):1670-1680

[85] Nimaichand S, Devi AM, Li and W.J. Direct plant growth-promoting ability of actinobacteria in grain legumes. In: Subramaniam G, Arumugam S, Rajendran V, editors. Plant Growth Promoting Actinobacteria: A New Avenue for Enhancing the Productivity and Soil Fertility of Grain Legumes. Singapore: Springer Nature; 2016. pp. 1-16

[86] Jacob S, Sudini HK. Indirect plant growth promotion in grain legumes: Role of Actinobacteria. In: Subramaniam G, Arumugam S, Rajendran V, editors. Plant Growth Promoting Actinobacteria: A New Avenue for Enhancing the Productivity and Soil Fertility of Grain Legumes. Singapore: Springer Nature; 2016. pp. 17-32

[87] El-Tarabily K, Hardy GEJ, Sivasithamparam K. Performance of three endophytic actinobacteriain relation to plant growth promotion and biological control of *Pythium aphanidermatum*, a pathogen of cucumber under commercial field production conditions in the United Arab Emirates. European Journal of Plant Pathology. 2010;**128**:527-539

[88] Hastuti RD, Lestari Y, Suwanto A, Saraswati R. Endophytic *Streptomyces* spp. as biocontrol agents of Rice bacterial leaf blight pathogen

(*Xanthomonas oryzae* pv. Oryzae). Hayati Journal of Biosciences. 2012;**19**(4):155-162

[89] Rungin S, Indananda C, Suttiviriya P, Kruasuwan W, Jaemsaeng R, Thamchaipenet A. Plant growth enhancing effects by a siderophore-producing endophytic streptomycetes isolated from a Thai jasmine rice plant (*Oryza sativa* L. cv. KDML105). Antonie Van Leeuwenhoek. 2012;**102**(3):463-472

[90] Goudjal Y, Toumatia O, Sabaou N, Barakate M, Mathieu F, Zitouni A. Endophytic actinobacteriafrom spontaneous plants of Algerian Sahara: Indole-3-acetic acid production and tomato plants growth promoting activity. World Journal of Microbiology and Biotechnology. 2013;**29**(10):1821-1829

[91] Purushotham N, Jones E, Monk J, Ridgway H. Community structure of endophytic Actinobacteria in a New Zealand native medicinal plant *Pseudowintera colorata* (horopito) and their influence on plant growth. Microbial Ecology. 2018;**76**(3):729-740

[92] Vardharajula S, Skz A, Vurukonda SSKP. Plant growth promoting endophytes and their interaction with plants to alleviate abiotic stress. Current Biotechnology. 2017;**6**:252-263

[93] Khamna S, Yokota A, Peberdy J, Lumyong S. Indole-3-acetic acid production by *Streptomyces* sp. isolated from Thai medicinal rhizosphere soils. Eur Asian Journal of Biosciences. 2010;**4**:23-32

[94] Palaniyandi S, Yang SH, Damodharan K, Suh JW. Genetic and functional characterization of culturable plant-beneficial actinobacteria associated with yam rhizosphere. Journal of Basic Microbiology. 2013;**53**(12):985-995

[95] Duca D, Lorv J, Patten CL, Rose D, Glick BR. Indole-3-acetic acid in plant–microbe interactions. Antonie Van Leeuwenhoek. 2014;**106**(1):85-125

[96] Passari AK, Mishra VK, Singh G, Singh P, Kumar B, Gupta VK, et al. Insights into the functionality of endophytic actinobacteria with a focus on their biosynthetic potential and secondary metabolites production. Scientific Reports. 2017;**7**(1):11809

[97] Madhurama G, Sonam D, Urmil PG, Ravindra NK. Diversity and biopotential of endophytic actinobacteriafrom three medicinal plants in India. African Journal of Microbiology Research. 2014;**8**(2):184-191

[98] Aldesuquy HS, Mansour FA, Abo-Hamed SA. Effect of the culture filtrates of *Streptomyces* on growth and productivity of wheat plants. Folia Microbiologica. 1998;**43**(5):465-470

[99] Tokala RK, Strap JL, Jung CM, Crawford DL, Salove MH, Deobald LA, et al. Novel plant-microbe rhizosphere interaction involving *Streptomyces lydicus* WYEC108 and the pea plant (*Pisum sativum*). Applied and Environmental Microbiology. 2002;**68**(5):2161-2171

[100] Verma VC, Singh SK, Prakash S. Bio-control and plant growth promotion potential of siderophore producing endophytic *Streptomyces* from *Azadirachta indica* A. Juss. Journal of Basic Microbiology. 2011;**51**(5):550-556

[101] Lin L, Xu X. Indole-3-acetic acid production by endophytic *Streptomyces* sp. En-1 isolated from medicinal plants. Current Microbiology. 2013;**67**(2):209-217

[102] Vijayabharathi R, Sathya A, Gopalakrishnan S. A Renaissance in Plant Growth Promoting and Biocontrol Agents by Endophytes. India: Springer; 2016. pp. 37-61

[103] Sreevidya M, Gopalakrishnan S, Kudapa H, Varshney RK. Exploring plant growth-promotion actinobacteria from vermicompost and rhizosphere soil for yield enhancement in chickpea. Brazilian Journal of Microbiology. 2016;**47**(1):85-95

[104] Abdallah ME, Haroun SA, Gomah AA, El-Naggar NE, Badr HH. Application of actinobacteria as biocontrol agents in the management of onion bacterial rot diseases. Archives of Phytopathology and Plant Protection. 2013;**46**(15):1797-1808

[105] Ahemad M, Kibret M. Mechanisms and applications of plant growth promoting rhizobacteria: Current perspective. Journal of King Saud University Science. 2014;**26**(1):1-20

[106] Hamdali H, Bouizgarne B, Hafidi M, Lebrihi A, Virolle MJ, Ouhdouch Y. Screening for rock phosphate solubilizing Actinobacteria from Moroccan phosphate mines. Applied Soil Ecology. 2008;**38**(1):12-19

[107] Ezawa TS, Smith SE, Smith FA. P metabolism and transport in AM fungi. Plant and Soil. 2002;**244**:221-230

[108] van der Heijden MGA, Bardgett RD, van Straalen NM. The unseen majority: Soil microbes as drivers of plant diversity and productivity in terrestrial ecosystems. Ecology Letters. 2008;**11**(3):296-310

[109] Jog R, Pandya M, Nareshkumar G, Rajkumar S. Mechanism of phosphate solubilization and antifungal activity of *Streptomyces* spp. isolated from wheat roots and rhizosphere and their application in improving plant growth. Microbiology. 2014;**160**(4):778-788

[110] Hamdali H, Hafidi M, Virolle MJ, Ouhdouch Y. Rock phosphate solubilizing actinomycetes: Screening for plant growth promoting activities. World Journal of Microbiology and Biotechnology. 2008;**24**:2565-2575

[111] Bothwell TH. Overview and mechanisms of iron regulation. Nutrition Reviews. 1995;**53**:237-245

[112] Francis I, Holsters M, Vereecke D. The gram-positive side of plantmicrobe interactions. Environmental Microbiology. 2010;**12**(1):1-12

[113] Sessitsch A, Kuffner M, Kidd P, Vangronsveld J, Wenzel WW, Fallmann K, et al. The role of plant-associated bacteria in the mobilization and phytoextraction of trace elements in contaminated soils. Soil Biology and Biochemistry. 2013;**60**:182-194

[114] De Oliveira MF, da Silva GM, Sand STVD. Anti-phytopathogen potential of endophytic actinobacteria isolated from tomato plants (*Lycopersicon esculentum*) in southern Brazil, and characterization of *Streptomyces* sp. R18(6), a potential biocontrol agent. Research in Microbiology. 2010;**161**(7):565-572

[115] Ghodhbane-Gtari F, Essoussi I, Chattaoui M, Chouaia B, Jaouani A, Daffonchio D, et al. Isolation and characterization of non-Frankia actinobacteria from root nodules of *Alnus glutinosa*, *Casuarina glauca* and *Elaeagnus angustifolia*. Symbiosis. 2010;**50**(1-2):51-57

[116] Glick BR, Penrose DM, Li J. A model for lowering of plant ethylene concentrations by plant growth promoting bacteria. Journal of Theoretical Biology. 1998;**190**:62-68

[117] Nascimento FX, Rossi MJ, Glick BR. Role of ACC deaminase in stress control of leguminous plants. In: Plant Growth Promoting Actinobacteria. Singapore: Springer; 2016. pp. 179-192

[118] Nascimento FX, Rossi MJ, Soares CRFS, McConkey BJ, Glick BR. New insights into 1-Aminocyclopropane-1-carboxylate

(ACC) deaminase phylogeny evolution and ecological significance. PLoS One. 2014;**9**(6):e99168

[119] Jaemsaeng R, Jantasuriyarat C, Thamchaipenet A. Molecular interaction of 1-aminocyclopropane-1-carboxylate deaminase (ACCD)-producing endophytic *Streptomyces* sp. GMKU 336 towards salt-stress resistance of *Oryza sativa* L. cv KDML105. Scientific Reports. 2018;**8**:1950

[120] Lee SO, Choi GJ, Choi YH, Jang KS, Park DJ, Kim CJ. Isolation and characterization of endophytic Actinobacteria from Chinese cabbage roots as antagonists to *Plasmodiophora brassicae*. Journal of Microbiology and Biotechnology. 2008;**18**:1741-1746

[121] Coombs JT, Michelsen PP, Franco CMM. Evaluation of endophytic actinobacteria as antagonists of *Gaeumannomyces graminis* var. *tritici* in wheat. Biological Control. 2004;**29**(3):359-366

[122] Cao L, Qiu Z, You J, Tan H, Zhou S. Isolation and characterization of endophytic *Streptomyces* strains from surface-sterilized tomato (*Lycopersicon esculentum*) roots. Letters in Applied Microbiology. 2004;**39**(5):425-430

[123] Arfaoui A, El Hadrami A, Daayf F. Pre-treatment of soybean plants with calcium stimulates ROS responses and mitigates infection by *Sclerotinia sclerotiorum*. Plant Physiology and Biochemistry. 2018;**122**:121-128

[124] Liu D, Yan R, Fu Y, Wang X, Zhang J, Xiang W. Antifungal, plant growth-promoting, and genomic properties of an endophytic Actinobacterium *Streptomyces* sp. NEAU-S7GS2. Frontiers in Microbiology. 2019;**10**:2077

[125] Kaur T, Rani R, Manhas RK. Biocontrol and plant growth promoting

potential of phylogenetically new *Streptomyces* sp. MR14 of rhizospheric origin. AMB Express. 2019;**9**(1):125

[126] Maggini V, De Leo M, Mengoni A, Gallo ER, Miceli E, Reidel RVB, et al. Plant-endophytes interaction influences the secondary metabolism in *Echinacea purpurea* (L.) Moench: An *in vitro* model. Scientific Reports. 2017;**7**(1):16924

[127] Wan M, Li G, Zhang J, Jiang D, Huang HC. Effect of volatile substances of *Streptomyces platensis* F-1 on control of plant fungal diseases. Biological Control. 2008;**46**(3):552-559

[128] Liang J, Xue Q, Niu X, Li Z. Root colonization and effects of seven strains of actinobacteriaon leaf PAL and PPO activities of *capsicum*. Acta Botanica Boreali-Occidentalia Sinica. 2005;**25**(10):2118-2123

[129] Ningthoujam DS, Sanasam S, Nimaichand S. Studies on bioactive actinobacteriain a niche biotope, Nambul River in Manipur, India. International Journal of Molecular and Cellular Medicine. 2011;**S6**:1-6

[130] Gopalakrishnan S, Humayun P, Vadlamudi S, Vijayabharathi R, Bhimineni RK, Rupela O. Plant growth-promoting traits of *Streptomyces* with biocontrol potential isolated from herbal vermicompost. Biocontrol Science and Technology. 2011;**22**(10):1199-1210

[131] Mingma R, Pathom-aree W, Trakulnaleamsai S, Thamchaipenet A, Duangmal K. Isolation of rhizospheric and roots endophytic actinobacteriafrom *Leguminosae* plant and their activities to inhibit soybean pathogen, *Xanthomonas campestris* pv. Glycine. World Journal of Microbiology and Biotechnology. 2014;**30**:271-280

[132] Jalaludeen SAM, Othman K, Ahmad RM, Abidin Z. Isolation and

characterization of actinomycets with *in vitro* antagonistic acvity against *fusarium oxysporum* from rhizosphere of chilli. International Journal of Environmental Science and Technology. 2014;**3**:54-61

[133] Ashokvardhan T, Rajithasri AB, Prathyusha P, Satyaprasad K. Actinobacteriafrom *Capsicum annuum* L. rhizosphere soil have the biocontrol potential against pathogenic fungi. International Journal of Current Microbiology and Applied Sciences. 2014;**3**(4):894-903

[134] Cheng G, Huang Y, Yang H, Liu F. *Streptomyces felleus* YJ1: Potential biocontrol agents against the Sclerotinia stem rot (*Sclerotinia sclerotiorum*) of oilseed rape. Journal of Agricultural Science. 2014;**6**(4):91-98

[135] Kunova A, Bondaldi M, Saracchi M, Pizzatti C, Chen X. Selection of *Streptomyces* against soil borne fungal pathogens standardized dual culture assay and evaluation of their effects on seed germination and plant growth. BMC Microbiology. 2016;**16**(1):272

[136] Singh SP, Gaur R. Evaluation of antagonistic and plant growth promoting activities of chitinolytic endophytic actinobacteria associated with medicinal plants against *Sclerotium rolfsii* in chickpea. Journal of Applied Microbiology. 2016;**121**(2):506-518

[137] Passari AK, Chandra P, Zothanpuia M, V.K., Leo, V.V., Gupta, V.K., Kumar, B. and Singh, B.P. Detection of biosynthetic gene and phytohormone production by endophytic actinobacteria associated with *Solanum lycopersicum* and their plant-growth-promoting effect. Research in Microbiology. 2016;**167**(8):692-705

[138] Hassan N, Nakasuji SE, Naznin HA, Kubota M, Ketta H,

Shimizu M. Biocontrol potential of an endophytic *Streptomyces* sp. strain MBCN152-1against *Alternaria brassicicola* on cabbage plug seedlings. Microbes and Environments. 2017;**32**(2):133-141

[139] Merrouche R, Yekkou RA, Lamari L, Zitouni A, Mathieu F, Sabaou N. Efciency of *Saccharothrix algeriensis* NRRL B-24137 and its produced antifungal dithiolopyrrolones compounds to suppress *fusarium oxysporum*-induced wilt disease occurring in some cultivated crops. Arabian Journal for Science and Engineering. 2017;**42**:2321-2327

[140] Patil HJ, Srivastava AK, Singh DP, Chaudhari BL, Arora DK. Actinobacteriamediated biochemical responses in tomato (*Solanum lycopersicum*) enhances bioprotection against *Rhizoctonia solani*. Crop Protection. 2011;**30**(10):1269-1273

[141] Liu Y, Guo J, Li L, Asem MD, Zhang Y, Mohamad OA, et al. Endophytic bacteria associated with endangered plant *Ferula sinkiangensis* K. M. Shen in an arid land: Diversity and plant growth-promoting traits. Journal of Arid Land. 2017;**9**(3):432-445

[142] Perez-Rosales E, Alcaraz-Melendez L, Puente ME, Vazquez-Juarez R, Quiroz-Guzmán E, Zenteno-Savín T. Isolation and characterization of endophytic bacteria associated with roots of jojoba (*Simmondsia chinensis* (link) Schneid). Current Science. 2017;**112**(2):396

[143] Conn VM, Walker AR, Franco CMM. Endophytic Actinobacteria induce defense pathways in *Arabidopsis thaliana*. Molecular Plant-Microbe Interactions. 2008;**21**(2):208-218

[144] Salla TD, Astarita LV, Santarém ER. Defense responses in plants of *eucalyptus* elicited by *Streptomyces* and challenged

with *Botrytis cinerea*. Planta. 2016;**243**(4):1055-1070

[145] Kurth F, Mailander S, Bonn M, Feldhahn L, Herrmann S, Grobe I, et al. *Streptomyces* induced resistance against oak powdery mildew involves host plant responses in defense, photosynthesis, and secondary metabolism pathways. Molecular Plant Microbe Interaction. 2014;**27**:891-900

[146] Beyer M, Diekmann H. The chitinase system of *Streptomyces* sp. ATCC 11238 and its significance for fungal cell wall degradation. Applied Microbiology and Biotechnology. 1985;**23**(2):140-146

[147] Taechowisan T, Peberdy JF, Lumyong S. Chitinase production by endophytic *Streptomyces aureofaciens* CMUAc130 and its antagonism against phytopathogenic fungi. Annals of Microbiology. 2003;**53**:447-461

[148] Xue L, Xue Q, Chen Q, Lin C, Shen G, Zhao J. Isolation and evaluation of rhizosphere actinobacteria with potential application for biocontrol of *Verticillium* wilt of cotton. Crop Protection. 2013;**43**:231-240

[149] Pal KK, McSpadden Gardener B. Biological control of plant pathogens. The Plant Health Instructor. 2006;**2**:1-15

[150] Prapagdee B, Kuekulvong C, Mongkolsuk S. Antifungal potential of extracellular metabolites produced by *Streptomyces hygroscopicus* against phytopathogenic fungi. International Journal of Biological Sciences. 2008;**4**(5):330

[151] Guo B, Wang Y, Sun X, Tang K. Bioactive natural products from endophytes: A review. Applied Biochemistry and Microbiology. 2008;**44**(2):136-142

[152] Zhang HW, Song YC, Tan RX. Biology and chemistry of endophytes. Natural Product Reports. 2006;**23**(5):753

[153] Priya MR. Endophytic actinobacteriafrom Indian medicinal plants as antagonists to some phytopathogenic fungi. Scientific Reports. 2012;**4**(1):259

[154] Castillo UF, Strobel GA, Ford EJ, Hess WM, Porter H, Jensen JB, et al. Munumbicins, wide-spectrum antibiotics produced by *Streptomyces* NRRL 30562, endophytic on Kennedianigriscansaa the GenBank accession number for the sequence determined in this work is AY127079. Microbiology. 2002;**148**(9):2675-2685

Section 2

Bioactive Metabolites from Actinobacteria

Chapter 4

Antimicrobials: Shift from Conventional to Extreme Sources

Aasif Majeed Bhat, Qazi Parvaiz Hassan
and Aehtesham Hussain

Abstract

Antimicrobials- the chemical substances that inhibit the growth of microorganisms and stop their multiplication are immensely useful in the context of pathogenic microorganisms where these substances either contain their growth by inhibiting them from growing (bacteriostatic) or killing them permanently (bacteriocidal). They may broadly be either antibiotics, antifungals, antivirals and antiparasitics. A major class of antimicrobials are antibiotics and almost half of the total percent of antibiotics driven from microbials are sourced from different taxonomic levels of actinomycetota (formerly actinobacteria), significantly from the genus Streptomyces. Adaptability and mechanisms to resist drug effects has outpushed the evolution of drug resisitant pathogenic microorganisms and outnumbered their growth vis a vis the discovery of new antimicrobials. Gone is the golden age of antibiotics: the tussle between antimicrobials to resist the growth of pathogens and the latter to contain the inhibitory effects of former has largely weighed on the pathogenic side- thanks to the inefficient and excessive use of antibiotics and their misapplication. Growth of drug (multi-drug) resistant pathogens coupled with inadequate antibiotics has set a dire need to explore new habitats-aquatic, terrestrial and microbiomes associated as endophytes in other plants and animals. The shift in habitat selection from conventional to extreme locations is met with convincingly successful outcomes. Researchers successfully explore the actinomycetota drug discovery potential of deep sea oceans, extreme high altitude Himalayas that remain capped with snow and glaciers round the year. The abyssopelagic and glaciated peaks both share similarity in that they are constrained by different pressure parameters. The environmental pressures associated with deep pelagic oceans are partial to complete exclusion of light, lack of photosynthesis and associated vegetation, limited nutrition and hydrostatic pressure by thounsands of pounds per square inch. Mountain peaks are glaciated, ice cold with limited nutrition and oligotrophic in nature. These temperature constraints in both the aquatic and terrestrial environments have activated the drug expression secondary metabolite machinary of actinomycetota to kill or inhibit other microorganisms and spare the already limited resources for their own growth. This antibiotic secretion paradigm also applies to actinomycetota living as endophytes in an interactive dynamic environments with insects and other organisms. The antibiotic potential hidden in these extreme selected sites is worthy of killing the microbial bugs and conatining the ever growing resistant pathogen load. Successful exploitation strategies should be hastened to garner the antimicrobial potential of these extreme sources.

Keywords: antimicrobials, antibiotics, drug discovery, Actinomycetota, Actinobacteria, extreme habitats, NRPS and PKS

1. Introduction

Antimicrobials are the substances or agents that kill, inhibit the growth and/or stop the spread of microorganisms. These are named based on the type of microorganism against which they act. Accordingly they are broadly of four different types [1]:

1. Antibiotics (Antibacterial antimicrobials): Prevent or treat infections by bacteria.

2. Antifungal antimicrobials: Prevent or treat infections by fungus.

3. Antiviral antimicrobials: Prevent or treat infections by viruses.

4. Antiparasitic antimicrobials: Prevent or treat infections by parasites.

These are suffixed as ~cidal or ~ static antimicrobials depend on whether they kill or inhibit the growth of microorganisms respectively. For instance a bactericidal antibiotic is an antimicrobial that kills the microorganism, e.g., Vancomycin, Rifampin, Pencillins & Cephalosporins, Aminoglycosides (at high doses), Quinolones, Isoniazid, Metronidazole, Polymyxins and Bacitracin;

However, a bacteriostatic antibiotic is an antimicrobial that stops microorganisms from growing and stalls the process of reproduction, without killing them necessarily e.g. Tetracyclines, Clindamycin, Chloramphenicol, Macrolides, Sulfonamides and Timethoprim. The major difference to differentiate between ~cidal or ~ static antimicrobials is that in the former case, upon removal at the decline phase of an antimicrobial having ~cidal effect, the growth curve of target microorganism continues to decline and never resumes while in later case, since the growth of microorganism is stalled and plateaus at stationary phase, removal of such antimicrobials resumes the growth of target microorganisms from stationary to log phase [2]. Though we discussed of antimicrobials above as source agents that kill or inhibit the spread of microorganisms, yet antimicrobial is a broad term that also includes agents applied to non living surfaces e.g. disinfectants like bleach, non pharmaceuticals like essential oils [3, 4], antimicrobial pesticides and pesticide products [5], and ozone [6] among other antimicrobial properties of metal and metal alloys [7, 8]. However here in this chapter we will limit our discussion on antimicrobials in connection with Actinomycetota (formerly called Actinobacteria) - a class of gram positive microorganisms high in Guanine Cytosine (GC) base pair composition in their DNA and evolutionary viewed as rich source of antimicrobials and FDA approved antibiotics among all the microbial taxa [9, 10]. The name change of phylum Actinobacteria to Actinomycetota is very recent and an innumerable number of research articles communicated still retain the word as Actinobacteria and researchers also use the term Actinobacteria very frequently, so here for the sake of brevity, we will use Actinobacteria and Actinomycetota interchangeably.

Although both ~cidal and ~ static antimicrobials display vital possibility to stop the spread of pathogen causing diseases but the antimicrobial crisis to unlock the potential to counter the menace of antimicrobial drug resistance grows continuously. Resistant pathogenic microorganisms employ one or a combination of following anti-bacterial resistance mechanisms to evade killing by approved antimicrobials;

1. Decreased membrane permeability by resistant microorganism and hence dearth of antimicrobial entry.

2. Antimicrobial drug removal by membrane Efflux transport system.

3. Drug receptor alteration that causes decreased affinity of an antimicrobial to a target receptor site.

4. Antibiotic inactivation by resistant microorganisms using different enzymes/protein inactivation systems.

With every new antibiotic discovery, microorganisms opt for one or the other resistance mechanisms to evade killing by an antimicrobial and sometimes execute it successfully-thus add to the already burdened drug resistance pathogenic load [11]. Thoroughly screened and verified efforts need to be searched to subdue antimicrobial drug resistance patterns.

From the **Table 1**, it is clear that different classes of antimicrobials-having different chemical moieties are being produced by varying Actinobacterial strains, majority of which belong to the genus Streptomyces. These antimicrobials range from majorly characterised chemical classes like tetracyclines, β – Lactams, macrolides, aminoglycosides, lactones, alkaloids, glycopeptides to less known drug moieties like peptides-including simple peptides and lipopeptides, esters and nucleosides. Majority of the broadly classified antimicrobials are a product of non-ribosomally synthesised bioactive chemicals by mega-enzyme complexes called NRPS Non-ribosomal peptide synthetases (NRPS) and polyketide synthetases (PKS) and hybrid NRPS-PKS complexes.

2. Shift from conventional to extreme terrestrial habitats - North Western Himalaya (NWH) to emerge a new hope for antibiotic drug discovery

In Golden era of antibiotics (1940–1962), discovery of new and novel antimicrobials was at its utmost peak. These antimicrobials called as "miracle drugs" reduced the mortality by pathogenic infections [12, 13]. But the selective pressure on these infectious agents by misuse and misapplication of miracle antibiotic drugs set in different resistance mechanisms [14, 15]. Infectious microorganisms tried to evade killing by antimicrobials by employing one or a combination of different resistance mechanisms as stated above. Subsequently search for new antimicrobials from already explored habitats met with limited success. These habitats thus became conventional and yielded diminished returns of drug discovery. To counter this, microbiologists shifted their focus from these conventional environments to extreme terrestrial and aquatic habitats. Actinomycetota from oligotrophic soils of high altitudes of North Western Himalaya serve as potential search sources to isolate bioactive actinobacteria of pharmaceutical importance. North Western Himalaya is unique terrestrial habitat in its sub-zero temperature, ice caped mountain peaks, oligotrophic nutrients and limited vegetation. These conditions create competitive environments for the isolation of novel Actinomycetota species and/or the production of novel biochemical scaffolds to be used as new antimicrobials. NWH also grow substantial prospective isolates of novel actinobacteria, some of which are pharmacologically active ones. In our laboratory at IIIM Jammu, Microbiological Researchers isolated hundreds of actinobacteria and screened successfully against different Gram positive, Gram negative and fungal pathogenic strains.

Antimicrobial/ Antibiotic	Producer strain	Chemical class	References
Teicoplanin	Actinoplanes teichomyceticus	Glycopeptide	https://doi.org/10.1099/mic.0.26507-0
Rifamycin	Amycolatopsis mediterranei	Ansamycins	https://doi.org/10.1021/cr030112j
Fortimicin	Micromonospora olivasterospora	Aminoglycoside	https://doi.org/10.7164/antibiotics.30.1064
Gentamycin	Micromonospora spp	Aminoglycoside	https://doi.org/10.1016/S0032-9592(99)00106-5
Cephamycin C	Nocardia lactamdurans	β – Lactam	https://doi.org/10.1016/j.biortech.2008.11.046
Vancomycin	Nocardia orientalis/ Amycolatopsis orientalis	Glycopeptide	https://doi.org/10.7164/antibiotics.39.694
Nocardicin	Nocardia uniformis	β – Lactam	https://doi.org/10.7164/antibiotics.29.492
Spiramycin	Streptomyces ambofaciens	Macrolide (PK)	doi: 10.1001/archopht.1961.00960010611029
Oleandomycin	Streptomyces antibioticus	Macrolide	https://doi.org/10.1099/00221287-136-8-1447
Tetracycline	Streptomyces aureofaciens	Naphthacene	Darken et al. 1960
Chlortetracycline	S. aureofaciens	Tetracycline	https://doi.org/10.1007/s11274-004-2778-z
Thienamycin	S. cattleya	β-Lactam Peptidoglycan	https://doi.org/10.7164/antibiotics.32.1
Clavulanic acid	Streptomyces clavuligerus	β – Lactam	https://doi.org/10.1128/AAC.11.5.852
Neomycin A, B and C	Streptomyces fradiae	Aminoglycoside	https://doi.org/10.1042/bj1200271
Fosfomycin	S. fradiae	Phosphoric acid	https://doi.org/10.1128/AAC.5.2.121
Streptomycin	S. griseus	Aminoglycoside	https://doi.org/10.1128/JB.00204-08
Kanamycin	Streptomyces kanamyceticus	Aminoglycoside	https://doi.org/10.1371/journal.pone.0181971
Fumaramidmycin	Streptomyces kurssanovii	Alkaloids	https://doi.org/10.7164/antibiotics.28.636
Lincomycinn	Streptomyces lincolnensis	Sugar—amide	https://doi.org/10.3390/molecules26154504
Novobiocin	S. neveus	Aminocoumarin	https://doi.org/10.3390/molecules26154504
Amphotericin B	S. nodosus	Polyene Macrolide	https://doi.org/10.1016/j.micres.2020.126623
Seromycin	S. orchidaceus	Peptide	https://doi.org/10.1107/S0365110X56002643
Daunorubicin	S. Peucetius	Peptide	https://doi.org/10.1128/jb.174.1.144-154.1992

Antimicrobial/ Antibiotic	Producer strain	Chemical class	References
Oxytetracycline	S. rimosus	Tetracycline	https://doi.org/10.17113/ ftb.55.01.17.4617
Daptomycin	S. rodeosporus	Lipopeptide	https://doi.org/10.1099/ mic.0.2008/020685-0
Nikkomycin	Streptomyces tendae	Nucleoside	https://doi. org/10.1111/j.1365-2672.1992. tb01823.x
Tobramycin	Streptoalloteichus tenebrarius	Aminoglycoside	https://doi.org/10.1016/ S0378-1097(03)00881-4
Puromycin	S.alboniger	Cinnamamido adenosine	https://doi. org/10.1099/00221287-131-11-2877
Tetracycline	S.antibioticus	Naphthacene	https://doi.org/10.1128/ AAC.44.5.1322-1327.2000
Avermectin	S.avermitilis	Lactone	https://doi.org/10.1128/ jb.169.12.5615-5621.1987
Cephalosporin	S.clavuligerus	β - Lactam	https://doi.org/10.1007/BF01950159
Erythromycin	S.erythraeus	Macrolide	https://doi.org/10.1128/ jb.164.1.425-433.1985
Actinomycin Z	S.fradiae	Chromopeptide lactone	https://doi.org/10.1021/np990416u
Dekamycin	S.fradiae	Aminoglycoside	DOI: 10.21276/ap.2017.6.1.3
Cycloserine (Seromycin)	S.garyphalus	Amino acid analogue	https://doi. org/10.1128/9781555817770.ch30
Clindamycin	S.lincolensis	Lincosamide	DOI: 10.21276/ap.2017.6.1.3
clindamycin	S.mediterranei	Macrolide	https://doi.org/10.1016/ S0006-291X(76)80072-1
Novobicin	S.niveus	Coumarin lactone	https://doi.org/10.1128/AAC.1.2.123
Nistatin A1,A2 and A3	S.noursei	Macrolide	https://doi.org/10.1128/ AAC.48.11.4120-4129.2004
Platenomycin	S.platensis	Macrolide	https://doi.org/10.7164/ antibiotics.28.770
Ribostamycin	S.ribosidificus	Aminoglycoside	https://doi.org/10.1021/ja00408a076
Paromomycin	S.rimosus	Aminoglycoside	https://doi.org/10.1186/ s12866-021-02093-6
	S.spectabilis	Aminoglycoside/ Aminocyclitol	https://doi.org/10.1007/ s00284-008-9204-y
Spectinomycin	S.spectabilis	Aminoglycoside/ Aminocyclitol	https://doi. org/10.1111/j.1365-2672.2008.03788.x
FK506	S.tubercidicus	Macrolide	https://doi. org/10.1111/1758-2229.12617
Chloramphenicol	S.venezuelae	Organochlorine Acetamide	https://doi.org/10.1128/ AAC.04272-14
Viomycin	S.vinaceus	tuberactinomycin	https://doi.org/10.1016/ S0378-1119(03)00617-6
Tetracycline	S.viridifaciens	Naphthacene	https://doi. org/10.1046/j.1365-2672.2001.01243.x

Antimicrobial/ Antibiotic	Producer strain	Chemical class	References
Anthramycin	Streptomyces refuineus	Benzodiazepine Alkaloid	https://doi.org/10.1016/j. chembiol.2007.05.009
Thermomycin	Streptomyces thermophilus	Polyketide Antibiotic	David et al. 1955
Pyridine-2,5-diacetamide	Streptomyces sp. DA3–7	pyridine alkaloid	https://doi.org/10.1016/j. micres.2017.11.012
1, 4-butanediol, adipic acid, & terephthalic acid	Thermomonospora fusca	aliphatic-aromatic copolyesters	https://doi.org/10.1021/bp020048b

Table 1.
Representative antimicrobials from actinobacteria, their producer strains and chemical class.

Pharmacological compound bio evaluation from these actinobacteria yielded anti-tuberculosis and anti-cancer antimicrobials [16–18]. Despite this other pharmacological active metabolites were also isolated from these bioactive strains [19–24]. Actinobacteria exploration from NWH can thus serve as an understudied reserve source for isolation and bio evaluation of pharmacologically potential antimicrobials to be used as next generation new antibiotics.

3. Actinobacteria from oceanic habitats

The chemical synthesis of bioactive molecules as derivatives of natural product secondary metabolites has added to the discovery of antimicrobials. The antimicrobial activity of these synthetic derivatives includes different classes like Quinolones, sulphonamides, anti-tuberculosis, anti-fungal and anti-viral antimicrobials. Despite antimicrobial addition by synthetic means, traditional approaches of culturable isolation of actinobacteria from unfathomed terrestrial and oceanic habitats, their natural product purification and pharmacological antimicrobial evaluation are fairly guerdoning [25]. Oceans are biodiversity rich environments [26] and the microbial biodiversity of oceanic habitats is understudied. Deep sea oceanic habitats are unique in its physical parameters like extreme pressure, hyper saline water, chilling temperatures. These extreme conditions have activated the transcription of gene clusters in actinobacteria to contain the growth of surrounding microorganisms, ensure maximum utilisation of already limited nutrients and enhance their survival in deep sea dynamic oceans [27]. Further the scientific expectations of continued and prolonged miracle drug discovery efforts as was witnessed in golden era of antibiotic discovery started fading away [12]. Two antimicrobial drug discovery strategic problems i.e. diminished returns of antimicrobials from well explored environments and the resistance rate outpacing the antibiotic drug discovery rate, had shaken the research thinking with an effort to reinvigorate the antibiotic pipelines [28]. Researchers started diving deep into the oceans to rediscover deep sea microbiology and search for potential antimicrobial producing actinobacteria species is being carried rapidly. Over time different antimicrobials were identified and successfully evaluated for their antimicrobial and pharmacological studies. Antibacterials like pseudonocardians, caerulomycins, abyssomimicins, Taromycin, Lynamimicin and Flustatin are verified to have been produced by independent isolates of different species of ocean dwelling actinobacteria. Moreover actinobacteria living in symbiotic association with other marine organisms have also been reported to produce different antibacterials like Arenjimysin, bendigoles, peptidolipins, solwaric acids, rifamycins, saccharothrixmicnes. These actinobacteria live as symbionts ranging from marine sponges, ascidians to molluscs [29]. Ocean microbiome is a dynamic repository of drug candidates and endeavours should be hastened for tapping such immense deep sea potential bioresources.

4. NRPS and PKS clusters as gene sources of antimicrobial secondary metabolites

Non-ribosomal peptide synthetases (NRPS) and polyketide synthetases (PKS) metabolic pathways encompass a cluster of multi domain subunits, where each subunit performs a separate enzymatic activity. The coordinated activity of these multi domain units in a mega synthetases complex performs the synthesis of Non ribosomal peptides (NRPs) and three different Polyketides (PKs)-secondary metabolites that exhibit clinically valuable biological activities as anti-microbial, anti-fungal, anti-tumour, anti-parasitic, and immunosuppressive agents [30]. The NRPs biosynthesis on NRPS enzyme complex is done through ordered arrangement and addition of amino acid monomers whereas the PKs biosynthesis on PKS enzyme complex follows the sequential addition of 2C ketide unit derived from thioester of acetate precursors or other short chain carboxylic acids [31]. These enzyme clusters are either modular (NRPS and modular type I PKS) or iterative (iterative type I PKS, type II PKS and type III PKS). In case of NRPS and modular type I PKS, each module is designed to hold an obligatory or a minimal core domain. The minimal core domain in NRPS module consists of an Adenylation domain (A) - for selective activation of amino acid from a pool of precursor amino acids, Condensation domain (C) for peptide bond formation and chain elongation Thiolation/Peptidyl carrier protein (T/PCP) domain with a phosphopantetheine group that transfer the starter monomer units or an extender growing chain to different catalytic sites in a mega enzyme complex. Likewise a modular type I PKS obligatory or a minimal core domain includes an Acyl transferase domain (AT) for starter/extender unit loading of acyl-CoA on acyl carrier protein (ACP) and a Ketoacyl synthase domain (KS) for condensation and decarboxylation of acyl CoA starter or extender units. In both cases of NRPs and PKs biosynthesis, the Thioesterase domain (TE) catalyses the release of full length NRPs and PKs [32–36]. There are few starter or extender units for biosynthesis of PKs however a larger pool of about 50 different amino acid precursors- natural or unnatural act as starter or extender units for biosynthesis of NRPs. Thus though the substrate specificity for PKS is not a complex process, the prediction of substrate specificity for NRPS is a challenging task [31]. The corresponding modules in NRPS and modular PKS are held together by short peptide chains called linkers that establish functional communication between modules [32]. In addition to core domains of NRPS and PKS, some non obligatory but essential auxiliary domains can be loaded mostly on elongation modules. These auxiliary domains include ketoreductase (KR), dehydratase (DH), or enoylacyl reductase (ER) enzymatic domains for partial and/or complete reduction of keto groups. These ketide chain length modifications enhance the structural complexity and increase diversity of mature PKs [37]. The auxiliary domains loaded on the modules of NRPS include cyclization of peptide chain into thiazoline or oxazoline rings, oxidation of thiazolines and oxazolines to thiazoles and oxazoles, reduction into thiazolidines and oxazolidines, amino acid epimerization into D isomers. Other processing modification of final NRPS chain peptide includes acylation, glycosylation, hydroxylation and halogenations [38–39]. Notably, it is reported that actinobacteria have a higher number of these biosynthetic genes [40]. These genes upon translation form modular NRPS and PKS, non modular iterative PKS and type III PKS. The modular genetic engineering of NRPS and PKS and biochemical and bioinformatic investigation of iterative PKS to unlock and discovery more iterative enzymes complexes of relative function are gaining attention. Addition or deletion of whole modules in an enzyme complex or most importantly an auxiliary domain addition or deletion in a module alters the chain length and modify the enzyme complex. This if executed successfully may give rise to diverse novel secondary metabolites, many of which could work as potential antimicrobials. Amalgamation of NRPS and PKS to form a Hybrid NRPS-PKS synthesised secondary metabolite are also successfully engineered [41–43].

The antiSMASH (antibiotics and secondary metabolites analysis shell) database is a handy tool in secondary metabolite gene cluster prediction analysis of bacterial genomes, it can however also be used against fungal and plant complete or draft genomes. Genome mining by antiSMASH gives an overview of the antimicrobial potential of different gene clusters along the genomic stretch of a given query organism (e.g. NRPS, different types of PKS, hybrid NRPS-PKS, lanthipeptides, siderophores, ectoines and terpenes). The antiSMASH results depict the type of gene cluster to which query is most similar to along with the percentage similarity. It searches a query sequence against the MIBiG database of different characterised gene clusters, selects the best possible hit, determines the start and stop origins or cluster coordinates along the genome length and percentage statistics of top hit to the query sequence.

5. Insect microbiome: symbiotic actinomycetota as antimicrobial sources

The mechanism of defensive symbiosis is employed by insects and this association with antimicrobial producing bacterial symbionts is critical for insect survival. Until recently soil microbiome was considered the only rich source of actinobacteria. Metagenomic analysis for actinobacteria from soil, fresh water, oceanic and insect associated microbiome revealed that the number of streptomyces reads per megabase (rpM) to be 172.72 rpM, 47.49 rpM, 24.65 rpM, 129.32 rpM- suggesting that insect microbiome also serve as the rich source of actinobacteria. Further when compared to other sources, the insect associated streptomyces exhibit higher inhibition against gram positive, gram negative and fungal microorganisms and insect streptomyces are inhibitory against antimicrobial resistant pathogens more than the soil streptomyces. Antimicrobial defensive symbiosis is shown in wasps, beetles, fungus growing ants where actinobacteria live in symbiosis with these insects; produce several antibacterial, antifungal and antimalarial substances akin to that used in human system. A discovery lead by Marc G. Chevrette et al. and published in 2019 exploited the insect microbiome diversity for antimicrobial detection. The studies described Cyphomycin-a new antimicrobial molecule against MDR pathogens. Genomic and metagenomic revelations show that the streptomyces from insect micro biota have immense potential to synthesise bioactive metabolites. The inhibitory secretions by Actinomycetota stop the spread of pathogenic microorganisms in insects and help them successfully flourish different microbe dwelling habitats [44]. Despite the above mentioned habitat sources for these predominant antimicrobial producing microorganisms, actinobacteria have also been found to grow in other extreme habitats like hyper saline and hyper alkaline marine and terrestrial regions, hyper arid deserts, volcanoes and glaciers. But for the sake of brevity we have limited our discussions to only the sources highlighted in this chapter. Current and future research on all extreme sources will delve deep into the bioactivity evaluation of these extremophilic actinobacteria and pave way for isolation and characterisation of new drugs from these still to be believed as golden drug reserves for next generation antibiotic discovery [45–48].

6. Conclusion

True that almost half of the antibiotic drugs isolated from microbiota of different habitats are being produced by different members of the phylum Actinomycetota, but the recent shortfalls in antibiotic discovery has shifted the focus of microbiologists to more extreme habitats- both terrestrial and aquatic. NWH- one of the world's high altitude and highly diverse ecosystems is an attractive location to uncover the understudied bioactive potential of these unexplored ice caped mountain ranges. Deep sea oceans also serve as parallel sources to augment microbial drug discovery efforts. Diving deep into the ocean floor and/or collecting samples from oceanic trenches are attractive selection sites for adding the phylogeny of Actinomycetota and unlocking the unfathomed antibiotic potential. To define taxonomic identity of undiscovered novel species, efforts should be made to consider two or more conserved genes along with 16S rRNA, like β subunit of bacterial RNA polymerase (rpoB), DNA gyrase subunit B (gyrB), 70 kilodalton heat shock proteins (hsp70 or DnaK), Tryptophan synthase beta chain (trpB), ATP-dependent DNA helicase (recG). Diversifying the conserved taxonomic molecular identifiers serves as an important methodology for accurate taxonomic classification. Looking into the success in drug discovery although not as expected from these extreme habitats, vigorous efforts should be made to diversify sample selection locations and outreach further northern Arctic and southern Antarctic. Genomes of Actinomycetota most specifically Streptomyces are highly encoded with Biosynthetic Gene Clusters (BGCs) like NRPS, different types of PKS, hybrid NRPS-PKS, other metabolite clusters such as siderophores, ectoines, terpenes, melanin, RiPP like, indoles and other secondary metabolite gene clusters. Many of these clusters are still uncharacterized and display structural similarity and homology to compounds of immense bioactive pharmacological activity. Sequencing more and more Actinomycetota genomes for presence of BGCs alongside their spectroscopic compound validation will augment the new insights into next generation drug discovery efforts. Parallel efforts to isolate Actinomycetota from extreme soil and water habitats and as symbionts in insects and other animals has the capacity to uncover the new domains of antibiotic drug discovery and unlock the bioactive potential hidden in these golden micro flora drug reserves.

Acknowledgements

We acknowledge the IntechOpen publishing house for providing the opportunity to write a chapter in the fascinating book titled Actinobacteria. Our special thanks to CSIR-IIIM (Council for Scientific and Industrial Research-Indian Institute of Integrative Medicine) for providing laboratory facilities to gain deep insights into the actinomycetota of North Western Himalayan altitudes and to UGC (University Grants Commission) for funding the research.

Conflict of interest

The authors declare no conflict of interest.

Note of thanks

Our special thanks to Director, CSIR-IIIM, Dr. D. Srinivasa Reddy for funding the laboratory.

Author details

Aasif Majeed Bhat[1,2], Qazi Parvaiz Hassan[2*] and Aehtesham Hussain[3]

1 Academy of Scientific and Innovative Research-AcSIR, India

2 Council for Scientific and Industrial Research, Indian Institute of Integrative Medicine-CSIR-IIIM, India

3 NCMR-National Centre for Cell Science, India

*Address all correspondence to: qphassan@iiim.ac.in

IntechOpen

References

[1] Strohl WR. Antimicrobials. Microbial Diversity and Bioprospecting. 2003:336-355

[2] Pankey GA, Sabath LD. Clinical relevance of bacteriostatic versus bactericidal mechanisms of action in the treatment of gram-positive bacterial infections. Clinical Infectious Diseases. 2004;**38**(6):864-870

[3] Smith-Palmer A, Stewart J, Fyfe L. Antimicrobial properties of plant essential oils and essences against five important food-borne pathogens. Letters in Applied Microbiology. 1998;**26**(2):118-122

[4] Kalemba DA, Kunicka A. Antibacterial and antifungal properties of essential oils. Current Medicinal Chemistry. 2003;**10**(10):813-829

[5] Astani A, Reichling J, Schnitzler P. Comparative study on the antiviral activity of selected monoterpenes derived from essential oils. Phytotherapy Research: An International Journal Devoted to Pharmacological and Toxicological Evaluation of Natural Product Derivatives. 2010;**24**(5):673-679

[6] Khadre MA, Yousef AE, Kim JG. Microbiological aspects of ozone applications in food: A review. Journal of Food Science. 2001;**66**(9):1242-1252

[7] Ingle AP, Duran N, Rai M. Bioactivity, mechanism of action, and cytotoxicity of copper-based nanoparticles: A review. Applied Microbiology and Biotechnology. 2014;**98**(3):1001-1009

[8] Sun D, Babar Shahzad M, Li M, Wang G, Xu D. Antimicrobial materials with medical applications. Materials Technology. 2015;**30**(suppl. 6):B90-B95

[9] Barka EA, Vatsa P, Sanchez L, Gaveau-Vaillant N, Jacquard C, Klenk HP, et al. Taxonomy, physiology, and natural products of Actinobacteria. Microbiology and Molecular Biology Reviews. 2016;**80**(1):1-43

[10] Anandan R, Dharumadurai D, Manogaran GP. An introduction to actinobacteria. In: Actinobacteria-Basics and Biotechnological Applications. London, UK: IntechOpen Limited; 2016

[11] Amabile-Cuevas CF, Cardenas-Garcia M, Ludgar M. Antibiotic resistance. American Scientist. 1995;**83**(4):320-329

[12] Schatz A, Bugle E, Waksman SA. Streptomycin, a substance exhibiting antibiotic activity against gram-positive and gram-negative bacteria. Proceedings of the Society for Experimental Biology and Medicine. 1944;**55**(1):66-69

[13] Kardos N, Demain AL. Penicillin: The medicine with the greatest impact on therapeutic outcomes. Applied Microbiology and Biotechnology. 2011;**92**(4):677

[14] Davies J, Davies D. Origins and evolution of antibiotic resistance. Microbiology and Molecular Biology Reviews. 2010;**74**(3):417-433

[15] Rice LB. Federal funding for the study of antimicrobial resistance in nosocomial pathogens: no ESKAPE. The Journal of Infectious Diseases. 2008;**197**:1079-1081

[16] Hassan QP, Bhat AM, Shah AM. Bioprospecting Actinobacteria for bioactive secondary metabolites from untapped ecoregions of the northwestern Himalayas. In: New and Future Developments in Microbial Biotechnology and Bioengineering. Oxford, United Kingdom: Susan Dennis Chemistry and Chemical Engineering

Books Publisher at Elsevier; 2019. pp. 77-85

[17] Shah AM, Hussain A, Mushtaq S, Rather MA, Shah A, Ahmad Z, et al. Antimicrobial investigation of selected soil actinomycetes isolated from unexplored regions of Kashmir Himalayas, India. Microbial Pathogenesis. 2017;**110**:93-99

[18] Hussain A, Rather MA, Shah AM, Bhat ZS, Shah A, Ahmad Z, et al. Antituberculotic activity of actinobacteria isolated from the rare habitats. Letters in Applied Microbiology. 2017;**65**(3):256-264

[19] Hussain A, Rather MA, Dar MS, Dangroo NA, Aga MA, Shah AM, et al. Streptomyces puniceus strain AS13., production, characterization and evaluation of bioactive metabolites: A new face of dinactin as an antitumor antibiotic. Microbiological Research. 2018;**207**:196-202

[20] Hussain A, Rather MA, Bhat ZS, Majeed A, Maqbool M, Shah AM, et al. In vitro evaluation of dinactin, a potent microbial metabolite against mycobacterium tuberculosis. International Journal of Antimicrobial Agents. 2019;**53**(1):49-53

[21] Hussain A, Dar MS, Bano N, Hossain MM, Basit R, Bhat AQ, et al. Identification of dinactin, a macrolide antibiotic, as a natural product-based small molecule targeting Wnt/β-catenin signaling pathway in cancer cells. Cancer Chemotherapy and Pharmacology. 2019;**84**(3):551-559

[22] Shah AM, Wani A, Qazi PH, Rehman SU, Mushtaq S, Ali SA, et al. Isolation and characterization of alborixin from Streptomyces scabrisporus: A potent cytotoxic agent against human colon (HCT-116) cancer cells. Chemico-Biological Interactions. 2016;**256**:198-208

[23] Singh VP, Sharma R, Sharma V, Raina C, Kapoor KK, Kumar A, et al. Isolation of depsipeptides and optimization for enhanced production of valinomycin from the North-Western Himalayan cold desert strain Streptomyces lavendulae. The Journal of Antibiotics. 2019;**72**(8):617-624

[24] Rather SA, Shah AM, Ali SA, Dar RA, Rah B, Ali A, et al. Isolation and characterization of Streptomyces tauricus from Thajiwas glacier—A new source of actinomycin-D. Medicinal Chemistry Research. 2017;**26**(9):1897-1902

[25] Axenov-Gribanov DV, Voytsekhovskaya IV, Tokovenko BT, Protasov ES, Gamaiunov SV, Rebets YV, et al. Correction: Actinobacteria isolated from an underground lake and moonmilk speleothem from the biggest conglomeratic karstic cave in Siberia as sources of novel biologically active compounds. PLoS One. 2016;**11**(3): e0152957

[26] Donia M, Hamann MT. Marine natural products and their potential applications as anti-infective agents. The Lancet Infectious Diseases. 2003;**3**(6):338-348

[27] Fenical W. Chemical studies of marine bacteria: Developing a new resource. Chemical Reviews. 1993;**93**(5):1673-1683

[28] Baltz RH. Marcel Faber roundtable: Is our antibiotic pipeline unproductive because of starvation, constipation or lack of inspiration? Journal of Industrial Microbiology and Biotechnology. 2006;**33**(7):507-513

[29] Dhakal D, Pokhrel AR, Shrestha B, Sohng JK. Marine rare actinobacteria: Isolation, characterization, and strategies for harnessing bioactive compounds. Frontiers in Microbiology. 2017;**8**:1106

[30] Salomon CE, Magarvey NA, Sherman DH. Merging the potential of

microbial genetics with biological and chemical diversity: An even brighter future for marine natural product drug discovery. Natural Product Reports. 2004;**21**(1):105-121

[31] Ansari MZ, Yadav G, Gokhale RS, Mohanty D. NRPS-PKS: A knowledge-based resource for analysis of NRPS/PKS megasynthases. Nucleic Acids Research. 2004;**32**(suppl_2): W405-W413

[32] Gokhale RS, Tuteja D. Biochemistry of polyketide synthases. Biotechnology Set. 2001;**10**:341-372

[33] Hopwood DA. Genetic contributions to understanding polyketide synthases. Chemical Reviews. 1997;**97**(7):2465-2498

[34] Staunton J, Weissman KJ. Polyketide biosynthesis: A millennium review. Natural Product Reports. 2001;**18**(4):380-416

[35] Khosla C, Gokhale RS, Jacobsen JR, Cane DE. Tolerance and specificity of polyketide synthases. Annual Review of Biochemistry. 1999;**68**(1):219-253

[36] Du L, Shen B. Biosynthesis of hybrid peptide-polyketide natural products. Current Opinion in Drug Discovery & Development. 2001;**4**(2):215-228

[37] Hertweck C. The biosynthetic logic of polyketide diversity. Angewandte Chemie International Edition. 2009;**48**(26):4688-4716

[38] Sieber SA, Marahiel MA. Learning from Nature's drug factories: Nonribosomal synthesis of macrocyclic peptides. Journal of Bacteriology. 2003;**185**(24):7036-7043

[39] Grünewald J, Marahiel MA. Chemoenzymatic and template-directed synthesis of bioactive macrocyclic

peptides. Microbiology and Molecular Biology Reviews. 2006;**70**(1):121-146

[40] Donadio S, Monciardini P, Sosio M. Polyketide synthases and nonribosomal peptide synthetases: The emerging view from bacterial genomics. Natural Product Reports. 2007;**24**(5):1073-1109

[41] Jenke-Kodama H, Dittmann E. Bioinformatic perspectives on NRPS/PKS megasynthases: Advances and challenges. Natural Product Reports. 2009;**26**(7):874-883

[42] Li Y, Weissman KJ, Müller R. Insights into multienzyme docking in hybrid PKS–NRPS megasynthetases revealed by heterologous expression and genetic engineering. Chem Bio Chem. 2010;**11**(8):1069-1075

[43] Beck C, Garzón JF, Weber T. Recent advances in re-engineering modular PKS and NRPS assembly lines. Biotechnology and Bioprocess Engineering. 2020;**1-9**:886-894

[44] Chevrette MG, Carlson CM, Ortega HE, Thomas C, Ananiev GE, Barns KJ, et al. The antimicrobial potential of Streptomyces from insect microbiomes. Nature Communications. 2019;**10**(1):1-1

[45] Trenozhnikova L, Azizan A. Discovery of actinomycetes from extreme environments with potential to produce novel antibiotics. Central Asian Journal of Globalization and Health. 2018;**7**(1):337

[46] Jose PA, Jebakumar SR. Unexplored hypersaline habitats are sources of novel actinomycetes. Frontiers in Microbiology. 2014;**5**:242

[47] Williams ST, Locci R, Beswick A, Kurtböke DI, Kuznetsov VD, Le Monnier FJ, et al. Detection and identification of novel actinomycetes.

Research in Microbiology.
1993;**144**(8):653-656

[48] Ningsih F, Yokota A, Sakai Y,
Nanatani K, Yabe S, Oetari A, et al.
Gandjariella thermophila gen. Nov., sp.
nov., a new member of the family
Pseudonocardiaceae, isolated from
forest soil in a geothermal area.
International Journal of Systematic and
Evolutionary Microbiology.
2019;**69**(10):3080-3086

Multiplicity in the Genes of Carbon Metabolism in Antibiotic-Producing Streptomycetes

Toshiko Takahashi, Jonathan Alanís, Polonia Hernández and María Elena Flores

Abstract

Streptomycetes exhibit genetic multiplicity, like many other microorganisms, and redundancy occurs in many of the genes involved in carbon metabolism. The enzymes of the glycolytic pathway presenting the greatest multiplicity were phosphofructokinase, fructose 1,6-bisphosphate aldolase, glyceraldehyde-3-phosphate dehydrogenase, and pyruvate kinase. The genes that encode citrate synthase and subunits of the succinate dehydrogenase complex are the ones that show the greatest multiplicity, while in the phosphoenolpyruvate-pyruvate-oxaloacetate node, only malic enzymes and pyruvate phosphate dikinase present two copies in some *Streptomyces*. The extra DNA from these multiple gene copies can be more than 50 kb, and the question arises whether all of these genes are transcribed and translated. As far as we know, there is few information about the transcription of these genes in any of this *Streptomyces*, nor if any of the activities that are encoded by a single gene could be limiting both for growth and for the formation of precursors of the antibiotics produced by these microorganisms. Therefore, it is important to study the transcription and translation of genes involved in carbon metabolism in antibiotic-producing *Streptomyces* growing on various sugars.

Keywords: gene multiplicity, carbon metabolism, *Streptomyces*, antibiotic, glycolytic pathway

1. Introduction

Gene multiplicity or redundancy is a characteristic of microorganisms and means there are two or more genes coding for proteins that perform the same function. Inactivation of any of these genes does not affect or has little relevance to the biological phenotype [1]. Gene redundancy has been observed in all organisms, including prokaryotes and eukaryotes, and is particularly important for actinobacteria that produce metabolites of industrial interest [2]. The most significant property of *Streptomyces* species is the production of several secondary metabolites (antibiotics and biologically active compounds) that can be generated at the industrial level. These metabolites are beneficial and indispensable for human and animal health [3]. Although the structures of secondary metabolites are diverse, they have been classified into at least five classes (sugar, polyketide, shikimate, amino acid, and terpene pathways) based on the precursor molecules incorporated during their biosynthesis. All precursors are derived from central primary metabolism, glycolysis (Embden–Meyerhof–Parnas EM pathway), the pentose phosphate pathway (PP), and the tricarboxylic acid (TCA) cycle. Glucose-6-phosphate, a precursor of aminoglycosides, is supplied from the early stage of the EM pathway, and acetyl-CoA and succinyl-CoA as precursors of polyketides are supplied from the final stage of the EM pathway and TCA cycle, respectively. Aromatic amino acids, as precursors of chloramphenicol, are supplied from the PP pathway, whereas other amino acids are supplied from the central metabolism to be precursors of peptide antibiotics. Acetyl-CoA, glyceraldehyde-3-phosphate, and pyruvate (PYR) are precursors of isopentenyl diphosphate and dimethylallyl diphosphate, which are the building blocks of terpenes. Malonyl-CoA, methylmalonyl-CoA, and ethylmalonyl-CoA are used as the extension units in macrolide antibiotic biosynthesis [4], and reduced cofactor (NADPH) is used during secondary metabolite biosynthesis and is generated from the PP pathway and TCA cycle. Accordingly, the dynamics of central metabolism, including the EM and PP pathways and the TCA cycle, which generate primary metabolites (as precursors for secondary metabolites) and cofactors, will influence the biosynthetic process of secondary metabolites [5].

One little-studied area is the carbon metabolism in these organisms. Few studies have examined the presence of genes that participate in the glycolytic pathway, TCA cycle, or phosphoenolpyruvate-pyruvate-oxaloacetate (PEP-PYR-OXA) node. In general, microorganisms metabolize glucose through the glycolysis and hexose monophosphate pathways [6]. Many of the intermediates in these metabolic pathways are used to synthesize other essential bacterial compounds (amino acids, polysaccharides, nucleic acids, lipids, fatty acids, and antibiotics).

The EMP pathway consists of nine reactions, in which the final product is pyruvate. The first reaction consists of the isomerization of glucose 6-phosphate to fructose 6-phosphate catalyzed by phosphoglucose isomerase. Another phosphate with adenosine triphosphate (ATP) as the donor is incorporated into fructose 6-phosphate by phosphofructokinase. The next step is the cleavage of fructose 1,6-diphosphate by aldolase, generating dihydroxyacetone phosphate and glyceraldehyde 3-phosphate, which can be interconverted by triose phosphate isomerase. Glyceraldehyde 3-phosphate is oxidized to 1,3-diphosphoglycerate by glyceraldehyde 3-phosphate dehydrogenase to generate NADH. The phosphoglycerate kinase-catalyzed reaction generates an ATP molecule, an example of substrate-level phosphorylation. The 3-phosphoglycerate is converted to 2-phosphoglycerate by phosphoglycerate mutase and is subsequently dehydrated by enolase. Phosphoenolpyruvate (PEP) is used to generate another molecule of ATP and PYR through a reaction catalyzed by pyruvate kinase. Then, four ATP molecules are generated, and two high-energy phosphate bonds are used; thus, the net gain is two ATPs per oxidized glucose molecule [6].

The TCA cycle is one of the most important metabolic pathways, not only as part of catabolism but also as an important intermediate for amino acid biosynthesis and synthesizing secondary metabolites. Generally, the citric acid cycle is the main oxidation pathway for carbon chains of carbohydrates, fatty acids, and many amino acids to CO_2 and water. At each turn of the cycle, two molecules of CO_2 are released. Most of the energy generated during oxidation is stored as NADH, FADH, or ATP (or GTP).

Because the intermediates of the TCA cycle are used as precursors in other pathways, they are replenished through anaplerotic reactions. Under normal conditions, the reactions that take intermediates in the cycle and those that replace them are kept in dynamic equilibrium; therefore, their concentrations remain constant. The most common anaplerotic reactions are those in which PYR or PEP are converted to Oxaloacetic acid (OXA) or malate. For example, this first reaction can be mediated by PEP carboxylase in some plants, yeasts, and bacteria or by the malic enzyme (ME), which is widely distributed in prokaryotes and eukaryotes [7].

2. Importance of *Streptomyces*

Streptomycetes are bacteria that produce the largest amount of commercially used antibiotics worldwide. To date, many *Streptomyces* genomes have been sequenced, and any approach that offers the possibility of increasing yields must be analyzed to increase production levels or produce more effective compounds against bacterial infections. Currently, there is great interest in the isolation of new *Streptomyces* strains from unusual environments as new sources of antibiotics [8].

Many antibiotics have precursors that act as intermediates for different metabolic pathways. Gunnarsson et al. (2004) described how central carbon metabolism is linked to producing many different antibiotics [9]; however, little has been achieved to improve the synthesis of inermediates and influence the biosynthesis of antibiotics or other commercially important compounds.

Genome sequencing of important antibiotic-producing *Streptomyces* has allowed the analysis of genes that encode the enzymes involved in carbon metabolism pathways. Gene multiplicity exists in the genomes of these organisms and up to four copies of the same gene can be found; however, the relevance of this fact has not been established. Gene multiplicity or redundancy is very common in the chromosomes of many microorganisms, especially *Streptomyces*.

3. Gene multiplicity in glycolytic pathway genes

There are many databases, such as KEGG, to identify how many genes code for the same activity. It is known that nine enzymes participate in the glycolytic pathway, as shown in **Table 1**, and it was found that three genes encode phosphofructokinase, which catalyzes the phosphorylation of fructose 6-phosphate in most antibiotic-producing *Streptomyces* species. Only *S. coelicolor* and *S. venezuelae* have two genes.

The conversion of fructose 6-phosphate to fructose 1,6-bisphosphate with the concomitant hydrolysis of adenosine triphosphate represents the first irreversible step specific to glycolysis. This reaction catalyzed by phosphofructokinase (PFK; EC 2.7.1.11) is subjected to tight control, thus rendering it a critical regulatory point of the glycolytic flux [10]. The genes that encode PFK present a high multiplicity, and in these 11 antibiotic-producing *Streptomyces*, 31 proteins have been noted. The PFKs have amino acids between 341 and 345, and only *S. griseus* and *S. venezuelae* have two copies, whereas the remainder has three genes. The 31 proteins have a very high identity with some identical regions along the sequence, and at the amino-terminal end, a highly conserved domain GGDCPGLNAVIR is present; 133 residues out of 350 are fully conserved, representing 38% identity. Despite the high resemblance, the phylogenetic tree is divided into two clades, one of which is split into two subgroups, as shown in the tree, each copy is distributed in one clade, and two of them are more closely related to each other. This clade includes *S. griseus* and *S. venezuelae*, which have only two copies of the PFK-coding genes, suggesting that retention of a third gene copy has not occurred in these species (**Figure 1**). Unlike *Escherichia coli*, which has two PFKs that do not have a common ancestor, *Streptomyces* does [11].

	SALS	SAVERM	SCLF	SCO	SGR	SHJGH	SKA	SLAV	SNOUR	SPRI	SVEN
Antibiotic	Salinomycin	Avermectin	Clavulanic acid	Actinorhodin, Undecylprodigiosine	Streptomycin	Rapamycin	Kanamycin	Streptothricin	Nystatine	streptogramin	Chloramphenicol
Phosphoglucomutase		SAVERM_803 (556)		SCO7443 (546)	SGR_6728 (547)	SHJGH_1760 (546)					
Phosphoglucose isomerase	SLNWT_5962 (550)	SAVERM_1770 (549) SAVERM_6302 (550)	BB341_22345 (550)	SCO1942 (551) SCO6659 (550)	SGR_5578 (552)	SHJGH_3162 (550) SHJGH_7334 (550)	CP970_33250 (550)	SLAV_27865 (553)	SNOUR_30680 (558)	SPRL_5587 (552)	SVEN_1571 (556)
Phosphofructokinase	SLNWT_1861 (341) SLNWT_5764 (345) SLNWT_7260 (341)	SAVERM_2822 (341) SAVERM_6083 (342) SAVERM_7123 (341)	BB341_07185 (341) BB341_21445 (342) BB341_25335 (341)	SCO1214 (341) SCO2119 (342) SCO5426 (341)	SGR_2110 (341) SGR_6306 (341)	SHJGH_2417 (341) SHJGH_3367 (342) SHJGH_6262 (341)	CP970_13335 (341) CP970_32070 (342) CP970_41110 (341)	SLAV_12675 (341) SLAV_26605 (343) SLAV_31335 (342)	SNOUR_05765 (341) SNOUR_14515 (341) SNOUR_29760 (342)	SPRL_2465 (341) SPRL_5380 (342) SPRL_6334 (341)	SVEN_0823 (341) SVEN_5078 (341)
Aldolase	SLNWT_0350 (285) SLNWT_3143 (343)	SAVERM_1445 (301) SAVERM_4523 (340)	BB341_12300 (283) BB341_15265 (340)	SCO3649 (343) SCO5852 (282)	SGR_285 (289) SGR_3418 (343)	SHJGH_2271 (286) SHJGH_5149 (340)	CP970_20370 (343) CP970_37965 (278)	SLAV_00910 (278) SLAV_20325 (340) SLAV_32740 (281) SLAV_38480 (278)	SNOUR_09120 (293) SNOUR_22110 (340)	SPRL_0508 (299) SPRL_3596 (340)	SVEN_3414 (340) SVEN_7329 (278)
Triose phosphate isomerase	SLNWT_5959 (259)	SAVERM_6298 (258)	BB341_22330 (258)	SCO0578 (259) SCO1945 (258)	SGR_5575 (258)	SHJGH_3166 (258)	CP970_33235 (261)	SLAV_27850 (262)	SNOUR_30660 (258)	SPRL_5584 (262)	SVEN_1574 (259)
Glyceraldehyde 3P-dehydrogenase	SLNWT_5957 (335)	SAVERM_2990 (334) SAVERM_6296 (335)	BB341_01000 (481) BB341_22320 (335)	SCO1947 (336) SCO7040 (481) SCO7511 (332)	SGR_5573 (336) SGR_936 (481)	SHJGH_1334 (337) SHJGH_1590 (332) SHJGH_1800 (481) SHJGH_3168 (335)	CP970_08935 (481) CP970_33225 (336)	SLAV_27840 (334) SLAV_31635 (481) SLAV_35550 (332)	SNOUR_30650 (334) SNOUR_35775 (481)	SPRL_5582 (336) SPRL_6547 (481)	SVEN_0459 (461) SVEN_1576 336 SVEN_7344 (331)

	SALS	SAVERM	SCLF	SCO	SGR	SHJGH	SKA	SLAV	SNOUR	SPRI	SVEN
Phosphoglycerate kinase	SLNWT_5958 (403)	SAVERM_6297 (403)	BB341_22325 (403)	SCO1946 (403)	SGR_5574 (403)	SHJGH_3167 (403)	CP970_33230 (403)	SLAV_27845 (403)	SNOUR_30655 (403)	SPRI_5583 (403)	SVEN_1575 (403)
Phosphoglycerate mutase	SLNWT_3230 (511)	SAVERM_3979 (253)	BB341_12505 (252) BB341_19285 (217)	SCO4209 (253) SCO6818 (511)	SGR_4005 (253)	SHJGH_4634 (253)	CP970_23555 (253)	SLAV_17715 (252)	SNOUR_07945 (218) SNOUR_18235 (253)	SPRI_4058 (253)	SVEN_3958 (252)
Enolase	SLNWT_1793 (427)	SAVERM_3533 (428)	BB341_17295 (426)	SCO3096 (426) SCO7638 (434)	SGR_4439 (426) SGR_6721 (434)	SHJGH_4327 (431)	CP970_25900 (428)	SLAV_05265 (435) SLAV_22580 (426)	SNOUR_24895 (426)	SPRI_0700 (432) SPRI_4452 (428)	SVEN_2899 (428) SVEN_6040 (433)
Pyruvate kinase	SLNWT_1865 (474) SLNWT_5886 (478)	SAVERM_2825 (476) SAVERM_6217 (478)	BB341_07200 (474) BB341_22070 (477)	SCO2014 (478) SCO5423 (476)	SGR_2113 (476) SGR_5516 (479)	SHJGH_3253 (478) SHJGH_6259 (457)	CP970_13350 (476) CP970_32715 (478)	SLAV_12690 (476) SLAV_27525 (475)	SNOUR_14530 (475) SNOUR_30335 (480)	SPRI_2468 (475) SPRI_5515 (474)	SVEN_1640 (475) SVEN_5075 (476)

Streptomyces lavendulae (SLAV), Streptomyces albus DSM 41398 (SALS), Streptomyces clavuligerus F613-1 (SCLF), Streptomyces pristinaespiralis HCCB 10218 (SPRI), Streptomyces kanamyceticus ATCC 12853 (SKA), Streptomyces noursei ATCC 11455 (SNOUR), Streptomyces hygroscopicus subsp. jinggagensis TL01 (SHJGH), Streptomyces coelicolor A3(2) (SCO), Streptomyces avermitilis MA-4680 (SAVERM), Streptomyces griseus subsp. griseus NBRC 13350 (SGR), Streptomyces venezuelae ATCC 10712 (SVEN). The number of amino acids of each encoded protein is shown in parentheses.

Table 1.
Gene multiplicity in glycolytic pathway genes.

```
slx_SLAV_12675      ---MRIGVLTAGGDCPGLNAVIRSVVHRAVVGHGDEVIGFEDGFKGLLDGHYRPLDINAV
sgr_SGR_2110        ---MRIGILTAGGDCPGLNAVIRSVVHRAVVGHGDEVIGFEDGFKGLLDGHFRPLDLNAV
sclf_BB341_07185    ---MRIGVLTAGGDCPGLNAVIRSVVHRAITGYGDEVIGFEDGFKGLLDGHYRPLDLNAV
spri_SPRI_2465      ---MRIGVLTAGGDCPGLNAVIRSVVHRAMTGHGDEVIGFEDGFKGLLDGHYRPLDLNAV
sve_SVEN_5078       ---MRIGVLTAGGDCPGLNAVIRSVVHRALTGHDDEVIGFEDGFKGLLDGHYRPLDLNAV
snr_SNOUR_14515     ---MRIGVLTAGGDCPGLNAVIRSVVHRALTGHGDEVIGFEDGFKGLLDGRFRKLDLDAV
sco_SCO5426         ---MRIGVLTAGGDCPGLNAVIRSVVHRAVDNYGDEVIGFEDGYAGLLDGRYRALDLNAV
sma_SAVERM_2822     ---MRIGVLTAGGDCPGLNAVIRSVVHRAVTMYGDEVIGFEDGYAGLLDGRYRALDLNAV
sho_SHJGH_6262      ---MRIGVLTAGGDCPGLNAVIRSVVHRAVAQYGDEVIGFEDGYRGLLDRHYRTLDLDAV
ska_CP970_13335     ---MRIGILTAGGDCPGLNAVIRSVVHRAVTHYGDEVIGFEDGYAGLLDGRYRPLDLNAV
sals_SLNWT_1861     ---MRIGVLTAGGDCPGLNAVIRSVVHRAVAQYGDEVIGFEDGYAGLLEGRYRPLDLNAV
sgr_SGR_6306        ---MRIGVLTSGGDCPGLNAVIRSVVHRAVVDHGDEVIGFHDGWKGLLECDYRKLDLDAV
sve_SVEN_0823       ---MRIGVLTSGGDCPGLNAVIRSVVHRAVVDHGDEVIGFHDGWKGLLECDYRKLDLEAV
sclf_BB341_25335    ---MRIGVLTSGGDCPGLNAVIRSVVHRAVVDHGDEVIGFHDGWRGLLECDYRKLDLDAV
spri_SPRI_6334      ---MRIGVLTSGGDCPGLNAVIRSVVHRAVVDHGDEVIGFHDGWKGLLECDYRKLDLDAV
snr_SNOUR_05765     ---MRIGVLTSGGDCPGLNAVIRSVVHRATADHGDEVIGFKDGWKGLLECDYRKLDLDAV
sho_SHJGH_2417      ---MRIGVLTSGGDCPGLNAVIRSVVHRAVADHGDEVIGFRDGWKGLLECDYLKLDLDAV
sco_SCO1214         ---MRIGVLTSGGDCPGLNAVIRSVVHRAVVDHGDEVIGFRDGWKGLLECDYLKLDLDAV
sma_SAVERM_7123     ---MRIGVLTSGGDCPGLNAVIRSVVHRAVVDHGDEVIGFHDGWKGLLECDYLKLDLDAV
ska_CP970_41110     ---MRIGVLTSGGDCPGLNAVIRSVVHRAVVDHGDEVIGFHDGWKGLLECDYRKLDLDAV
sals_SLNWT_7260     ---MRPMGVLTSGGDCPGLNAVIRSVVHRAVVDHGDDVIGFHDGWKGLLECDYRKLDLDAV
slx_SLAV_31335      ---MRIGVLTSGGDCPGLNAVIRSVVHRAVVDHGDEVIGFHDGWRGLLECDYRKLDLDAV
slx_SLAV_26605      ---MKVGVLTGGGDCPGLNAVIRAVVRKGTQEYGYGFVGFKDGWRGPVEGDAVPLGIPAV
spri_SPRI_5380      ---MRVGVLTGGGDCPGLNAVIRGLVRKGVQEYGYEFVGYRDGWRGPLEGDTVPLDIPTV
sma_SAVERM_6083     ---MRVGVLTGGGDCPGLNAVIRGVVRKGVQEYGYDFVGFRDGWRGPLEGDAVRLDIPAV
ska_CP970_32070     ---MRVGVLTGGGDCPGLNAVIRGVVRKGVQEYGYDFVGFRDGWRGPLEGDTVQLDIPAV
snr_SNOUR_29760     ---MRVGVLTGGGDCPGLNAVIRGIVRKGTQEYGYDFVGFRDGWRGPLEGRSMRLDIPAV
sho_SHJGH_3367      ---MRVGVLTGGGDCPGLNAVIRGIVRKGVQEYGYDFVGFRDGWRGPLEGDTVRLDIPAV
sco_SCO2119         ---MKVGVLTGGGDCPGLNAVIRAVVRKGVQEYGYDFTGFRDGWRGPLEGDTVPLDIPAV
sclf_BB341_21445    ---MRVGVLTGGGDCPGLNAVIRGAVRKGVQEYGYEF1GYRDGWRGPLEGDTLPLAIPNV
sals_SLNWT_5764     MSVTRIGVLTGGGDCPGLNAVIRAVVRKGVQSHGYEFTGFRDGWRGTLDGRTVKLDIPAV
                       ::*:**.************.  *::.    :.   . *:.**:  *  ::     *  :  *
```

A

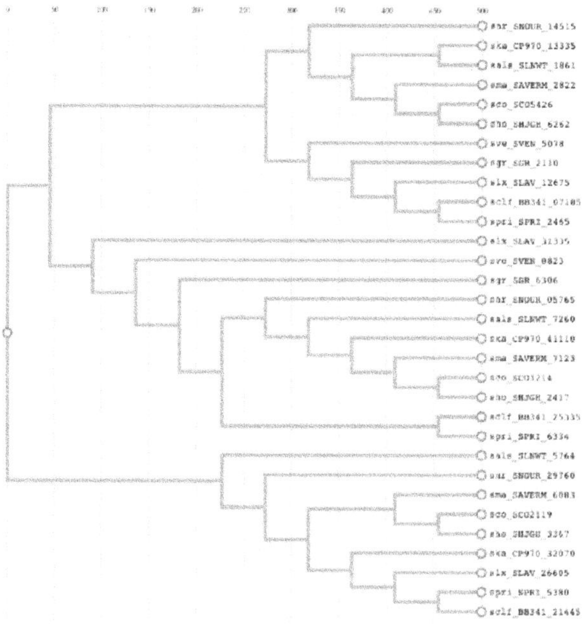

B

Figure 1.
Partial multiple alignments of fosfofructokinase proteins from antibiotic producing Streptomyces *(A) and derived bootstrapped tree (B).* slx, *Streptomyces lavendulae,* sals, *Streptomyces albus DSM 41398,* sclf, *Streptomyces clavuligerus F613-1,* spri, *Streptomyces pristinaespiralis HCCB 10218,* ska, *Streptomyces kanamyceticus ATCC 12853,* snr, *Streptomyces noursei ATCC 11455,* sho, *Streptomyces hygroscopicus subps. jinggagensis TL01,* sco, *Streptomyces coelicolor A3(2),* sma, *Streptomyces avermitilis MA-4680,* sgr, *Streptomyces griseus subsp. griseus NBRC 13350,* sve, *Streptomyces venezuelae ATCC 10712.*

The next reaction is performed using aldolase (EC 4.1.2.13), which catalyzes the conversion of fructose 1-6-diphosphate to glyceraldehyde 3-phosphate and dihydroxy-acetone phosphate. Most *Streptomyces* have two genes that code for this enzyme, except for *S. lavendulae*, which has four genes. The SLAV_00910 and SLAV_38480 proteins have 100% identity; therefore, are coded by duplicated genes.

Triosephosphate isomerase (EC 5.3.1.1) is an enzyme that converts dihydroxy-acetone phosphate and glyceraldehyde-3-phosphate. Of all the antibiotic-producing *Streptomyces*, only *S. coelicolor* had two proteins for this activity with an identity of 31% between the two copies. The average molecular weight of all the triose phosphate isomerases is 27 kDa with an identity between them greater than 85%, except for SCO0578.

Another *Streptomyces* gene that showed genetic redundancy was the one coding for glyceraldehyde 3-phosphate dehydrogenase. Glyceraldehyde 3P-dehydrogenase enzyme shows two classes of proteins, some with 331–335 amino acids (approx. 36.1 kDa) and others with a higher molecular weight, the majority of 481 residues (52.9 kDa), except for *S. venezuelae* with 461 amino acids (50.1 kDa). Only *S. albus* has a single small protein and the rest of the microorganisms had two or three, one large and one or two small. *S. hygroscopicus* has four paralogues, three of approximately 335 amino acids and one of 481 residues. Alignment of the amino acid sequences of glyceraldehyde 3P-dehydrogenase showed that the larger proteins (SLAV_31635, SPRI_6547, SVEN_0459, CP970_08935, SNOUR_35775, SHJGH_1800, SCO7040, SGR_936, and BB341_01000) had an extended N-terminal region of approximately 126 residues. The highly conserved regions GFGRIGR, ASCTTNA, PRVPV, and WYDNEXG were near or into the NAD-binding and C-terminal domains of the enzyme (**Figure 2**). In the phylogenetic analysis, the sequences are divided into two lineages. The first clade is formed by small-sized proteins, whereas the other included all proteins with large-size copies. Small protein clades are split into two subgroups. One is formed by a single organism with the copy with less divergence, and the other subgroup is divided into two clades, each containing a single homolog of each organism; therefore, the copies diverged to the point of being more related between species than between duplicates of the same species (**Figure 2**).

In all antibiotic-producing *Streptomyces,* the only reaction whose enzyme is encoded by one gene is 2,3-bisphosphoglycerate-dependent phosphoglycerate kinase (EC 2.7.2.3), which catalyzes the reversible conversion of 1,3-diphospho-glycerate to 3-phosphoglycerate with the generation of an ATP molecule. All proteins were 403 amino acids long and had high identity with each other. The phosphoglycerate kinase of *S. clavuligerus*, an overproducer of clavulanic acid, was detected in the proteome of this microorganism grown in tryptic soy broth, indicating that its gene was expressed in this culture medium [12].

Phosphophoglycerate mutase (EC 5.4.2.11) performs the internal transfer of a phosphate group from the C-3 carbon to the C-2 carbon, resulting in the isomerization of 3-phosphoglycerate to 2-phosphoglycerate. Only two genes are involved in this activity in *S. griseus, S. clavuligerus, S. coelicolor*, and *S. noursei*. Most of the proteins are 252–253 amino acids in length, which are very similar (identity greater than 90%), and there are only two proteins with 511 amino acids long, SLNWT_3230 and SCO6818 also, with 82% identity and 91.6% similarity between them. One of the two *S. noursei* mutases (SNOUR_07945) is smaller (218 amino acid residues), with 22% identity to 253 amino acid mutases and 92% identity to histidine phosphatases of various *Streptomyces*. Then probably, this mutase is wrongly annotated in its genome.

Enolase (EC 4.2.1.11), also known as phosphopyruvate hydratase, is a metalloenzyme responsible for converting 2-phosphoglycerate to PEP. *S. coelicolor*,

```
slx_SLAV_27840    ------MTIRVGINGFGRIGRNYFRALLEQGAD-------IEIVGVNDLTDNATLVHLLK
snr_SNOUR_30650   ------MTIRVGINGFGRIGRNYFRALLEQGAD-------IEIVGVNDLTDNATLVHLLK
ska_CP970_33225   ------MTIRVGINGFGRIGRNYFRALLEQGAD-------IEIVAVNDLGDTATTAHLLK
sgr_SGR_5573      ------MTIRVGINGFGRIGRNYFRALLEQGAD-------IEIVAVNDLGDTATTAHLLK
sco_SCO1947       ------MTIRVGINGFGRIGRNYFRALLEQGAD-------IEIVAVNDLGDTATTAHLLK
spri_SPRI_5582    ------MTIRVGINGFGRIGRNYFRALLEQGAK-------IEIVAVNDLGDTATTAHLLK
sve_SVEN_1576     ------MTIRVGINGFGRIGRNYFRALLDQGAD-------IEIVAVNDLGDTATTAHLLK
sclf_BB341_22320  ------MTIRVGINGFGRIGRNYFRALLEQGAD-------IEIVAVNDLGDTATTAHLLK
sma_SAVERM_6296   ------MTIRVGINGFGRIGRNYFRALLEQGAD-------IEIVAVNDLGDTATTAHLLK
sals_SLNWT_5957   ------MTIRVGINGFGRIGRNYFRALLEQGAD-------IEIVAVNDLGDTATTAHLLK
sho_SHJGH_3168    ------MTIRVGINGFGRIGRNYFRALLEQGAD-------IEVVAVNDLGDTATTAHLLK
sma_SAVERM_2990   ------MTVRVGINGFGRIGRNVFRAAATRGAD-------LEIVAVNDLGDVATMAHLLA
slx_SLAV_35550    -------MTRIAINGFGRIGRNVLRALLERDSD-------LDVVAVNDLTEPATLARLLA
sve_SVEN_7344     -------MTRIAINGFGRIGRNVLRALLERDTE-------LEVVAVNDLTEPTALARLLA
sho_SHJGH_1590    -------MTRIAINGFGRIGRNVLRALLERDSA-------LEIVAVNDLTEPATLARLLA
sco_SCO7511       -------MTRIGINGFGRIGRNVLFALLERDTK-------LEVVAVNDLTEPATLARLLA
sho_SHJGH_1334    ------MTVRVGINGFGRIGRTYLRAALDRAEAGT---QDVHVVAINDIAPPATIAHLLE
slx_SLAV_31635    NKIERGAGRDVVLYGFGRIGRLVARLLIEKAGSGNGLRLRAIVVRGGGEADLVKRASLLR
spri_SPRI_6547    NKIERGDGRDVVLYGFGRIGRLVARLLIEKSGSGNGLRLRAIVVRGGGEQDLVKRASLLR
sve_SVEN_0459     NKIERRDGRDVVLYGFGRIGPLVARLLIEKAGSGNGLRLRAIVVRGGGDQDLVKRASLLR
ska_CP970_08935   NKIERCAPRDVVLYGFGRIGRLVARLLIEKAGSGNGLRLRAVVVRRGGEQDLVKRASLLR
snr_SNOUR_35775   NKIECRQGRDVVLYGFGRIGRLVARLLIEKTGSGNGLRLRAIVVRQGGEQDIVKRASLLR
sho_SHJGH_1800    DKIERREPRDVVLYGFGRIGRLVARLLIEKAGSGNGLRLRAIVVRGGGAQDLVKRASLLR
sco_SCO7040       NKIDRREGRDVVLYGFGRIGRLVARLLIEKAGSGNGLRLRAIVVRGGGEQDLVKRASLLR
sgr_SGR_936       NKIERRASRDVVLYGFGRIGRLIARLLIEKAGSGNGLRLRAIVVRRGSGQDIVKRASLLR
sclf_BB341_01000  NKIERREGRDVVLYGFGRIGRLLARLLIEKAGSGNGLRLRAIVVRKGAGQDLVKRASLLR
                  : :  *******   *     :        :*.   .  . .**
```

A

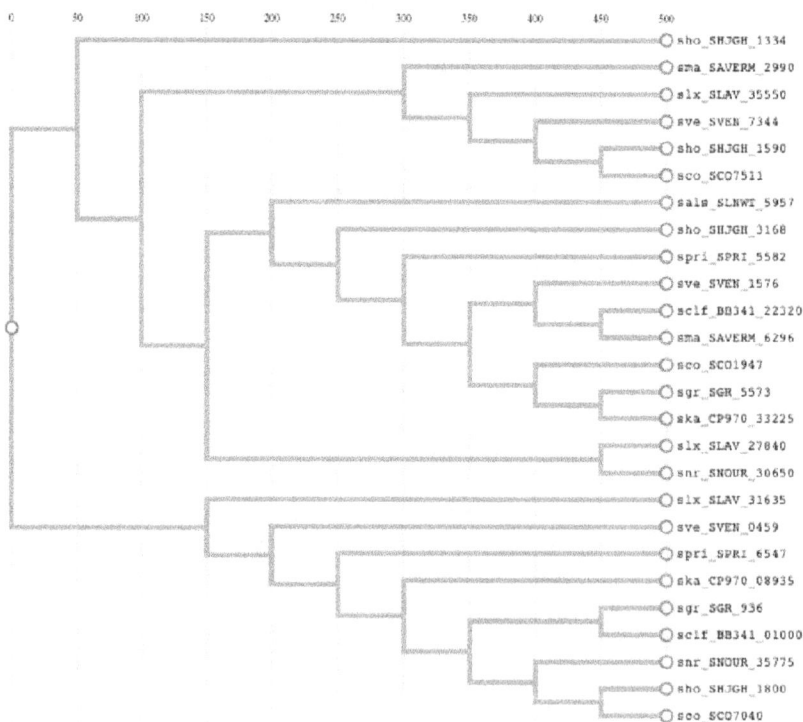

B

Figure 2.
Partial multiple alignment of glyceraldehyde 3-phosphate dehydrogenase proteins (A) in antibiotic producing Streptomyces *and derived bootstrapped tree (B).* slx, Streptomyces lavendulae, sals, Streptomyces albus DSM 41398, sclf, Streptomyces clavuligerus F613-1, spri, Streptomyces pristinaespiralis HCCB 10218, ska, Streptomyces kanamyceticus ATCC 12853, snr, Streptomyces noursei ATCC 11455, sho, Streptomyces hygroscopicus *subps.* jinggagensis TL01, sco, Streptomyces coelicolor A3(2), sma, Streptomyces avermitilis MA-4680, sgr, Streptomyces griseus *subsp.* griseus NBRC 13350, sve, Streptomyces venezuelae ATCC 10712.

S. griseus, *S. lavendulae*, *S. pristinaspiralis*, and *S. venezuelae* have two enolases, one with 426–428 amino acids and the other with 432–435 residues. The rest have only one enolase, which in most of these *Streptomyces* has 426 amino acids. Between the two proteins of different sizes, there is approximately 56% identity; however, between the small proteins of the different microorganisms, there is an identity greater than 90%, and between the larger ones, there is an identity greater than 83%.

Pyruvate kinase (EC 2.7.1.40) is a key enzyme involved in the last step of glycolysis that catalyzes the transfer of a phosphate group from PEP to ADP, yielding one molecule of PYR and one molecule of ATP. There are two types: type I and type II, and both enzymes show positive cooperative effects concerning PEP. The type I enzyme is activated by fructose 1,6-bisphosphate (F1,6BP) and the type II by AMP [13]. According to the amino acid sequences of PYKF and PYKA enzymes from *E. coli*, the smaller *Streptomyces* PYR kinases are type I, and the others are type II. Most proteins have 474–479 amino acids regardless of whether they are class I or class II and have greater than 60% identity between them.

4. TCA cycle

The TCA cycle is a central metabolic pathway of all aerobic organisms and synthesizes many important precursors and molecules [14] and eight reactions are performed for complete glucose oxidation.

The first reaction is citrate synthase (CS, EC 2.3.3.1), which catalyzes the irreversible conversion of OXA and acetyl-CoA into citrate. The proteins are classified into two types: type I and type II, and are encoded by four genes in eight of the selected antibiotic-producing *Streptomyces*, three in *S. venezuelae*, two in *S. noursei*, and *S. pristinaespiralis*. Their coding genes show the greatest redundancy among all the genes involved in carbon metabolism in most *Streptomyces* species (**Table 2**). Sals_SLNWT_1427 (*S. albus*) and ska_CP970_28100 (*S. kanamyceticus*) are the largest, with 439 and 451 amino acid residues, respectively. In the amino acid sequence alignment, two groups of CSs are distinguished: small ones ranging from 363 to 395 amino acid residues and large ones ranging from 416 to 451 residues. Small or large types of either type I or II have been found, implying that type does not depend on size. As shown in **Figure 3**, a highly conserved histidine is embedded in the conserved region GPLHGXA. Another conserved amino acid is arginine in the DPR conserved amino acid sequence and aspartic or glutamic acid in the conserved NVD/E. All of these were previously reported as essential residues involved in interactions with the substrate OXA in *Streptomyces* [15]. All CSs have a common ancestor but are separated into two main lineages. One of them is split into two subgroups, which include CS classified as types 1 and 2, like the proteins of *S. lavendulae* (type 1) and *S. pristinaespiralis* (type 2) with approximately 429–433 amino acid residues. The other clade was also split into two subgroups, including proteins with 366 and 390 residues. The CSs of *S. coelicolor* encoded by *sco2736*, *sco4388* (both classified as type 2), and *sco5832* (unclassified) are grouped in one of the main clades and distributed in the two subgroups. The sequence encoded by *sco5831* is grouped into the other principal clade with another seven proteins. The % of identity among the proteins included in each subgroup is 87%–94% and similarity 91%–98%. These differences are mainly observed at the amino terminus. Taking *S. coelicolor* CSs as an example, the identity between the CSs found in each subgroup, was low, between 24.6% and 28.1%, and the similarity between 46% and 49%.

Aconitase (EC 4.2.1.3), isocitrate dehydrogenase (EC 1.1.1.42), the E1 component of 2-oxoglutarate dehydrogenase (EC 1.2.4.2), and malate dehydrogenase (EC 1.1.1.37) are encoded by unique genes in all *Streptomyces*, probably because of the

	SALS	SAVERM	SCLF	SCO	SGR	SHJGH	SKA	SLAV	SNOUR	SPRI	SVEN
Citrate synthase	SLNWT_1427 (439)	SAVERM_2427 (388)	BB341_05380 (375)	SCO2736 (429)	SGR_1691 (381)	SHJGH_4003 (433)	CP970_11090 (395)	SLAV_10725 1 (383)	SNOUR_17670 (367)	SPRI_4216 (366)	SVEN_2535 (433)
	SLNWT_1428 (422)	SAVERM_2428 (418)	BB341_05385 (432)	SCO4388 (366)	SGR_1692 (432)	SHJGH_4518 (221)	CP970_11095 (417)	SLAV_10730 2 (416)	SNOUR_26625 (433)	SPRI_4812 (429)	SVEN_4205 (366)
	SLNWT_4294 (369)	SAVERM_3859 (366)	BB341_11620 (363)	SCO5831 (421)	SGR_3076 (367)	SHJGH_6697 (421)	Cp970_24400 (366)	SLAV_17155 I (368)			SVEN_5584 (377)
	SLNWT_5026 (434)	SAVERM_5330 (429)	BB341_18800 (429)	SCO5832 (390)	SGR_4826 (432)	SHJGH_6698 (388)	CP970_28100 (451)	SLAV_24055 I (433)			
Aconitase	SLNWT_0836 (904)	SAVERM_2258 (905)	BB341_04765 (908)	SCO5999 (904)	SGR_1506 (911)	SHJGH_6830 (905)	CP970_10270 (904)	SLAV_10035 (904)	SNOUR_12030 (904)	SPRI_1818 (905)	SVEN_5812 (910)
Isocitrate dehydrogenase	SLNWT_6831 (739)	SAVERM_7214 (739)	BB341_23905 (739)	SCO7000 (739)	SGR_1224 (740)	SHJGH_7521 (739)	CP970_35840 (739)	SLAV_07110 (739)	SNOUR_1735 (406) SNOUR_33110 (740)	SPRI_3067 (409) SPRI_6612 (739)	SVEN_0436 (741)
2-oxoglutarate dehydrogenase E1	SLNWT_2008 (1276)	SAVERM_2972 (1276)	BB341_08105 (1287)	SCO5281 (1272)	SGR_2226 (1267)	SHJGH_6150 (1170)	CP970_14080 (1294)	SLAV_13205 (1305)	SNOUR_15075 (1262)	SPRI_2599 (1272)	SVEN_4966 (1266)
2-oxoglutarate dehydrogenase E2 (dihydrolipoamide succinyltransferase)	SLNWT_2008		BB341_08105	SCO1268 (372) SCO7123 (417)	SGR_2226	SHJGH_6150		SLAV_13205	SNOUR_15075	SPRI_2599	SVEN_4966
2-oxoglutarate dehydrogenase E2 (dihydrolipoamide dehydrogenase)	SLNWT_5161 (467)	SAVERM_2154 (478) SAVERM_6024 (462)	BB341_21130 (462)	SCO2180 (486)	SGR_5330 (468)	SHJGH_6150	CP970_31745 (462) CP970_41240 (467)	SLAV_26305 (462) SLAV_35195 (467)	SNOUR_29490 (462) SNOUR_37990 (466)	SPRI_0819 (466) SPRI_1688 (469) SPRI_5319 (462)	SVEN_1842 (501) SVEN_3731 (467)

	SALS	SAVERM	SCLF	SCO	SGR	SHJGH	SKA	SLAV	SNOUR	SPRI	SVEN
2-oxoglutarate ferredoxin oxidoreductase subunit beta	SLNWT_4515 (373)	SAVERM_4876 (359)	BB341_10730 (333)	SCO4594 (352) SCO6269 (350)	SGR_2930 (357)	SHJGH_5497 (353)	CP970_17695 (363)	SLAV_16235 (356) SLAV_39690 (467)	SNOUR_22995 (364)	SPRI_3245 (349)	SVEN_4303 (353)
2-oxoglutarate/2-oxoacid ferredoxin oxidoreductase subunit alpha	SLNWT_4516 (644)	SAVERM_4877 (642)	BB341_10725 (597) BB341_25540 (597)	SCO4595 (645) SCO6270 (630)	SGR_2929 (654)	SHJGH_5498 (642)	CP970_17690 (643)	SLAV_16230 (649) SLAV_39695 (621)	SNOUR_23000 (650)	SPRI_3244 (648)	SVEN_4304 (653)
Succiny CoA synthetase subunit β	SLNWT_4730 (392)	SAVERM_1818 (376) SAVERM_3452 (392)	BB341_10035 (391)	SCO4808 (394) SCO6585 (383)	SGR_2723 (393)	SHJGH_5672 (392) SHJGH_7288 (376)	CP970_05375 (374) CP970_16770 (392)	SLAV_15400 (392)	SNOUR_17190 (392)	SPRI_1230 (375) SPRI_3081 (393)	SVEN_4485 (424)
Succiny CoA synthetase subunit α	SLNWT_4731 (305)	SAVERM_1817 (299) SAVERM_3451 (294)	BB341_10030 (294)	SCO4809 (294) SCO6586 (308)	SGR_2722 (294)	SHJGH_5673 (294) SHJGH_7289 (293)	CP970_05370 (290) CP970_16765 (294)	SLAV_15395 (295)	SNOUR_17185 (294)	SPRI_1229 (296) SPRI_3080 (294)	SVEN_4486 (294)
Succinate dehydrogenase, iron-sulfur protein	SLNWT_4174 (248) SLNWT_4779 (244)	SAVERM_3182 (258) SAVERM_3398 (257) SAVERM_7309 (249)	BB341_01735 (248) BB341_08890 (255) BB341_09850 (252)	SCO0922 (248) SCO4855 (257) SCO5106 (259)	SGR_2690 (255) SGR_705 (248)	SHJGH_5739 (259) SHJGH_5963 (257) SHJGH_7644 (249)	CP970_03035 (248) CP970_15135 (256) CP970_16460 (255)	SLAV_06165 (246) SLAV_14090 (260) SLAV_15205 (252)	SNOUR_15960 (265) SNOUR_16955 (257) SNOUR_30130 (247)	SPRI_0753 (248) SPRI_2773 (256) SPRI_3031 (252)	SVEN_4531 (252) SVEN_4752 (257) SVEN_6837 (249)
Succinate dehydrogenase, flavoprotein subunit	SLNWT_4173 (652) SLNWT_4780 (584)	SAVERM_3181 (667) SAVERM_3397 (584) SAVERM_7308 (649)	BB341_08885 (667) BB341_09845 (584)	SCO0923 (649) SCO4856 (584) SCO5107 (653) SCO7109 (576)	SGR_2689 (584) SGR_706 (649)	SHJGH_5740 (584) SHJGH_5964 (651) SHJGH_7643 (649)	CP970_03040 (652) CP970_15130 (656) CP970_16455 (584)	SLAV_06170 (650) SLAV_14085 (633) SLAV_15200 (584)	SNOUR_15955 (640) SNOUR_16950 (584) SNOUR_30135 (648)	SPRI_0754 (647) SPRI_2772 (647) SPRI_3030 (584)	SVEN_4532 (584) SVEN_4753 (636) SVEN_6836 (648)

	SALS	SAVERM	SCLF	SCO	SGR	SHJGH	SKA	SLAV	SNOUR	SPRI	SVEN
Succinate dehydrogenase hydrophobic membrane anchor protein	SLNWT_4782 (161)	SAVERM_3396 (160)	BB341_09840 (158)	SCO4857 (160)	SGR_2688 (160)	SHJGH_5741 (160)	CP970_16450 (163)	SLAV_15195 (160)	SNOUR_16945 (154)	SPRI_3029 (158)	SVEN_4533 (159)
succinate dehydrogenase cytochrome β-556 subunit	SLNWT_4172 (223)	SAVERM_3395 (126)	BB341_01745 (207)	SCO0924 (243)	SGR_707 (234)	SHJGH_5742 (110)	CP970_03045 (235)	SLAV_06175 (241)	SNOUR_16940 (126)	SPRI_0755 (234)	SVEN_4534 (126)
	SLNWT_4781 (110)	SAVERM_7307 (235)	BB341_09835 (144)	SCO4858 (126)	SGR_2687 (126)	SHJGH_7642 (223)	CP970_16445 (126)	SLAV_15190 (126)	SNOUR_30140 (278)	SPRI_3028 (126)	SVEN_6835 (208)
Fumarate hydratase Class I	SLNWT_2307 (554)	SAVERM_3218 (558)	BB341_08995 (562)	SCO5044 (558)	SGR_2481 (558)	SHJGH_5903 (534)	CP970_15380 (558)	SLAV_14315 (559)	SNOUR_16110 (558)	SPRI_2823 (555)	SVEN_4713 (556)
Fumarate hydratase, class II	SLNWT_2312 (473)	SAVERM_3221 (467)		SCO5042 (461)	SGR_2491 (471)	SHJGH_5902 (461)	CP970_15390 (470)	SLAV_14340 (467)	SNOUR_16150 (464)		SVEN_4708 (469)
Malate dehydrogenase	SLNWT_4746 (329)	SAVERM_3436 (329)	BB341_09975 (329)	SCO4827 (329)	SGR_2711 (329)	SHJGH_5712 (329)	CP970_16700 (329)	SLAV_15340 (329)	SNOUR_17130 (329)	SPRI_3065 (329)	SVEN_4498 (329)

Streptomyces lavendulae (SLAV), Streptomyces albus DSM 41398 (SALS), Streptomyces pristinaespiralis HCCB 10218 (SPRI), Streptomyces kanamyceticus ATCC 12853 (SKA), Streptomyces noursei ATCC 11455 (SNOUR), Streptomyces hygroscopicus subps. jinggagensis TL01 (SHJGH), Streptomyces clavuligerus F613-1 (SCLF), Streptomyces coelicolor A3(2) (SCO), Streptomyces avermitilis MA-4680 (SAVERM), Streptomyces griseus subsp. griseus NBRC 13350 (SGR), Streptomyces venezuelae ATCC 10712 (SVEN). The number of amino acids of each encoded protein is shown in parentheses.

Table 2.
Gene multiplicity in TCA cycle genes.

```
slx_SLAV_10725      ------------------------------------------------------------
sclf_BB341_05380    ------------------------------------------------------------
sgr_SGR_1691        ------------------------------------------------------------
sals_SLNWT_1427     ----------------MRRLRTEPFALPPRPEHRQGINMSINSTADAGSQPRGTASGA
ska_CP970_11090     ------------------------------------------------------------
sma_SAVERM_2427     ------------------------------------------------------------
sho_SHJGH_6698      ------------------------------------------------------------
sco_SCO5832         ------------------------------------------------------------
ska_CP970_24400     ------------------------------------------------------------
snr_SNOUR_17670     ------------------------------------------------------------
slx_SLAV_17155      ------------------------------------------------------------
sve_SVEN_4205       ------------------------------------------------------------
spri_SPRI_4216      ------------------------------------------------------------
sgr_SGR_3076        ------------------------------------------------------------
sals_SLNWT_4294     ------------------------------------------------------------
sco_SCO4388         ------------------------------------------------------------
sma_SAVERM_3859     ------------------------------------------------------------
sclf_BB341_11620    ------------------------------------------------------------
sho_SHJGH_4518      ------------------------------------------------------------
spri_SPRI_4812      ----------------------------------MSDNSVVLRYADGEYTYPVVESTVGDK
sve_SVEN_2535       -------------------------------MRDDVSDNSVVLRYADGEYTYPVVESTVGDK
sclf_BB341_18800    ----------------------------------MSDNSVVLRYADGEYTYPVVESTVGDK
slx_SLAV_24055      -------------------------------MRDDVSDNSVVLRYADGEYTYPVVESTVGDQ
ska_CP970_28100     -----------MVETVRAPRRRGTTSVRDDVSENANNGVVLRYGDGEYTYPVIESTVGDK
sma_SAVERM_5330     ---------------------------------MSDNSVVLRYGDGEYTYPVIDSTVGDK
sals_SLNWT_5026     ---------------------------MSDNTQNNGAVVVRYGDGEYSYPVVDSTVGDK
sgr_SGR_4826        -------------------------------MSEHTNNAVVLRYGDDEYTYPVIDSTVGDK
sho_SHJGH_4003      -------------------------------MREDVSDNSVVLRYGDGEYTYPVVDSTVGDK
sco_SCO2736         -------------------------------MSDNSVVLRYGDGEYTYPVIDSTVGDK
snr_SNOUR_26625     -------------------------------MRDDVSDNSVVLRYGDGEYSYPVVDSTVGDK
sve_SVEN_5584       ------------------------------------------------------------
slx_SLAV_10730      --MSDHAADVPPAPEGTRRLTTQEAARLLGVKPATVYAYVSRGQLSSRRDPVGRGSSFDA
sclf_BB341_05385    ---------MTDHAAEAERLTTREAAEKLGVKPATLYAYVSRGLLLTSRRGPGGRGSTFDP
sgr_SGR_1692        -----MTDRSAPDEQGPPRLTTREAAELLGVKPATVYAYVSRGQLSSARAPGGRGSTFDA
sho_SHJGH_6697      --MAPMRDHDTAPHRPGRRLSTKETAELLGVKPETVYAYVSRGLLSSRREPGRGSSTFEA
sma_SAVERM_2428     -----MTDHEAAPTHPDRRLSTKEAAELLGVKPETVYAYVSRGQLSSRREPGGRGSTFDA
ska_CP970_11095     -------MTDQGTPREGRRLSTKEAAELLGVKPETVYAYVSRGQLSSSRGAGSRGSTFDA
sco_SCO5831         ----MRDHEPDHPADSERRLTTREAAELLGVKPETVYAYVSRGQLGSSRRAPGSRGSTFDA
sals_SLNWT_1428     MTDAQHPQRTQDGGDAERRLSTREAADLLGVKPETVYAYVSRGQLSSRRGGGGRGSTFDA
```

```
slx_SLAV_10725      ---------MNTTAEVPRGLAGVVVTETQLGDVRGLEGFYHYREYSAVELAESRTFEDVW
sclf_BB341_05380    ---------------MPRGLAGVVVTDTGLGDVRGREGFYHYRQYSAVELAESRTFEDVW
sgr_SGR_1691        ---------MTTTPAVPRGLAGVVVTDTALGDVRGREGFYHYRQYSAIELARTRGFEDVW
sals_SLNWT_1427     GGADGGAASGPGAADAPRGLAGVVVTRTALGDVRGREGFYHYRQFSAIELARTRSFEDVW
ska_CP970_11090     MPINGSAATPAEPVEVPRGLAGVVVTETELGDVRGYEGFYHYRQYSAVELAQSRSFEDVW
sma_SAVERM_2427     MPIN---RAASTSVDVPRGLAGVVVTDTSLGDVRGLEGFYHYRQYSAVELAQTRGFEDVW
sho_SHJGH_6698      ---MAVNMPATPLIDVPRGLAGVVVAETEVGDVRGREGFYHYRQYPAVDLARTRSFEDVW
sco_SCO5832         -MSVNRTAAAATPVEVPRGLAGVVVADTEVGDVRGLEGFYHYRQYSAVELARSRGFEDVW
ska_CP970_24400     -------------MSDFVPGLEGVVAFETEIAEPDKEGGALRYRGVDIEDLVGHVSFGNVW
snr_SNOUR_17670     -------------MSDFVPGLEGVVAFETEIAEPDKEGGALRYRGVDIEDLVGHVSFENVW
slx_SLAV_17155      -------------MSDFVPGLEGVVAFETEIAEPDKEGGSLRYRGVDIEDLVGHVSFGNVW
sve_SVEN_4205       -------------MSDFVPGLEGVVAFETEIAEPDKEGGALRYRGVDIEDLVGHVSFGNVW
spri_SPRI_4216      -------------MSDFVPGLEGVVAFETEIAEPDKEGGSLRYRGVDIEELVGHVSFGNVW
sgr_SGR_3076        -------------MSDFVPGLEGVVAFETEIAEPDKEGGSLRYRGVDIEDLVGHVSFGNVW
sals_SLNWT_4294     -------------MSDFVPGLEGVVAFETEIAEPDKEGGALRYRGVDIEDLVGHVSFGNVW
sco_SCO4388         -------------MSDFVPGLEGVVAFETEIAEPDKEGGALRYRGVDIEDLVGHVSFGNVW
sma_SAVERM_3859     -------------MSDFVPGLEGVIAFETEIAEPDKEGGALRYRGVDIEDLVGHVSFGNVW
sclf_BB341_11620    -------------MSDFVPGLEGVVAFETEIAEPDKEGGALRYRGVDIEDLVGRVPFGQVW
sho_SHJGH_4518      ------------------------------------------------------------
spri_SPRI_4812      GFDIGKLRAQTGLVTLDSGYGNTAAYKSAITYLDGEQGILRYRGYPIEQLAERSTFLEVA
sve_SVEN_2535       GFDIGKLRAQTGLVTLDSGYGNTAAYKSAITYLDGEQGILRYRGYPIEQLAERSTFLEVA
sclf_BB341_18800    GFDIGKLRAQTGLVTLDSGYGNTAAYKSAITYLDGERGILRYRGYPIEQLAERSTFLEVA
slx_SLAV_24055      GFDISKLRAQTGLVTLDSGYGNTAAYKSAITYLDGEQGILRYRGYPIEQLAERSTFIEVA
ska_CP970_28100     GFDIGKLRAQTGLVTLDSGYGNTAAYKSAITYLDGEQGILRYRGYPIEQLAERSTFTEVA
sma_SAVERM_5330     GFDIGKLRAQTGLVTLDSGYGNTAAYKSAITYLDGEQGILRYRGYPIEQLAERSTFLEVA
sals_SLNWT_5026     GFDIGKLRAQTGLVTLDSGYGNTAAYKSAITYLDGEQGILRYRGYPIEQLAEGSTFLEVA
sgr_SGR_4826        GFDIGKLRANTGLVTLDSGYGNTAAYKSAITYLDGEQGILRYRGYPIEQLAERSSFLEVA
sho_SHJGH_4003      GFDIGKLRAQTGLVTLDSGYGNTAAYKSAITYLDGEAGILRYRGYPIEQLAERSTFLEVA
sco_SCO2736         GFDIGKLRAQTGLVTLDSGYGNTAAYKSAITYLDGEAGILRYRGYPIEQLAERSSFVEVA
```

A

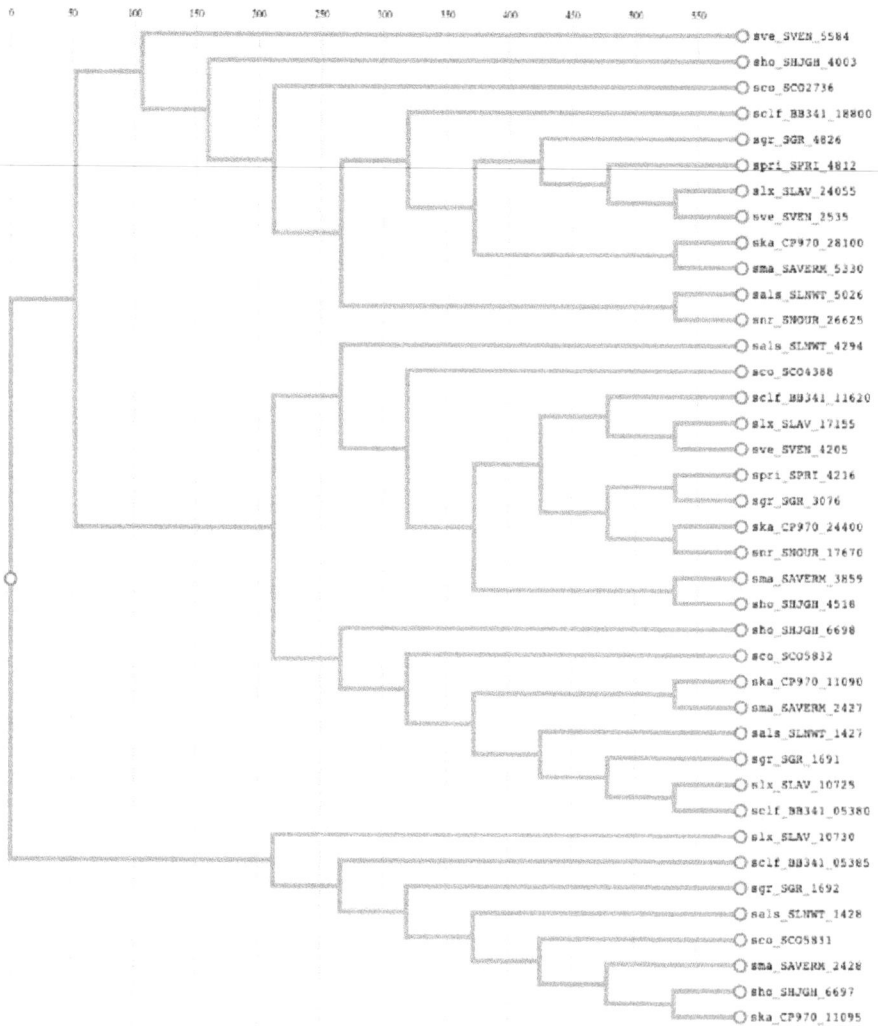

B

Figure 3.
Partial multiple alignment of citrate synthase proteins (A) in antibiotic producing Streptomyces *and derived bootstrapped tree (B). slx,* Streptomyces lavendulae, *sals,* Streptomyces albus *DSM 41398, sclf,* Streptomyces clavuligerus *F613-1, spri,* Streptomyces pristinaespiralis *HCCB 10218, ska,* Streptomyces kanamyceticus *ATCC 12853, snr,* Streptomyces noursei *ATCC 11455, sho,* Streptomyces hygroscopicus *subps. jinggagensis TL01, sco,* Streptomyces coelicolor *A3(2), sma,* Streptomyces avermitilis *MA-4680, sgr,* Streptomyces griseus *subsp. griseus NBRC 13350, sve,* Streptomyces venezuelae *ATCC 10712.*

importance of the reactions they catalyze. The molecular weight of the aconitases of these microorganisms is 97 kDa with an identity between them greater than 92%. Isocitrate dehydrogenases have a molecular weight of 79 kDa in all these *Streptomyces*, except *S. noursei* and *S. pristinaspirales*, which also have another protein (SNOUR_17135 and SPRI_3067) that are smaller (45 kDa). The identity between the large proteins and between the small ones is about 85% and 77%, respectively. On the other hand, all MDHs have 329 amino acids with an identity between 87 and 93%.

The 2-oxoglutarate dehydrogenase complex is a central enzyme in aerobic metabolism that catalyzes the oxidative decarboxylation of oxoglutarate, generating

NADH [16]. 2-oxoglutarate dehydrogenase is composed of three subunits, E1 (EC 1.2.4.2), E2 (EC 2.3.1.61, dihydrolipoamide succinyltransferase), and E3 (EC 1.8.1.4; dihydrolipoamide dehydrogenase). As previously mentioned, there is a single gene coding for component E1 in all selected antibiotic-producing *Streptomyces*. The E2 component is encoded by one gene, except in *S. coelicolor*, which has three. *S. pristinaespiralis, S. albus, S. clavuligerus, S. coelicolor, S. griseus*, and *S. hygroscopicus* genomes have only one gene coding for E3 component, while the remaining five microorganisms have two copies with a low identity ($\leq$ 38%) between them suggesting different origins.

Multiple alignments of the amino acid sequences of the E2 subunit of 2-oxoglutarate dehydrogenase showed close resemblance, with many highly conserved amino acids within the conserved domains, such as the YDHR region, which is part of the 2-oxoacid dehydrogenase acyltransferase catalytic domain. The proteins encoded by *sco7123* and *sco1268* of *S. coelicolor* were smaller than the rest, with sequences of 372 and 417 amino acids, respectively. In contrast, the one encoded by *sco5281* had 1272 residues, similar to the proteins of the rest of the *Streptomyces* studied. This characteristic was reflected in the phylogenetic tree, where the former two proteins were separated from the others in one clade. Large proteins were grouped in the second clade, with SCO5281 being the less related group member (**Figure 4**).

There is an alternative way to perform the synthesis of 2-oxoglutarate via 2-oxoglutarate ferredoxin oxidoreductase (EC 1.2.7.11) formed by two subunits called a and b. *S. avermitilis, S. griseus, S. kanamyceticus*, and *S. noursei* have only one copy of this pair of genes, whereas *S. albus* has two genes for subunit b and one for a, and *S. clavuligerus* and *S. noursei* have two for subunit a and one for b. In contrast, *S. coelicolor* has two copies of each, generated by gene duplication because the proteins have 99% identity. *S. lavendulae* also has two copies of each gene, with an identity of 38.4%.

The next reaction is catalyzed by succinyl-CoA synthetase (EC 6.2.1.5), which is also composed of two protein subunits: α and β. *S. albus, S. clavuligerus, S. griseus, S. lavendulae, S. noursei*, and *S. venezuelae* have a single copy of the genes that code for each subunit. Two copies are present in the genomes of *S. avermitilis, S. coelicolor, S. pristinaespiralis, S. hygroscopicus*, and *S. kanamyceticus*; a subunit is smaller (290–299 residues) than b (376–394), one of the copies being around 376 amino acids and the other larger in all cases. In the first four microorganisms, the genes for the a and b subunits are physically together for both pairs of genes, whereas in the latter, they are separated by five genes in both cases (CP970_05375/CP970_5370; CP970_16770/ CP970_16765). The identity and similarity between the two proteins subunits a or b in each of the *Streptomyces* range from 57.9% to 78.6% and 74.8% to 86.2%, respectively.

Succinate dehydrogenase (SDH; EC 1.3.5.1, EC 1.3.5.4), which catalyzes the oxidation of succinic acid to generate fumaric acid is a four-subunit multimeric enzyme (iron-sulfur protein, flavoprotein subunit, hydrophobic membrane anchor protein, and cytochrome b-556 subunit). The number of genes that code for each subunit varies, and even the same genome *Streptomyces* can have different number of genes for each subunit. As shown in **Table 2**, the *S. albus* genome contains two genes for the SDH flavoprotein subunit, two for the iron-sulfur protein, one for the SDH hydrophobic membrane anchor protein, and two for the SDH cytochrome b-556 subunit. *S. avermitilis* has three genes for flavoprotein subunit, while *S. coelicolor* has four. On the other hand, all of these microorganisms have one copy of the gene coding the hydrophobic membrane anchor protein and two for the cytochrome b-556 subunit.

The genes that code for iron-sulfur protein have great redundancy, finding three copies in all antibiotic-producing *Streptomyces* except *S. albus*, which has two. In total, 28 proteins were found that have very similar molecular weights, however, according to the identities between them, three groups can be distinguished. The

```
sho_SHJGH_6150    --------------------------------------------------------------
sco_SCO5281       ------------MSPQSPSNSSISTDTDQAGTAGKNPAAAFGANEWLVDEIYQQYLQDP
ska_CP970_14080   MPAVHHLVKRKRAVTAVSSQSPSSSSISTDQDGQGKDPAAAFGPNEWLVDEIYQQYLQDP
sma_SAVERM_2972   ---------------MSPQSPSNSSISTDDQAGKNPAAAFGPNEWLVDEIYQQYLQDP
sals_SLNWT_2008   -------MSPQSPSNSTSPGLKPGGTPTTDQDGQGKNPAAAFGPNEWLVDEIYQQYLQDP
sclf_BB341_08105  -------MSSQSPSNSSSPSLRPGGAPTTDESAQGAGPAAAFGPNEWLVDEIYQQYLQDP
sve_SVEN_4966     -------MSSQSPSTPSSP---------TDQDGQGQNPATATGANEWLVDEIYQQYLQDP
spri_SPRI_2599    -------MSSQSPSNSSIS---------TDQAGQGQNPAAAFGANEWLVDEIYQQYLQDP
sgr_SGR_2226      -------MSSQSPSNSSIS---------TDQAGPGTNPAAAFGANEWLVDEIYQQYLQDP
snr_SNOUR_15075   ----------------MSLQSPNSASVSTEQGAQGKNPAAAFGSNEWLVDEIYQQYLQDP
slx_SLAV_13205    ----------------MSPQSPSNPSTTTEAAEGGKTPASGFGANEWLVDEIYQQYLQDP
sco_SCO7123       ---------MTVSVTLPALGESVTEGTVTRWLKQVGDRVEADEPLLEVSTDKVDTEIPSPA
sco_SCO1268       ---------MIGIELPTLNTNDSSYTLVEWLVPDGGSIEGEEPLAVVETSKASEEIESTG
```

```
sho_SHJGH_6150    --------------------------------------------------------------
sco_SCO5281       NSVDRAWWDFFADYKPGAAATPTAAGTVP-----------------------TDAGS
ska_CP970_14080   NSVDRAWWDFFADYKPGVGASVAAAPAKP-----------------------AGDAA
sma_SAVERM_2972   NSVDRAWWDFFADYKP-----GAAAASAP-----------------------AGTAA
sals_SLNWT_2008   NSVDRAWWDFFADYKPGGGATQAKQTAAP-----------------------AAQAP
sclf_BB341_08105  NSVDRAWWDFFADYKPGAQNTAATGAEP------------------------TA
sve_SVEN_4966     NSVDRAWWDFFADYKPGASETPTAPAPT------------------------TQ
spri_SPRI_2599    NSVDRAWWDFFADYKPGAAATADAPAAG------------------------AK
sgr_SGR_2226      SSVDRAWWDFFADYKPGASGTADKPVPG------------------------VA
snr_SNOUR_15075   NSVDRAWWDFFADYKPGGTGAADAAPPAA----------------------P----
slx_SLAV_13205    NSVDRAWWDFFADYKPGGAVAPVKADGPGKSTTTDGASAQAATSAAAPQAPVAAATGAAR
sco_SCO7123       AGVLLEILAAEDETVEVGAGLGIIGAPDTAP------------------------AAP
sco_SCO1268       PGILHVLVGGGQECRPGQTIAYLFESAEE---------------------------
```

A

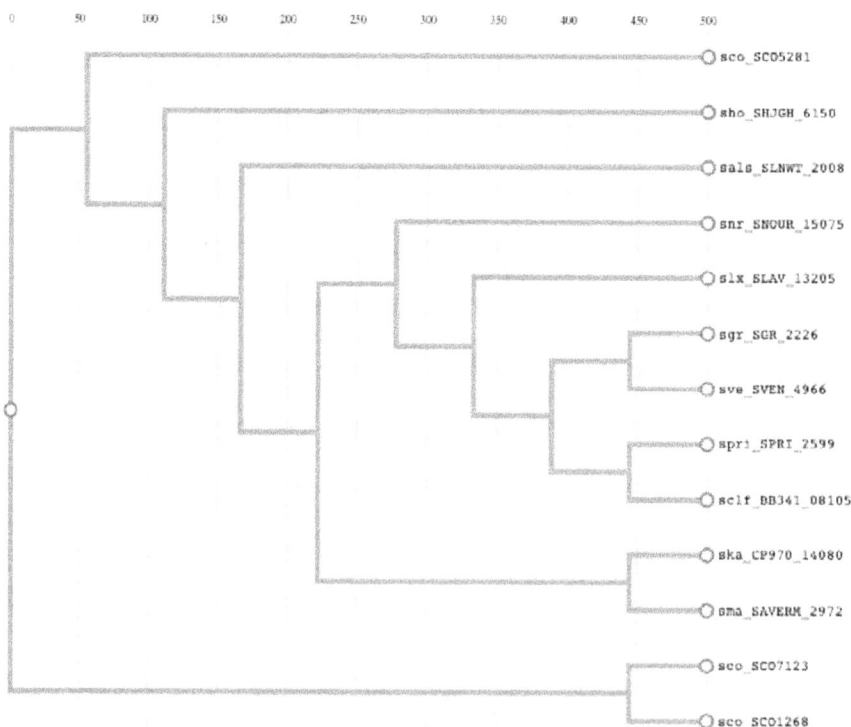

B

Figure 4.
Multiple alignment of 2-oxoglutarate dehydrogenase component E2 (A) in antibiotic producing Streptomyces and derived bootstrapped tree (B). slx, Streptomyces lavendulae, sals, Streptomyces albus DSM 41398, sclf, Streptomyces clavuligerus F613-1, spri, Streptomyces pristinaespiralis HCCB 10218, ska, Streptomyces kanamyceticus ATCC 12853, snr, Streptomyces noursei ATCC 11455, sho, Streptomyces hygroscopicus subps. jinggagensis TL01, sco, Streptomyces coelicolor A3(2), sma, Streptomyces avermitilis MA-4680, sgr, Streptomyces griseus subsp. griseus NBRC 13350, sve, Streptomyces venezuelae ATCC 10712.

```
sho_SHJGH_5712     MTRTPVNVTVTGAAGQIGYALLFRIASGQLLGADVPVRLRLLEITPALKAAEGTAMELDD
sco_SCO4827        MTRTPVNVTVTGAAGQIGYALLFRIASGQLLGADVPVKLRLLEITPALKAAEGTAMELDD
sma_SAVERM_3436    MTRTPVNVTVTGAAGQIGYALLFRIASGQLLGADVPVKLRLLEITPALKAAEGTAMELDD
sals_SLNWT_4746    MTRTPVNVTVTGAAGQIGYALLFRIASGQLLGADVPVKLRLLEITPALKAAEGTAMELDD
sclf_BB341_09975   MTRTPVNVTVTGAAGQIGYALLFRIASGHLLGPDVPVKLRLLEITPALGAAQGTAMELDD
ska_CP970_16700    MTRTPVNVTVTGAAGQIGYALLFRIASGHLLGADVPVNLRLLEITPALKAAEGTAMELDD
snr_SNOUR_17130    MTRTPVNVTVTGAAGQIGYALLFRIASGHLLGADVPVKLRLLEIPQGLKAAEGTAMELDD
slx_SLAV_15340     MTRTPVNVTVTGAAGQIGYALLFRIASGHLLGADVPVKLRLLEIPQGMKAAEGTAMELDD
sve_SVEN_4498      MTRTPVNVTVTGAAGQIGYALLFRIASGHLLGADVPVKLRLLEIPQGVKAAEGTAMELDD
spri_SPRI_3065     MTRTPVNVTVTGAAGQIGYALLFRIASGHLLGADVPVKLRLLEIPQGLKAAEGTAMELDD
sgr_SGR_2711       MTRTPVNVTVTGAAGQIGYALLFRIASGHLLGPDVPVNLRLLEIPQGLKAAEGTAMELDD
                   ************************************:***.****.******. .: **:********

sho_SHJGH_5712     CAFPLLQGIDITDDPNVAFDGANVALLVGARPRTKGMERGDLLEANGGIFKPQGKAINDH
sco_SCO4827        CAFPLLQGIEITDDPNVAFDGANVALLVGARPRTKGMERGDLLEANGGIFKPQGKAINDH
sma_SAVERM_3436    CAFPLLQGIDITDDPNVAFDGTNVGLLVGARPRTKGMERGDLLSANGGIFKPQGKAINDN
sals_SLNWT_4746    GAFPLLQSIEISDDPNVAFDGTNVALLVGARPRAKGMERGDLLEANGGIFKPQGKAINDH
sclf_BB341_09975   CAFPLLRGIDITDDPNVAFDGANVALLVGARPRTKGMERGDLLEANGGIFKPQGKAINDH
ska_CP970_16700    CAFPLLNSIEISDDPNVAFDGANVALLVGARPRTKGMERGDLLEANGGIFKPQGKAINDN
snr_SNOUR_17130    CAFPLLRGIDITDDPNVAFAGANVALLVGARPRTKGMERGDLLEANGGIFKPQGKAINDH
slx_SLAV_15340     CAFPLLKGIDIFDDPNKGFEGANVALLVGARPRTAGMERGDLLSANGGIFKPQGAAINAH
sve_SVEN_4498      CAFPLLKGIDIFDDPNQGFEGANVALLVGARPRTKGMERGDLLAANGGIFKPQGKAINDH
spri_SPRI_3065     CAFPLLKGIEITDDPNVGFDGANVALLVGARPRTKGMERGDLLSANGGIFKPQGKAINDN
sgr_SGR_2711       CAFPLLRGIEITDDPNVGFAGANVALLVGARPRTKGMERGDLLAANGGIFKPQGKAINDH
```

A

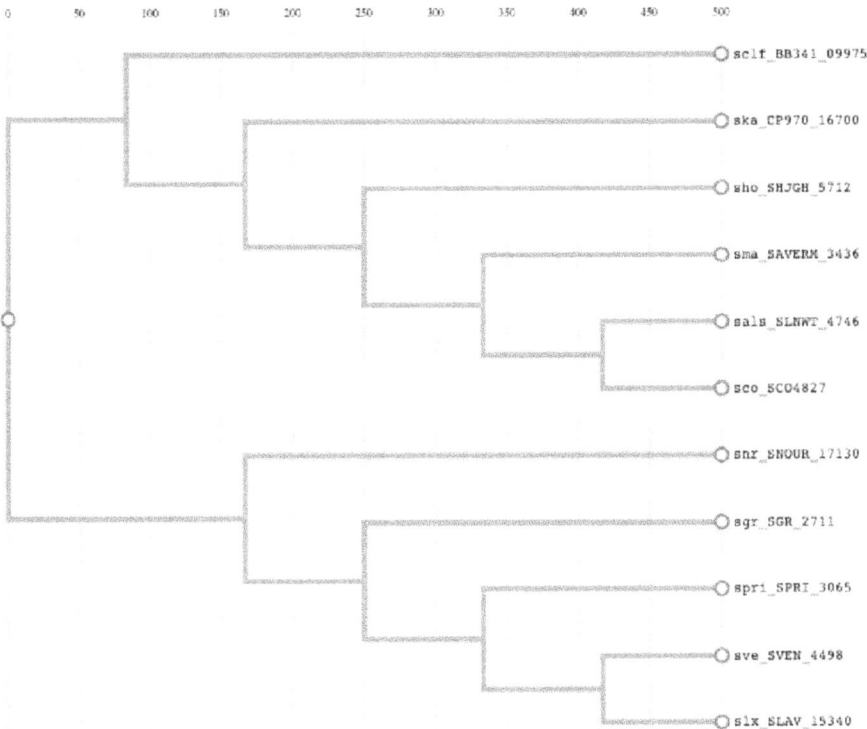

B

Figure 5.
Partial multiple alignment of malate dehydrogenase proteins (A) in antibiotic producing Streptomyces *and derived bootstrapped tree (B).* slx, *Streptomyces lavendulae,* sals, *Streptomyces albus DSM 41398, sclf, Streptomyces clavuligerus F613-1,* spri, *Streptomyces pristinaespiralis HCCB 10218, ska, Streptomyces kanamyceticus ATCC 12853, snr, Streptomyces noursei ATCC 11455, sho, Streptomyces hygroscopicus subps. jinggagensis TL01, sco, Streptomyces coelicolor A3(2), sma, Streptomyces avermitilis MA-4680, sgr, Streptomyces griseus subsp. griseus NBRC 13350, sve, Streptomyces venezuelae ATCC 10712.*

first includes proteins of 246–249 amino acids and that have an identity between them greater than 88% and is one of the copies in all these microorganisms. In another group are those with 252 amino acids and identities greater than 94%. The last group includes proteins with 256–267 amino acids with identities greater than 85%. All these data indicate that although the subunits have very similar molecular weights, they are actually not so similar, suggesting that they had a common ancestor but that they have diverged a lot over time.

It has been found that there are three different flavoprotein proteins, some with 649–652 amino acids with identities greater than 90%, another group of proteins with 584 amino acids and identities between 91 and 93%, and the last group with 633–667 residues and identities greater than 80%. The flavoproteins SCO7109 and SHJGH_5964 do not resemble each other or the proteins of the previous groups.

The hydrophobic membrane anchor proteins are smaller than the previous ones, around 17 kDa, and present identities between 77% and 82%, while there are two types of cytochrome b subunit, one of around 223–243 amino acids with lower identities than the previous ones, between 58% and 71%. The second type includes proteins of smaller size and identity between 74% and89%.

The hydration reaction of fumarate to generate malate is catalyzed by fumarate hydratase (EC 4.2.1.2; fumarase). There are two classes of proteins: fumarase classes I and II. Most antibiotic-producing *Streptomyces* have a single gene for fumarase Class I, while only seven of these *Streptomyces* have fumarase class II (**Table 2**). The class I proteins are larger (60.2 kDa) than the class II proteins (50.3 kDa). The identity among class I fumarases and class II is 74% and 75%, respectively.

The final TCA cycle reaction is catalyzed by malate dehydrogenase, which catalyzes the reversible conversion of malate to OXA using NAD^+ or $NADP^+$ as the coenzyme [17], which is only encoded by a single gene in all microorganisms that produce different antibiotics. The multiple sequence alignment of this enzyme showed a high degree of conservation between them, distributed in two main clades from a common ancestor, which was then subdivided into eight subclades (**Figure 5**). The identity is higher than 90%.

5. PEP-PYR-OXA node

The PEP-PYR-OXA node is a major branch of central carbon metabolism and acts as a connection point between glycolysis, gluconeogenesis, and the TCA cycle [7]. A large variety of enzymes involved in the node have been reported, such as PEP carboxylase (EC 4.1.1.31), pyruvate carboxylase (EC 6.4.1.1), PEP carboxykinase (EC 4.1.1.32), malic enzymes (EC 1.1.1.38), pyruvate kinase (EC 2.7.1.40), and pyruvate phosphate dikinase (EC 2.7.9.1). These enzymes are indispensable for distributing PEP, PYR, and OXA in *Streptomyces*. The enzymes involved in this node vary among microorganisms, and their activity depends on the culture conditions [2, 18].

This anaplerotic pathway does not present multiplicity in any of the genes that encode the enzymes participating in the PEP-PYR-OXA node, except for malic enzymes and pyruvate phosphate dikinase as shown in **Table 3**. PYR carboxylase is an enzyme that is present only in *S. albus*, *S. coelicolor*, *S. hygroscopicus*, and *S. pristinaspiralis*, with only one copy of the gene. The molecular weight is 121 kDa and with 77%–88% identity and 88%–95% similarity among all proteins.

PEP carboxylase, an enzyme that catalyzes the carboxylation of PEP to OXA, is encoded by a single gene and is present in all antibiotic-producing *Streptomyces*. The

	SALS	SAVERM	SCLF	SCO	SGR	SHJGH	SKA	SLAV	SNOUR	SPRI	SVEN
Pyruvate carboxylase	SLNWT_3899 (1124)			SCO0546 (1124)		SHJGH_7997 (1124)				SPRI_1970 (1124)	
Phosphoenol pyruvate carboxylase	SLNWT_4462 (912)	SAVERM_3566 (910)	BB341_17140 (909)	SCO3127 (911)	SGR_4379 (909)	SHJGH_4361 (910)	CP970_25710 (917)	SLAV_22415 (920)	SNOUR_24730 (921)	SPRI_4413 (909)	SVEN_2951 (909)
NAD-Malic enzyme	SLNWT_2497 (476)	SAVERM_1514 (587) SAVERM_3870 (573) SAVERM_5126 (477)	BB341_17990 (476)	SCO2951 (471)	SGR_4581 (478)	SHJGH_4187 (471) SHJGH_4528 (570)	CP970_26965 (477)	SLAV_23225 (474)	SNOUR_25740 (477)	SPRI_4590 (477)	SVEN_2718 (476)
NADP-Malic enzyme	SLNWT_2020 (407)	SAVERM_2981 (407)	BB341_08155 (412)	SCO5261 (409)	SGR_2236 (413)	SHJGH_6136 (407)	CP970_14130 (407)	SLAV_13290 (400)	SNOUR_15125 (409)	SPRI_2612 (403)	SVEN_4951 (403)
Phosphoenol pyruvate carboxykinase	SLNWT_4964 (609)	SAVERM_3287 (607)	BB341_09370 (621)	SCO4979 (609)	SGR_2556 (608)	SHJGH_5842 (607)	CP970_15680 (623)	SLAV_14715 (613)	SNOUR_16390 (605)	SPRI_2904 (606)	SVEN_4658 (607)
pyruvate phosphate dikinase	SLNWT_5381 (902)	SAVERM_5654 (916)	BB341_01330 (608) BB341_19830 (905)	SCO0208 (898) SCO2494 (909)	SGR_5048 (903)	SHJGH_3733 (906) SHJGH_8556 (895)	CP970_29700 (896)	SLAV_24965 (933)	SNOUR_27780 (911)	SPRI_5017 (903)	SVEN_2286 (934)

Streptomyces lavendulae (SLAV), Streptomyces albus DSM 41398 (SALS), Streptomyces clavuligerus F613-1 (SCLF), Streptomyces pristinaespiralis HCCB 10218 (SPRI), Streptomyces kanamyceticus ATCC 12853 (SKA), Streptomyces noursei ATCC 11455 (SNOUR), Streptomyces hygroscopicus subps. jinggagensis TL01 (SHJGH), Streptomyces coelicolor A3(2) (SCO), Streptomyces avermitilis MA-4680 (SAVERM), Streptomyces griseus subsp. griseus NBRC 13350 (SGR), Streptomyces venezuelae ATCC 10712 (SVEN). The number of amino acids of each encoded protein is shown in parentheses.

Table 3.
Gene multiplicity in PEP-PYR-OXA node.

proteins are very similar; they have 909 and 921 amino acids in and identity and similarity between 87.3–90.2% and 90.5–94.5%, respectively.

PEP carboxykinase, an enzyme with the opposite action to the previous ones, decarboxylates OXA to form PEP. This enzyme could be considered gluconeogenic; however, it is also part of the PEP-PYR-OXA node for distributing the carbon flux between the different central pathways of metabolism [19]. Its molecular weight is approximately 67 kDa, some of which are a few amino acids long. In the same way as the previous enzymes, *Streptomyces* carboxykinases have an identity greater than 84%.

MEs catalyzes the oxidative decarboxylation of L-malate to produce PYR and CO_2, coupled with the reduction of $NAD(P)^+$ cofactors (EC 1.1.1.38, EC 1.1.1.40). In general, NAD^+- dependent MEs function to provide PYR for the TCA cycle, whereas $NADP^+$- dependent MEs function to generate NADPH for anabolic reactions. Bacterial ME isoforms are comparatively understudied and collectively demonstrate greater structural and functional diversity (ranging from "minimal" 40 kDa subunits to much larger 85 kDa multidomain proteins). The greater bacterial ME complexity arises from the need for allosteric regulation owing to the non-compartmentalization of the bacterial cell [20].

The genes coding for MEs in these antibiotic-producing *Streptomyces* are those that present multiplicity. All these microorganisms have two MEs, one dependent on NAD^+ and the other on $NADP^+$, with molecular weights of approximately 48 kDa and 42 kDa, respectively. Three proteins (SAVERM_1514, SAVERM_3870, and SHJGH_4528) from *S. avermitilis* and *S. hygroscopicus* have higher molecular weight, with 570 and 583 amino acid residues, and less than 25% identity with the NAD- and NADP-dependent MEs. Its molecular weight is approximately 61–63 kDa, however, BLAST analysis showed that these enzymes are common in many other *Streptomyces*.

Pyruvate phosphate dikinase (EC 2.7.11.32) converts PEP, inorganic pyrophosphate, and AMP into PYR, inorganic phosphate, and ATP. This protein is a gluconeogenic and anaplerotic enzyme and, in some bacteria, it plays other important roles. This protein is associated with virulence in *Brucella ovis*, and in *Mycobacterium tuberculosis,* it is indispensable for its growth as a part of the node [21, 22]. The gene encoding pyruvate phosphate dikinase in these antibiotic-producing *Streptomyces* species is present in all of them, in some cases with two copies, for example, *S. coelicolor*, *S. clavuligerus*, *S. griseus*, *S. hygroscopicus*, and *S. lavendulae*, with molecular weights ranging from 98 to 102 kDa. One of the *S. clavuligerus* proteins is smaller (61.7 KDa), with 25.9% identity to the amino acid sequence of the protein encoded by SCO0208; however, as in the previous case, BLAST analysis found many other different *Streptomyces* species that have this enzyme of different size. The expression of the gene that encodes pyruvate phosphate dikinase has been little studied, however, Llamas et al. (2020) reported that *sco0208* which codes for this enzyme in *S. coelicolor*, was maximum at 36 h of growth in minimal medium with casamino acids as carbon and nitrogen sources.

Finally, the genes that encode all enzymes of the glycolytic pathway, the TCA cycle, and the PEP-PYR-OXA node are distributed throughout the genome and represent up to 50 kb that might not be needed. Although many genes are found in the core, others are also found in the arms, and the question arises as to whether all genes are expressed and translated. For example, four genes that encode CS are transcribed to generate functional proteins? Viollier et al. (2001) reported that the deletion of the *citA* gene of *S. coelicolor* (*sco2736*), which codes for one of the four CS, generated glutamate auxotrophy, indicating that it is the main enzyme in this microorganism for the condensation reaction to form six-carbon citrate [23]. Similarly, Takahashi and Flores (unpublished results) found that in *S. coelicolor* grown on glucose as a carbon source, *sco2736* mRNA levels were higher than *sco5832*, whereas *sco4388* and *sco5831* mRNA levels were very low, confirming that one of the four CS is predominant under these growth conditions. In *Saccharopolyspora erythraea*, another actinobacterium, strong relative transcription of *gltA-2* was observed only during the early exponential phase and declined thereafter, whereas the other two genes *citA* and *citA4* exhibited relatively low transcript levels during the early exponential phase, and transcription was gradually increased and reached a maximum level during the early stationary phase [24]. The expression of different genes for the same activity likely occurs under different growth conditions or carbon sources. If this is the case, genetic multiplicity allows *Streptomyces* to adapt and be robust, which may drive the expansion of primary metabolic capability [2].

As mentioned before, antibiotics are synthesized from precursors that are intermediates from carbon metabolism. By providing an overview of the gene multiplicity and determining which enzymes are encoded by a single gene, it will be possible to design strategies aimed at the sufficient biosynthesis of precursors for the metabolism of microorganisms to satisfy the demand for these same compounds to form antibiotics.

On the other hand, the growth of *Streptomyces* could be limited by insufficient synthesis of some of the enzymes involved in glucose metabolism. For example, the E1 component of 2-oxoglutarate dehydrogenase is encoded by a single gene in *Streptomyces*, whereas gene multiplicity exists in the other two components to form a complex that performs the conversion reaction of 2-oxoglutarate to succinate. In addition, this protein is a part of the PYR dehydrogenase complex, therefore, the question arises as to whether the expression of this protein could limit the growth of glucose as a carbon source. To date, no evidence has been reported indicating that it may or maybe not limiting.

6. Conclusions

Genetic multiplicity is present in all antibiotic-producing *Streptomyces* included here. The number of base pairs representing the redundant DNA was approximately 50 kb. The enzymes of the glycolytic pathway presenting the greatest multiplicity were phosphofructokinase, fructose 1,6-bisphosphate aldolase, glyceraldehyde 3 phosphate dehydrogenase and PYR kinase. The TCA cycle enzymes with the most gene copies were CS, both subunits of succinyl-CoA synthetase, the iron-sulfur protein subunit, flavoprotein subunit, and cytochrome b-556 subunit of succinate dehydrogenase, and fumarase. The MEs genes of the PEP-PYR-OXA node were the only ones that presented multiplicity. More research is required on the transcription of these genes and also on translation in order to establish their importance in these microorganisms.

Acknowledgements

This study was partially supported by grant IN214116 (DGAPA-UNAM).

Conflict of interest

The authors declare no conflict of interest.

Author details

Toshiko Takahashi[1], Jonathan Alanís[1], Polonia Hernández[2] and María Elena Flores[1*]

1 Departamento de Biología Molecular y Biotecnología, Universidad Nacional Autómoma de México, Instituto de Investigaciones Biomédicas, CDMX, México

2 Universidad Nacional Autónoma de México, Programa Jóvenes hacia la Investigación, México

*Address all correspondence to: mflores@iibiomedicas.unam.mx

IntechOpen

References

[1] Nowak MA, Boerlijst MC, Cooke J, Smith JM. Evolution of genetic redundancy. Nature. 1997;**388**:167-171

[2] Schniete JK, Cruz-Morales P, Selem-Mojica N, Fernández-Martínez LT, Hunter IS, Barona-Gómez F, et al. Expanding primary metabolism helps generate the metabolic robustness to facilitate antibiotic biosynthesis in. MBio. 2018;**9**

[3] Pye CR, Bertin MJ, Lokey RS, Gerwick WH, Linington RG. Retrospective analysis of natural products provides insights for future discovery trends. Proceedings of the National Academy of Sciences. 2017; **114**:5601

[4] van Keulen GS, Dijkhuizen J. Central carbon metabolic pathways in *Streptomyces*. In: P D *Streptomyces: Molecular Biology and Biotechnology*. Norfolk, UK: Caister Academic Press; 2011. pp. 105-124

[5] Doi S, Komatsu M, Ikeda H. Modifications to central carbon metabolism in an engineered Streptomyces host to enhance secondary metabolite production. Journal of Bioscience and Bioengineering. 2020; **130**:563-570

[6] Kim BH, Gadd GM. Glycolysis. In: Kim BH, Gadd GM, editors. Prokaryotic Metabolism and Physiology. 2nd ed. Cambridge: Cambridge University Press; 2019. pp. 58-79

[7] Llamas-Ramírez R, Takahashi-Iñiguez T, Flores ME. The phosphoenolpyruvate-pyruvate-oxaloacetate node genes and enzymes in Streptomyces coelicolor M-145. International Microbiology. 2020;**23**: 429-439

[8] Quinn GA, Banat AM, Abdelhameed AM, Banat IM.

Streptomyces from traditional medicine: Sources of new innovations in antibiotic discovery. Journal of Medical Microbiology. 2020;**69**:1040-1048

[9] Gunnarsson N, Eliasson A, Nielsen J. Control of fluxes towards antibiotics and the role of primary metabolism in production of antibiotics. Advances in Biochemical Engineering/Biotechnology. 2004;**88**:137-178

[10] Webb BA, Forouhar F, Szu FE, Seetharaman J, Tong L, Barber DL. Structures of human phosphofructokinase-1 and atomic basis of cancer-associated mutations. Nature. 2015;**523**:111-114

[11] Cabrera R, Baez M, Pereira HM, Caniuguir A, Garratt RC, Babul J. The crystal complex of phosphofructokinase-2 of Escherichia coli with fructose-6-phosphate: Kinetic and structural analysis of the allosteric ATP inhibition. The Journal of Biological Chemistry. 2011;**286**:5774-5783

[12] Ünsaldı E, Kurt-Kızıldoğan A, Voigt B, Becher D, Özcengiz G. Proteome-wide alterations in an industrial clavulanic acid producing strain of. Synthetic and Systems Biotechnology. 2017;**2**:39-48

[13] Muñoz ME, Ponce E. Pyruvate kinase: Current status of regulatory and functional properties. Comparative Biochemistry and Physiology. Part B, Biochemistry & Molecular Biology. 2003;**135**:197-218

[14] Ferraris DM, Spallek R, Oehlmann W, Singh M, Rizzi M. Structures of citrate synthase and malate dehydrogenase of Mycobacterium tuberculosis. Proteins. 2015;**83**:389-394

[15] Ge Y, Cao Z, Song P, Zhu G. Identification and characterization of a novel citrate synthase from

Streptomyces diastaticus No. 7 strain M1033. Biotechnology and Applied Biochemistry. 2015;**62**:300-308

[16] Ambrus A, Nemeria NS, Torocsik B, Tretter L, Nilsson M, Jordan F, et al. Formation of reactive oxygen species by human and bacterial pyruvate and 2-oxoglutarate dehydrogenase multienzyme complexes reconstituted from recombinant components. Free Radical Biology and Medicine. 2015;**89**: 642-650

[17] Ge Y-D, Cao Z-Y, Wang Z-D, Chen L-L, Zhu Y-M, Zhu G-P. Identification and biochemical characterization of a thermostable malate dehydrogenase from the mesophile Streptomyces coelicolorA3(2). Bioscience, Biotechnology, and Biochemistry. 2014; **74**:2194-2201

[18] Sauer U, Eikmanns BJ. The PEP-pyruvate-oxaloacetate node as the switch point for carbon flux distribution in bacteria. FEMS Microbiology Reviews. 2005;**29**:765-794

[19] Papagianni M. Recent advances in engineering the central carbon metabolism of industrially important bacteria. Microbial Cell Factories. 2012; **11**:50

[20] Harding CJ, Cadby IT, Moynihan PJ, Lovering AL. A rotary mechanism for allostery in bacterial hybrid malic enzymes. Nature Communications. 2021;**12**:1228

[21] Basu P, Sandhu N, Bhatt A, Singh A, Balhana R, Gobe I, et al. The anaplerotic node is essential for the intracellular survival of. The Journal of Biological Chemistry. 2018;**293**:5695-5704

[22] Vizcaíno N, Pérez-Etayo L, Conde-Álvarez R, Iriarte M, Moriyón I, Zúñiga-Ripa A. Disruption of pyruvate phosphate dikinase in Brucella ovis PA CO. Veterinary Research. 2020; **51**:101

[23] Viollier PH, Minas W, Dale GE, Folcher M, Thompson CJ. Role of acid metabolism in Streptomyces coelicolor morphological differentiation and antibiotic biosynthesis. Journal of Bacteriology. 2001;**183**:3184-3192

[24] Liao CH, Yao L, Ye B-C, Ye BC. Three genes encoding citrate synthases in Saccharopolyspora erythraea are regulated by the global nutrient-sensing regulators GlnR, DasR, and CRP. Molecular Microbiology. **94**:1065-1084

Chapter 6

Metabolites from Actinobacteria for Mosquito Control

Pathalam Ganesan and Savarimuthu Ignacimuthu

Abstract

Arthropods like mosquitoes are well-known vectors which are mainly involved in the transmission of pathogens to different human and vertebrate diseases. Most of the pathogens like viruses and nematodes are transmitted by mosquitoes. Controlling vector populations by using actinobacteria can be particularly very effective. Actinobacteria which contain also non filamentous forms of bacteria which produce a large number of biologically active secondary metabolites. Even though many antibiotics have been developed from actinobacteria, not much work have been conducted in the field of pest control. The actinobacteria and their metabolites effectively control mosquito populations and the transmission of diseases by them. The microbial metabolites have many advantages over synthetic chemicals because many of them are host-specific and safe for beneficial organisms. Due to this species-specific effect, microbial pesticides are more reliable to control mosquito populations. These types of metabolites have to be evaluated for the development of novel insecticides for vector control. Some studies have reported the mosquitocidal effects of actinobacterial metabolites like tetranectin, avermectins, spinosad, macrotetrolides, etc; they have less or no residual effect in the environment. This chapter focuses on the mosquitocidal effects of actinobacteria and their metabolites.

Keywords: Actinobacteria, *Streptomyces*, *Aedes aegypti*, *Culex quinquefasciatus*, *Anopheles stephensi*, vector-borne diseases, microbial pesticides

1. Introduction

Arthropods are the most important organisms in relation to humans and environment in many ways. Most insects are beneficial to environment, humans and other animals; some of them are dangerous to humans and mammals. Insects which act as vectors can cause several devastating diseases to human beings and other mammals. Mosquitoes are the most harmful vectors among hematophagous insects [1]. They transmit harmful pathogens which cause millions of death every year; they produce a great impact on public health, labour outputs and economics [2, 3]. Mosquitoes mainly transmit the diseases like *Japanese encephalitis*, filariasis, dengue, malaria, dengue haemorrhagic fever, yellow fever, Zika and chikungunya [4–7].

World Health Organization declared mosquito as the 'public enemy number one' in 1996 [8]. Millions of people are dying every year due to mosquito-borne diseases. Mosquitoes can grow in different aquatic habitats such as ponds, overhead tanks, brackish water, sewage waters, freshwater pools, paddy fields, and even in stagnant rainwater in small containers [9]. Mosquitoes are important etiological agents not only to human beings but also to other native faunas [10]. Mosquito-borne diseases are becoming more extreme and spreading due to ecological and environmental changes like urbanization.

1.1 The most common mosquitoes as disease vectors

Five vector-borne diseases are considered as most dreadful diseases in India. They are malaria (transmitted by *Anopheles* spp.), dengue, chikungunya (*Aedes* spp.), filaria and *Japanese encephalitis*(*Culex* spp.) [11, 12].

1.1.1 Aedes spp.

The genus *Aedes* is the most important vector responsible for chikungunya and dengue, which are mainly found in the temperate regions of the world. Nearly 30–50% of the world population has been affected by dengue virus [13, 14]. *Aedes aegypti* is a tropical species that grows in fresh water in and around human dwelling areas. *Aedes polynesiensis* and *Aedes scutellaris* are in the western Pacific region and *Aedes mediovittatus* is in the Caribbean. *Aedes* is an important vector in Southeast Asia and it has spread to the Mediterranean rim, Americas, and western Africa [15, 16]. In 2006, several parts of southern India confirmed the re-emergence of chikungunya infection [17]. Chikungunya outbreak affected 1.25 million people from about 150 districts in eight states of India; the causative viral agent was spread by *Aedes aegypti* [18]. Mosquitoes not only transmit diseases to humans but also to other mammals like dogs, horses and cats. They cause diseases like West Nile fever, dog heartworm and Eastern equine encephalitis (EEE). Dog heartworm (*Dirofilaria immitis*) is a dreadful disease for canines.

1.1.2 Anopheles spp.

Malaria is transmitted by *Anopheles* spp. mostly in urban areas. *Anopheles* mosquitoes mainly breed in clean and rainwater storage amenities. Out of 59 Anopheline species in the world, nine occur in India as vectors. In Ethiopia, malaria is the main disease responsible for a large number of deaths. It is one of the significant interferences to financial enhancement as the important transmission time frames agree with top farming and collecting period [19–21]. In Orissa, a state which is located in the eastern part of India, a large number of malaria cases and malaria-related deaths were recorded [22]. Malaria outbreaks are common in Indonesia, India, Bangladesh, Myanmar, Thailand and Sri Lanka [23, 24]. In India, *Anopheles subpictus* was reported in the state of Rajasthan and it is identified to transmit malaria and filariasis [25, 26].

1.1.3 Culex spp.

Culex is a vector of many diseases and an important genus of mosquito which transmits diseases like filariasis, St. Louis encephalitis, *Japanese encephalitis* and avian West Nile fever. The adult *Culex* mosquito can size up to 0.16–0.4 inches [27]. It is a major house-dwelling mosquito in many tropical areas. *Culex* is an annoying mosquito to humans and main vector of filariasis in some countries. These mosquitoes mainly breed in polluted waters close to human residences. *Culex* mosquitoes are known to carry the nematode worm *Wuchereria bancrofti* which is responsible for causing lymphatic filariasis. More than 146 million people were affected all over the world [12, 28]. *Japanese encephalitis* virus (JEV) belongs to the family of *Flaviviridae* and is the primary pathogen of viral encephalitis. In earlier times, JEV was efficiently controlled primarily by vaccination [11, 29]. *Culex* spp.

are night-biting mosquitoes, with highest activities after 1 h of darkness. They are mainly exophilic and commonly stay indoors after feeding on blood. *Culex* bite causes sensitive responses like skin irritation, urticaria and angioedema [30].

With increasing human activities and climate change, several vector-borne diseases are emerging in the world. Humans are fighting to prevent mosquito-borne diseases for many centuries but still they are unable to find any definite way. To control mosquito-borne diseases we are following mainly two ways namely controlling mosquito population and protecting from bite. In eradication programmes, mosquitoes are killed at their immature and adult stage. Control of adult mosquitoes is mainly done in malaria control programmes, and larval control is done to eradicate filariasis, dengue and encephalitis [31–33]. In order to prevent mosquito-borne diseases, it is important to eradicate mosquito population for improving public health. Currently, eradication of mosquito programme is suffering because of the ever-increasing detrimental effects of synthetic chemicals on non-target organisms, development of pesticide resistance and environmental and public health concerns. The increased costs of insecticides and greater public concern over ecological pollution have required a continued search for alternative vector-control approaches, which would be naturally safer and specific in their action [12, 34–38]. Controlling of mosquito population at juvenile stages is done by direct application of insecticides in their breeding sites. Early insecticides such as DDT, BHC and methoxychlor were found to be effective in the beginning; after some years, due to the development of resistance by insect pests, their effectivity declined. Manmade chemicals such as chlorpyrifos, diflbenzuron, petroleum oils, pyriproxyfen, permethrin, malathion, methoprene, temephos and resmethrin are used to control mosquito population at the immature and adult stages [36, 39].

Applications of synthetic insecticides in the field have created a number of environmental problems, like resistance development in insect pests, environmental imbalance, and detrimental effects on animals. Repellents from synthetic chemicals have harmful impacts of toxic effects, undesirable effects like unpleasant odour and unpleasantly sticky skin; toxic reactions under some situations to different age groups, toxicity against the skin, nervous and immune systems usually occur when the product is used incorrectly or in the long term [40, 41]. Harmful effects of malathion are twitching of voluntary muscles, paralysis, ultimately death, incoordination, headaches, nausea, convulsions, blurred vision and pupil constriction, slowed heartbeat, respiratory depression, paralysis and coma [42]. Axonic poisons from pyrethroids cause toxicity to the nerve fibres which results in continuous nerve stimulation and suffering from headache, dizziness, nausea and diarrhoea. Different groups of pyrethroids like fenvalerate, sumithrin, d-trans allethrin and permethrin have some undesirable effects like disruption of hormonal pathways, reproductive dysfunction, developmental impairment and cancer [43].

Constant use of insecticides in the field and water bodies leads to serious hazards to the soil microorganisms and other beneficial organisms present in the environment. Highly sensitive reactions were observed when using malathion and carbendazim to *Hyphessobrycon erythrostigma*, *Colossoma macropomum*, *Nannostomus unifasciatus*, *Otocinclus affinis* and *Paracheirodon axelrodi*, one crustacean (*Macrobrachium ferreirai*), three insects (Hydrophilus sp., *Buenoa unguis*, and *Palustra laboulbeni*) and one freshwater snail (*Pomacea dilioides*) [44–46]. Thus, there is a continual need for developing biologically active molecules from natural resources which are toxic to insect pests but beneficial to environment. Insecticides derived from natural resources are generally preferred because of their less harmful nature to non-target organisms and innate biodegradability. Pesticides from natural sources are effective alternatives to such synthetic insecticides to control mosquito population [47–51]. In this chapter, the metabolites derived from actinobacteria which are toxic to mosquitoes, beneficial to non-target organisms, and their role in eco-friendly mosquito control programmes at present and in the future are discussed.

2. Chemical insecticides in mosquito control

Synthetic insecticides play an important role in controlling vector-borne diseases [52]. Methoprene is an important hormone that was first registered by EPA that acts as an insect growth-regulating hormone and inhibits the normal development of immature stages of insects. It mimics the insect growth hormone; it was successfully developed as a biorational insecticide based on understanding the physiology of insects [53–55]. Temephos is an organophosphate (OP) class of pesticides, and it was first registered by EPA in 1965. It was mainly used for controlling immature stages of mosquitoes and also other insect pests. Application of temephos is mainly on standing water, swamps, marshes, shallow ponds and intertidal zones to kill mosquito life stages [56]. Monomolecular films are prepared from renewable plant oils. Presently, two types of monomolecular surface film products Agnique® MMF and Agnique® MMF are available in the market for controlling the population of mosquitoes [57]. Using oils in vector control programme is the best way to control the population of insect pests. Oils are mainly derived from crude petroleum and several petroleum products. They act as contact poisons, with effective mosquitocidal efficiency [58]. Pyrethroid, resmethrin and chlorpyrifos have the ability to kill insects quickly mainly in mature stages, flying mosquitoes and other insect pests. Organophosphates and Malathion are mostly used for controlling both larvae and adults. DEET provides long-lasting protection against a wide range of insect pests, and it has been documented in several reports especially for preventing mosquito biting [59]. DEET is available in various commercial formulations such as lotions, gels, creams, aerosols, solutions, sticks, and impregnated towelettes [41]. Synthetic chemicals are preferred over natural because of their quick and strong efficacy to control vector population. However, chemicals derived from nature to control vectors are given preferences by government and non-government agencies to reduce the use of synthetic chemicals.

3. Ecofriendly management of mosquitoes

Predatory organisms mostly act as the main agents that are hazardous to mosquitoes but favourable to human beings. Without insecticides, predators can control mosquito population to some extent. Dragonflies consume the larvae and bats, and birds eat adult mosquitoes. Electrified coils are used to control insect pest population. They do not release toxic chemicals into the environment, but they can kill beneficial insects also and may cause danger to children and pets (http://homeguides.sfgate.com/ecofriendly-mosquito-killer-lawn-81948.html). Larvivorous fishes have been widely used all over the world for controlling mosquito population. In 1905, larvivorous fish *Gambusia affinis* was purposely introduced from its native Texas to Hawaiian Islands, to Italy and Spain during1920s and later to 60 other countries [60]. *Poecilia reticulata*, a native of South America, was introduced to control malaria in British India and many other countries [61]. Sound traps which are used to control mosquitoes attract mosquitoes from long distances [62]. Using the lure and killing is the best method to control mosquitoes, especially in genus *Anopheles* and other arthropods [62, 63]. Balancing ecosystem is the most important thing in the living world. Every organism in the ecosystem is dependent on some other organism.

4. Microbes as insecticides

Eradication of insect pest population through pesticides derived from microorganisms is highly effective and generally has benefits over synthetic insecticides. The metabolites derived from microbes are host specific and there is no detrimental effect on the non-target organism and surrounding environs. Around the world, only 5% of fungi and 0.1% of bacteria have been described [64]. Different types of microorganisms like fungi, bacteria, nematodes and viruses are biologically toxic to insect pests [65, 66]. Microorganisms from different sediments are important sources of bioactive components for antibiotics; many bioactive secondary metabolites are used for biotechnology and pharmacological studies [67].

Larvicides derived from microbes, especially bacteria have been used to eradicate mosquito population for the past few years. Bacteria like *Bacillus sphaericus* are widely used for potential biolarvicide in mosquito control programmes worldwide; it exhibited effective larvicidal activity against larvae of several mosquito species. Commercial larvicides from active strain of *B. sphaericus* are used to control various types of insects which act as vectors [68–70]. Toxins Bin and Mtxs are produced during the sporulation and vegetative stages of *B. sphaericus*, and some of the toxic strains have been extensively used for controlling the populations of mosquito [71]. Larvicides developed from *B. sphaericus* against mosquitoes have led to development of resistance [72]. *B. sphaericus* biolarvicide is limited in India due to the resistance development in the target mosquito. In the early stage, *An. stephensi* had developed resistance against *B. sphaericus* [73, 74].

Metabolites from *Bacillus thuringiensis var. israelensis* were toxic to the larvae and pupae of *Cx. quinquefasciatus* [75]. *Bacillus thuringiensis* is naturally present in the soil and normally it is used as a pest-control microorganism. Different types of *Bacillus thuringiensis* have been used to control insect pests. It is the only insecticide extensively used in all parts of the world. δ-endotoxin produced from *Bacillus thuringiensis* is toxic to various insect species. The toxin initiates growth of a lytic protein in the midgut epithelial membrane, which leads to cell lysis, termination of feeding, and leads to death of the larva. They produce two different types of toxin proteins such as Cry and Cyst proteins [76–78]. *Bacillus sphaericus* and *Bacillus thuringiensis* var. *israelensis* H-14, *Bacillus amyloliquefaciens* and *Bacillus amyloliquefaciens* were highly effective against different species of mosquitoes [79–82]. Secondary metabolites derived from various types of fungal species like *Beauveria bassiana*, *Chrysosporium tropicum*, *Aspergillus niger*, *Cochliobolus lunatus*, *Fusarium oxysporum*, *Chrysosporium lobatum*, *Trichophyton ajelloi*, *Fusarium moniliforme*, *Trichophyton mentagrophytes*, *Paecilomyces carneus*, *Paecilomyces marquandii*, *Isaria fumosorosea*, *Metarhizium anisopliae*, *Penicillium* sp., *Paecilomyces lilacinus* and *Evlachovaea kintrischic* are also used to control the immature stages of mosquito population [83–89]. In recent years, the research on microscopic organisms has increased to identify microbial agents for various biological uses.

5. Actinobacteria for mosquito control

Actinobacteria are a group of filamentous bacteria; they are Gram-positive, dwelling in the soil, marine, and some of them are endophytic; they produce a large number of secondary metabolites. In the pharmaceutical industries, 75–80% of antibiotics are derived from microbes like actinobacteria [90–92]. Microorganisms present in different environments serve as an important natural resource for novel antibiotics, antitumor agents, and other therapeutic substances. Antibiotics such as erythromycin, vancomycin, and streptomycin are used for various pharmacological purposes [93–95]. The molecules isolated from microbes like actinobacteria are extremely toxic to insects like mosquitoes and have low toxicity to other beneficial organisms and environment. The use of secondary metabolites from actinobacteria may be a good approach for environment-friendly insect pest management [96].

5.1 Isolation of actinobacteria

Actinobacteria are present in different habitats, and they have a capacity to produce a large amount of various active secondary metabolites. Between the diversity of microbes, actinobacteria produce an enormous amount of secondary metabolites which have been used for various biological and biotechnological activities like anticancer drugs, antibiotics, pesticides, immunosuppressors and enzyme inhibitors [97–100]. Isolation of actinobacteria from different environments like cold, halophilic and alkaliphilic is possible; some of them present in high temperatures are called extremophiles. More than 61% of the secondary metabolites are isolated from actino-bacteria genus *Streptomyces*; some of the metabolites have been isolated from rare acti-nobacteria. The samples collected from extreme environments — particularly in places which are isolated from human dwellings yield good results [101–104]. Generally, microorganism is isolated using serial membrane filter technique, dilution method and direct inoculation technique [105]. The collected samples are spread on different types of media used for actinobacteria isolation, such as actinomycetes isolation agar (AIA), humic acid vitamin agar (HVA), Starch casein agar, Glycerol-asparagine agar, Bennet's agar (BA) medium, Gause`s No.1 medium, Complex HV Agar, HV agar, humic acid vitamin agar, ZSSE (Zhang' Starch Soil Extract Agar, Kuster's agar, inor-ganic salt starch agar, starch nitrate agar, glycerol glycine agar, chitin agar, soil extracts agar and Glycerol-asparagine agar etc. [106–108]. The endophytic actinobacteria are isolated using the recommended method of Otoguro et al., [109]. Based on the colony morphology, the actinobacteria are selected and purified on ISP-2 (International *Streptomyces* project medium No. 2) for further bioactive studies [110].

5.2 Pre-treatment and selection of actinbacteria

Samples collected from different places are pre-treated using different procedures to remove the fungi, bacteria and other unwanted microbes. Pre-treatments of the collected samples encourage or enrich the growth of the actinobacteria, especially rare actinobacteria. In one of the pre-treatments, $CaCO_3$ was used to treat the soil samples to decrease the number of other unwanted bacteria, and it allowed the excess actinobacteria spores cells to survive [111]. In Physico-chemical treatment, the soil sample (1 g) was suspended in 10 ml of normal saline, and the sample was heated for 1 h at 120°C to increase and encourage the growth of the actino-bacteria [112]. Samples were treated with 1.5% phenol for 30 min at 30°C by the recommended method by Hayakawa et al., [112]. To decrease the growth of other unwanted microbes, the growth media were added with nalidixic acid (100 mg/l) and ketoconazole (30 mg/l) [113]. To increase the number of actinobacteria, the soil sample is treated with peptone (6%) and sodium lauryl sulphate (0.05%) at 50°C for 10 min [114]. Soil samples are added to 10 ml of sterilized distilled (Wet heat) water and heated in water bath at 30–50°C for 2–6 min and allowed to cool before serial dilution; without distilled water samples are heated (Dry heat) in hot air oven at 50–70°C [112, 115]. In other treatment, the soil sample is added to sterile water and centrifuged at 10,000 rpm for 30 min and used for isolation of actinobacteria [116]. Soil samples are treated with sodium dodecyl sulphate as per the prescribed method by Janaki et al., [117]. Sample is also treated in microwave oven as per the recommended method by Bulina et al., [118].

5.3 Identification of actinobacteria

Different physical, chemical and molecular methods are available for identification of actinobacteria species. Generally, actinobacteria are identified in the petriplate based on the aerial and substrate mycelia, melanin pigments, pigment production, elevation and surface of each culture on the media [119]. Kelly [120] designated the colony arrangement of the different types of actinobacteria. Sporulation arrangement and Spores structures of the actinobacteria species were examined microscopically [121]. To check the cultural characterization of actinobacteria, different strains were streaked in different optimized growth media. Physiological and biochemical characters were done using the streaking of the culture in different media gelatin agar plates (gelatin hydrolysis) and starch agar plates (starch hydrolysis and sodium chloride resistance) etc. Isolated actinobacteria were streaked in the Petri plates and incubated at different temperatures for 7 days to check the optimal temperature for maximum growth through visual analysis [122]. Gel-diffusion and Fluorescent antibody (FA) procedures were used to identify *Actinomyces* species [123]. *Actinomyces israelii* was identified by the method of Slack et al., [124]. Spores' arrangements and Mycelium of the actinobacteria were carried out using scanning electron microscope (SEM). For genetic level identification, 16s rRNA was used for different actinobacteria [125].

5.4 Secondary metabolites from actinobacteria for mosquito control

The secondary metabolites isolated from actinobacteria are highly toxic to mosquitoes and have low toxicity to nontarget organisms. It is a good source for eco-friendly control of immature stages of mosquitoes [96]. Metabolites from actinobacteria were tested against mosquito life stages: three actinobacteria were reported to have ovicidal activity and 35 strains of actinobacteria had larvicidal activity; two *Streptomyces* sp. and one *Paecilomyces* sp. showed potent activity against tested mosquitoes. Aqueous solutions of actinobacteria presented potent larvicidal activity [126]. Karthik et al., [11] isolated the extract of *S. gedanensis* and tested it against the larvae of *Cx. gelidus* and *Cx.tritaeniorhynchus*. The results exhibited promising activity with LC_{50} values of 108.08 ppm and 609.15 ppm. Crude extracts of *S. gedanensis* and *S. roseiscleroticus* also revealed repellent activity at 1,000 ppm against *Cx. tritaeniorhynchus* and *Cx. gelidus*. Govindarajan et al., [127] reported that four *Streptomyces* sp. (A14, A21, A49 and A63) revealed potent larvicidal activity. Tanvir et al., [128] isolated twenty-one endophytic actinobacteria from plants. Among them, 10 actinobacteria species exposed strong larvicidal activity. Kekuda et al., [129] isolated extract from *Streptomyces* sp. and it showed 100% larvicidal activity against *Ae. aegypti* mosquito larvae after 24 h of treatment. Dhanasekaran et al., [130] reported that 35 actinobacteria isolated from different samples exhibited good activity against mosquitoes. Deepika et al., [131] reported 100% larval mortality for extract from *Streptomyces* sp. against *Cx. quinquefasciatus*. The marine actinobacterium (LK1) was isolated and crude extract was purified using reversed-phase high-pressure liquid chromatography. The extract presented good larvicidal activity against *An. stephensi* and *Cx. tritaeniorhynchus* with LC_{50} values of 31.82 ppm and 26.62 ppm, respectively, at tested concentrations [132]. Seven *Streptomyces* sp. isolated from marine sediments of South China produced siderophores, which acted as biocontrol agents and inhibited the growth of *Vibrio* spp. [133]. Gomes et al., [134] reported that five *Streptomyces* spp. were very efficient signifying their potential as biocontrol agents. Dhanasekaran et al., [135] isolated some actinobacteria and tested them for insecticidal activity. Totally four isolates showed strong larvicidal (100%) activity against larvae of *Anopheles* mosquito. Marine actinobacteria extracts had larvicidal, repellent and ovicidal activity against *Culex gelidus* and *Culex tritaeniorhynchus* [11].

Anwar et al., [136] collected different soil samples from various sites in salt range of Kalar Kahar, Pakistan and isolated 41 actinobacteria cultures. Among them, three actinobacteria: *Streptomyces minutiscleroticus*, *Streptomyces rochei* and *Streptomyces phaeoluteigrisseus* presented 100% larvicidal activity against *Cx. quinquefasciatus*. Vijayakumar et al., [137] tested the actinobacteria extract in different concentrations. The results presented that the isolates CC11 and SH22 (20%), CC110 and SH23 (16%), SH15 (12%), CC19 and S22 (8%), and S21 (4%) had good activity at 3 h against *Anopheles* mosquito. Filtrates of *Streptomyces citreofluorescens* presented good activity against *A. stephensi* and *Cx. quinquefasciatus* with LC50 values of 122.6 and 60.0 μl/ml, respectively [113]. Sanjenbam and Kannabiran, [138] isolated *Streptomyces* sp. VITPK9 from soil sample and tested for mosquitocidal activity. Ethyl Acetate extract gave good mortality against *Cx. tritaeniorhynchus, An. subpictus* and *Cx. gelidus* with LC50 values of 489.21, 831.78, and151.29 at 1000 ppm concentration.

5.5 Compounds from actinobacteria for mosquito control

Pure compounds isolated from actinobacteria like faerifungin, tetranectin, avermectins, flavonoids and macrotetrolides were found to be toxic against immature stages of vector mosquito and other insect pests. Actinobacteria like *Streptomyces* sp., *Streptomyces griseus*, *Streptomyces avermitilis*, *Streptosporangium albidum* and *Streptomyces aureus* produce these kinds of toxic metabolites that kill mosquitoes. Different types of genera were found to be producing toxic metabolites against mosquitoes; they are Actinomadura, Sreptoverticillium, Actinoplanes, Micropolyspora, Nocardiopsis, Thermomonospora, Oerskonia, Micromonospora, and Chainia [139–143].

Three new alpha class milbemycins (named milbemycins alpha28, alpha29, and alpha30)were isolated from *Streptomyces bingchenggensis*. They exhibited potent acaricidal and nematocidal activities [144]. Ichthyomycin, a compound isolated from *Streptomyces* sp. (strain 1107) was checked against larvae of *Culex pipiens autogenicus*, and the results exhibited that mortality of larvae was concentration-dependent [143]. Deepika et al., [131] isolated *Streptomyces* sp. VITDDK3 which produced the compound (2S,5R,6R)-2-hydroxy-3,5,6-trimethyloctan-4-one and which was tested for acaricidal and larvicidal activities against blood-sucking parasites. A compound trioxacarcin A (2a) and D (2d), isolated from the extract of *Streptomyces* sp. (B8652), influenced particularly high antiplasmodial activity against *Plasmodium falciparum* [145]. Prumycin, isolated from *Streptomyces* sp. showed antimalarial activity against drug-resistant Plasmodia [146]. Actinobacteria like *Streptomyces spinosa* have been reported to have a high level of activity against phytophagous insects and insects impacting public and animal health [147]. Metacycloprodigiosin, bafilomycin A1, and spectinabilin, isolated from *Streptomyces spectabilis* (BCC 4785) showed strong in vitro activity against *P. falciparum* [148]. The compound, salinosporamide A, isolated from marine actinobacteria, *Salinispora tropica*, presented strong inhibitory activity against Plasmodium growth [149]. Isolation of 10 new nine-membered bislactones, splenocins A–J (1–10) from organic extract of *Streptomyces* species (strain CNQ431) presented potent biological activities [150].

The compounds tetranectin, avermectins, faerifungin, and macrotetrolides were isolated from *Streptomyces aureus*, *S. avermitilis*, *Streptomyces albidum* and *Streptomyces griseus* which showed insecticidal activity [139–141, 143]. Salinosporamide A and Depsipeptides derived from actinobacteria were used as antimalarial compounds [149, 151]. Avermectin family of 16-membered macrocyclic lac Streptomycestones isolated from *Streptomyces avermectinius* had antihelminthic activity [152]. Saurav et al. [153] isolated the pure compound, 5-(2, 4-dimethylbenzyl) pyrrolidin-2-one, from *Streptomyces* VITSVK5 sp. which exhibited strong activity against *R. (B.) microplus*, *An. stephensi*, and *Cx. tritaeniorhynchus*. Faeriefungin, a polyol polyene macrolide lactone was isolated from the mycelium of *S. griseusvar. autotrophicus*. It showed 100% larval mortality of *A. aegypti* [154]. The compound aculeximycin, isolated from the *Streptosporangium albidum*, exhibited strong larvicidal activity against mosquito larvae as well as antimicrobial activities [126].

The compounds 5-azidomethyl-3-(2-ethoxy carbonyl-ethyl)-4-ethoxycarbon-ylmethyl-1H-pyrrole-2-carboxylic acid, ethyl ester (18.2%) 2; and akuammilan-16-carboxylic acid, 17-(acetyloxy)-10-methoxy, methyl ester (16R) (53.3%) were isolated from *Streptomyces* VITSTK7 sp. which had mosquitocidal activity [155]. Antonio et al., [156] reported that spinosad is a mixture of two tetracyclic macro-lides produced during the fermentation of soil actinobacteria, and it was used for controlling dengue vector, *A. aegypti*.

6. Spinosad as a microbial pesticide

Insect control metabolite spinosad was isolated from soil bacterium *Saccharopolyspora spinosa*. It exhibited high toxicity to the insect pest compared to the formerly developed chemical insecticides. Environmental Protection Agency of the United States gave permission to use Spinosad against various insect pests [157, 158]. It is a combination of two tetracyclic macrolide neurotoxins, spinosyns A and D. Insecticide from spinosad targets the nervous system of pest which contains nicotinic acetyl-choline and GABA receptors leading to immobilization and death. Due to its specific toxicity and its favourable nontarget organism and ecological profile, spinosad is considered by IPM practitioners as a significant new-generation pesticide [159]. The pesticide developed from spinosad is currently used against different insect orders like dipteran, lepidopteran, thysanopteran, and some coleopteran. Recently, research reports have recognized that spinosads are used to control several important mosquito species which act as vectors [160, 161]. Some of the insect orders reported that spinosad acts as stomach poison with direct contact poison and it is most active against Diptera, Lepidoptera, some Coleoptera, ants, termites and thrips [162]. Spinosad presented effective controlling of the popula-tion, particularly during the development in immature aquatic stages of mosquito vectors such as *Ae. albopictus*, *Ae. gambiae*,*Ae. aegypti*, *Ae. pseudopunctipennis*, *Cx. pipiens*, *Ae.albimanus*and *Cx. quinquefasciatus* [163]. Spinosad has been found to eradicate mosquito population all over the world. Eradication of mosquitoes from water jars in Thailand, Mexico cemetery water containers, septic tanks in Turkey, field microcosms in California, flooded fields in Egypt, street drains, cesspits, and disused wells in India, plots in Florida, water tanks in India, basins in Connecticut USA has been reported [56, 147, 164–173].

7. Future prospects of actinobacteria for mosquito control

In this chapter, we have stated that actinobacteria are used to control mos-quito population in immature stages like egg, larvae, pupae and mature adults. Insecticides from secondary metabolites derived from actinobacteria are an impor-tant component in controlling vector-borne diseases by controlling the population of mosquitoes. Identification of the compound present in the secondary metabolites paves the way to preparing an effective insecticide to control insect pests, especially mosquito. Metabolites from actinobacteria show species-specific target activity and nontoxicity to other animals and humans. The feasibility of pesticides in field appli-cation is considered as the most consistent agent for controlling immature stages of mosquitoes. Preparation of mosquito coil, cream, repellent and evaporator from isolated compounds of actinobacteria give great impact on mosquito population control. Only a limited number of research have been done on the actinobacteria to control mosquitoes. In future, researchers should focus on actinobacteria to identify novel compounds to effectively control mosquitoes. An efficient mosquitocide prepared from mixing of different compounds eluted from actinobacteria acts as a best alternative to synthetic insecticides to control mosquito-borne diseases without adverse residual effects. Government and private sectors should give priority to these kinds of research to promote mosquito control programmes.

8. Conclusion

Pesticides from actinobacteria are reliable mosquito control agents; they are in wider use in field applications to control various insect populations. Several research report that compounds from actinobacteria exhibit promising activity against mosquito population. Insecticides from natural resources like actinobacteria

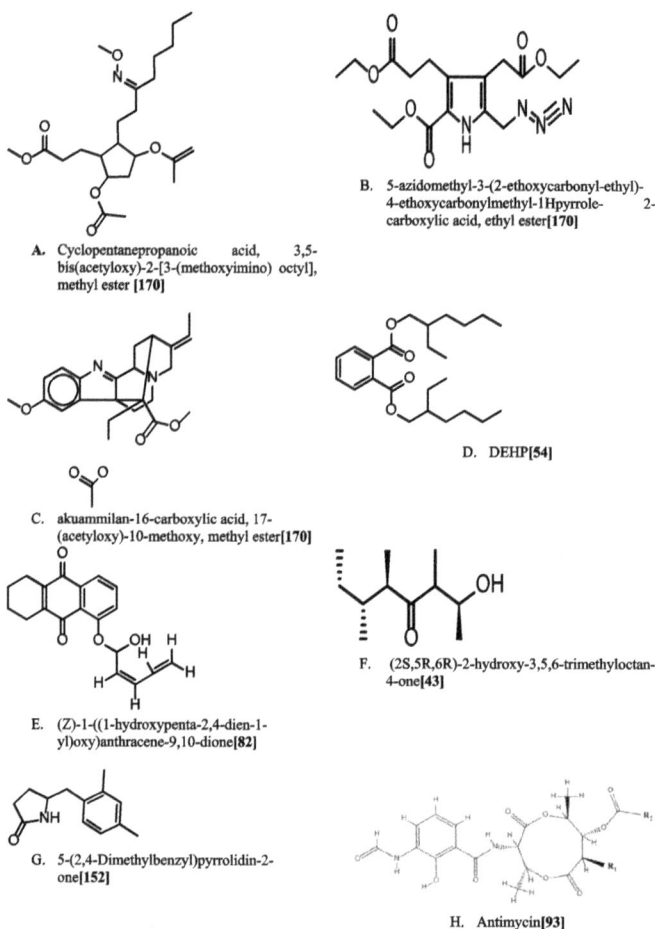

A. Cyclopentanepropanoic acid, 3,5-bis(acetyloxy)-2-[3-(methoxyimino) octyl], methyl ester [170]

B. 5-azidomethyl-3-(2-ethoxycarbonyl-ethyl)-4-ethoxycarbonylmethyl-1Hpyrrole- 2-carboxylic acid, ethyl ester[170]

C. akuammilan-16-carboxylic acid, 17-(acetyloxy)-10-methoxy, methyl ester[170]

D. DEHP[54]

E. (Z)-1-((1-hydroxypenta-2,4-dien-1-yl)oxy)anthracene-9,10-dione[82]

F. (2S,5R,6R)-2-hydroxy-3,5,6-trimethyloctan-4-one[43]

G. 5-(2,4-Dimethylbenzyl)pyrrolidin-2-one[152]

H. Antimycin[93]

Figure 1.
Some of the compounds isolated from actinobacteria for mosquito control. (A) Cyclopentanepropanoic acid, 3,5-bis(acetyloxy)-2-[3-(methoxyimino) octyl], methyl ester [2]; (B) 5-azidomethyl-3-(2-ethoxycarbonyl-ethyl)-4-ethoxycarbonylmethyl-1Hpyrrole- 2-carboxylic acid, ethyl ester [2]; (C) akuammilan-16-carboxylic acid, 17-(acetyloxy)-10-methoxy, methyl ester [2]; (D) DEHP [80]; (E) (Z)-1-((1-hydroxypenta-2,4-dien-1-yl)oxy)anthracene-9,10-dione [20]; (F) (2S,5R,6R)-2-hydroxy-3,5,6-trimethyloctan-4-one [53]; (G) 5-(2,4-Dimethylbenzyl)pyrrolidin-2-one [123]; and (H) Antimycin [64].

metabolites are easily produced in large quantities without disturbing other animals and ecosystems. In future, more research should focus on the isolation of compounds from actinobacteria with significant mosquito control metabolites from various natural sources such as desert, forest, marine and mangrove environments to control vector populations (**Figure 1** and **Table 1**).

Name of actinobacteria	Effective against Mosquito species	Reference
Streptomyces cacaoi	*Aedes aegypti*	[117]
Nocardiopsis sp. KA25-A	*Culex quinquefasciatus*	[174]
Streptomyces citreofluorescens	*Anopheles stephensi, Culex quinquefasciatus* and *Aedes aegypti*	[113]
Nocardia alba KC710971	*Anopheles stephensi, Culex quinquefasciatus* and *Aedes aegypti*	[175]
Streptomyces zaomyceticus Oc-5 and *Streptomyces pseudogriseolus* Acv-11	*Culex pipiens*	[176]
Streptomyces rochei *Streptomyces rimosus* *Streptomyces enissocaesilis* *Streptomyces enissocaesilis* *Streptomyces plicatus* *Streptomyces bungoensis* *Streptomyces ghanaensis* *Streptomyces vinaceus* *Streptomyces bungoensis* *Streptomyces vinaceusdrappus* *Streptomyces bungoensis*	*Aedes aegypti* *Anopheles stephensi* *Culex quinquefasciatus*	[177]
Streptomyces strain AN120537	*Aedes aegypti*	[178]
Streptomyces rimosus	*Culex quinquefasciatus*	[179]
Streptomyces fungicidicus, S. griseus, S. albus, S. alboflavus and *S. rochei*	*Aedes aegypti* and *Anopheles stephensi*	[180]
Streptomyces vinaceusdrappus	*Culex quinquefasciatus, Anopheles stephensi,* and *Aedes Aegypti*	[91]
Streptomyces sp. VITDDK3	*Anopheles subpictus* Grassi and *Culex quinquefasciatus* Say	[131]
Streptomyces capillispiralis Ca-1, *Streptomyces zaomyceticus* Oc-5, and *Streptomyces pseudogriseolus* Acv-11	*Culex pipiens*	[181]
Streptomyces sp. VITJS4	*Anopheles stephensi, Aedes aegypti* and *Culex quinquefasciatus*	[182]
Streptomyces rochei, Streptomyces minutiscleroticus and *Streptomyces phaeoluteigrisseus*	*Culex quinquefasciatus*	[136]
Streptomyces sp. PA9	*Culex quinquefasciatus*	[183]
Saccharomonospora spp. (LK-1), *Streptomyces roseiscleroticus* (LK-2), and *Streptomyces gedanensis* (LK-3)	*Culex tritaeniorhynchus* and *Culex gelidus,*	[11]
Streptomyces sp. and *Streptosporangium* sp.	*Anopheles*	[137]
Actinomycetes strain LK1	*Anopheles stephensi* and *Culex tritaeniorhynchus*	[132]

Streptomyces sp.VITPK9	*Anopheles subpictus, Culex tritaeniorhynchus, Culex gelidus,*	[138]
Streptomyces VITSTK7 sp.	*Anopheles subpictus; Culex quinquefasciatus*	[155]
Streptomyces VITSVK5 sp.	*Anopheles stephensi,* and *Culex tritaeniorhynchus*	[153]
Streptomyces albovinaceus and *Streptomyces badius*	*Culex quinquefasciatus*	[128]
Streptomyces sp. M25	*Anopheles subpictus, Culex quinquefasciatus* and *Aedes aegypti*	[184]
Streptomyces fungicidicus, Streptomyces griseus, Streptomyces albus, Streptomyces rochei, Streptomyces violaceus, Streptomyces alboflavus and *Streptomyces griseofuscus*	*Culex pipiens*	[185]
Streptomyces collinus	*Culex quinquefasciatus Aedes aegypti*	[7]

Table 1.
Some of the actinobacteria species used for mosquito control.

Acknowledgements

The authors thank Xavier Research Foundation, St. Xavier's College, Palayamkottai, Tirunelveli, Tamilnadu-627 002, INDIA for financial support and facilities.

Author details

Pathalam Ganesan and Savarimuthu Ignacimuthu*
Xavier Research Foundation, St. Xavier's College, Tirunelveli, Tamilnadu, India

*Address all correspondence to: xrfsxc@gmail.com

IntechOpen

References

[1] Rai MM, Rathod MK, Padole A, Khurad AM. Mosquitoes menace to humanity. In: William SJ, editor. Defeating the Public Enemy, the Mosquito: A Real Challenge. Loyola Publications; 2007. pp. 398-403

[2] World Health Organization. Lymphatic filariasis: The disease and its control, fifth report of the WHO Expert Committee on Filariasis [meeting held in Geneva from 1 to 8 October 1991]. World Health Organization; 1992

[3] WHO (World Health Organization). Global Malaria Programme. Geneva: WHO; 2013

[4] Anderson RL. Toxicity of fenvalerate and permethrin to several non-target aquatic invertebrates. Environmental Entomology. 1982;**9**:436-439

[5] Borah R, Kalita MC, Kar A, Talukdar AK. Larvicidal efficacy of *Toddalia asiatica* (Linn.) Lam against two mosquito vector *Aedes aegypti* and *Culex quinquefasciatus*. African Journal of Biotechnology. 2010;**9**:2527-2530

[6] Rahuman AA, Bagavan A, Kamaraj C, Saravanan E, Zahir AA, Elango G. Efficacy of the larvicidal botanical extracts against *Culex quinquefasciatus* Say (Dipetera: Culicidae). Parasitology Research. 2009;**104**:1365-1372

[7] Reegan AD, Kumar PS, Asharaja AC, Devi C, Jameela S, Balakrishna K, et al. Larvicidal and ovicidal activities of phenyl acetic acid isolated from Streptomyces collinus against Culex quinquefasciatus Say and Aedes aegypti L.(Diptera: Culicidae). Experimental Parasitology. 2021;**2021**:108120

[8] World Health Organization. The World Health Report 1996: Fighting Disease, Fostering Development. Geneva: WHO; 1996. p. 48

[9] Paulraj MG, Kumar PS, Ignacimuthu S, Sukumaran D. Natural insecticides from actinomycetes and other microbes for vector mosquito control. Herbal Insecticides, Repellents and Biomedicines: Effectiveness and Commercialization. 2016:85-99

[10] Snow RW, Guerra CA, Noor AM, Myint HY, Hay SI. The global distribution of clinical episodes of *Plasmodium falciparum* malaria. Nature. 2005;**434**:214-217

[11] Karthik L, Gaurav K, Rao BKV, Rajakumar G, Rahuman AA. Larvicidal, repellent, and ovicidal activity of marine actinobacteria extracts against *Culex tritaeniorhynchus* and *Culex gelidus*. Parasitology Research. 2011;**108**(6):1447-1455

[12] WHO Expert Committee on Vector Biology and Control, World Health Organization. Vector resistance to pesticides: Fifteenth report of the WHO Expert Committee on Vector Biology and Control [meeting held in Geneva from 5 to 12 March 1991]. World Health Organization. 1992

[13] Hales S, Wet ND, Maindonald J, Woodward A. Potential effect of population and climate changes on global distribution of dengue fever: An empirical model. Lancet. 2002;**360**:830-834

[14] Yang T, Liang L, Guiming F, Zhong S, Ding G, Xu R, et al. Epidemiology and vector efficiency during a dengue fever outbreak in Cixi, Zhejiang province, China. Journal of Vector Ecology. 2009;**34**:148-154

[15] Harrington LC, Scott TW, Lerdthusnee K. Dispersal of the dengue vector *Aedes aegypti* within and between rural communities. The American Journal of Tropical Medicine and Hygiene. 2005;**72**:209-220

[16] Honorio NA, Da Silva WC, Leite PJ, Gonçalves JM, Lounibos LP, Lourenço-de-Oliveira R. Dispersal of *Aedes aegypti* and *Aedes albopictus* (Diptera: Culicidae) in an urban endemic dengue area in the State of Rio de Janeiro. Brazil Memorias do Instituto Oswaldo Cruz. 2003;**98**:191-198

[17] WHO. WHO Director-General summarizes the outcome of the Emergency Committee regarding clusters of microcephaly and Guillain-Barré syndrome. 2016. http://www.who.int/mediacentre/news/statements/2016/emergencycommittee-zika-microcephaly/en/2016. [Accessed: September 22, 2016]

[18] Pialoux G, Gaüzère M, Jauréguiberry S, Strobel M. Chikungunya, an epidemic arbovirus. The Lancet Infectious Diseases. 2007;**7**:319-327

[19] Brower J, Chalk P. The Global Threat of New and Reemerging Infectious Diseases: Reconciling US National Security and Public Health Policy. Rand Corporation; 2003

[20] Karunamoorthi K, Bekele M. Prevalence of malaria fromperipheral blood smears examination: A 1-year retrospectivestudy from the Serbo Health Center, Kersa Woreda, Ethiopia. Journal of Infection and Public Health. 2009;**2**(4):171-176

[21] Mittal PK, Adak T, Subbarao SK. Inheritance of resistance to *Bacillus sphaericus* toxins in a laboratory selected strain of *A. stephensi* (Diptera: Culicidae) and its response to *Bacillus thuringiensis var. israelensis*. Current Science. 2005;**89**:442-443

[22] Sharma SK, Upadhyay AK, Haque MA, Tyagi PK, Raghavendra K, Dash AP. Wash-resistance and field evaluation of alpha cypermethrin treated long-lasting insecticidal net (Interceptor) against malaria vectors *Anopheles culicifacies* and *Anopheles fluviatilis* in a tribal area of Orissa, India. Acta Tropica. 2010;**116**(1):24-30

[23] Najera JA, Kouznetsov RL, Delacollette C. Malaria Epidemics Detection and Control Forecasting and Prevention. WHO; 1998

[24] Nájera JA, Kouznetsov RL, Delacollette C. Malaria epidemics, detection and control, forecasting and prevention. Division of Control of Tropical Diseases, WHO: Geneva; 1998

[25] Hoedojo PF, Atmosoedjono S, Purnomo TT. A study on vectors of *Bancroftian filariasis* in West Flores, Indonesia. The Southeast Asian Journal of Tropical Medicine and Public Health. 1980;**11**(3):399-404

[26] Tyagi BK, Yadav SP. Bionomics of malaria vectors in two physiographically different areas of the epidemic-prone Thar Desert, north-western Rajasthan (India). Journal of Arid Environments. 2001;**47**:161-172

[27] Molaei G. Host feeding pattern of *Culex quinquefasciatus* (Diptera: Culicidae) and its role in transmission of West Nile Virus in Harris County, Texas. The American Journal of Tropical Medicine and Hygiene. 2007;**77**(1):73-81

[28] Holder P. The mosquitoes of New Zealand and their animal disease significance. Surveillance. 1999;**26**:12-15

[29] Ghosh D, Basu A. *Japanese encephalitis*—a pathological and clinical perspective. PLoS Neglected Tropical Diseases. 2009;**3**(9):437

[30] Karunamoorthy K, Ilango K, Murugan K. Laboratory evaluation of traditionally used plant-based insect repellents against the malaria vector *Anopheles arabiensis* Patton. Parasitology Research. 2010;**106**:1217-1223

[31] Mulla SM. Biological control of mosquitoes withentomopathogenic bacteria. Chinese Journal of Entomology. 1991;**1991**:93-104

[32] Nerio LS, Olivero-Verbel J, Stashenko E. Repellent activity of essential oils: A review. Bioresource Technology. 2010;**101**:372-378

[33] Nerio Pascual M, Ahumada JA, Chaves LF, et al. Malaria resurgence in East African highlands: Temperature trends revisited. Proceedings of the National Academy Science USA. 2006;**A103**:5829-5834

[34] Coats JR. Risks from natural versus synthetic insecticides. Annual Review of Entomology. 1994;**39**:489-515

[35] Khan AR, Selman BJ. Microsporidian pathogens of mosquitoes and their potential of control agents. Agricultural Zoological Review. 1996;**7**:303-335

[36] Paulraj MG, Reegan AD, Ignacimuthu S. Toxicity of Benzaldehyde and Propionic acid against immature and adult stages of *Aedes aegypti* (Linn.) and *Culex quinquefasciatus* (Say) (Diptera: Culicidae). Journal of Entomology. 2011;**8**:539-547

[37] Peng Y, Song J, Tian G, Xue Q, Ge F, Yang J, et al. Field evaluations of *Romanomermis yunanensis* (Nematoda: Mermithidae) for control of Culicinae mosquitoes in China. Fundamental and Applied Nematology. 1998;**21**:227-232

[38] Sivanandhan S, Pathalam G, Antony S, Michael GP, Samuel R, Kedike B, et al. Mosquitocidal effect of monoterpene ester and its acetyl derivative from Blumea mollis (D. Don) Merr against *Culex quinquefasciatus* (Diptera: Culicidae) and their insilico studies. Experimental Parasitology. 2021;**223**:108076

[39] Brattsten LB, Hamilton GC, Sutherland DJ. Insecticides Recommended for Mosquito Control in New Jersey. New Jersey: New Jersey Agricultural Experiment Station; 2009

[40] Briassoulis G. Toxic encephalopathy associated with use of DEET insect repellents: A case analysis of its toxicity in children. Human & Experimental Toxicology. 2001;**20**:8-14

[41] Govere J, Durrheim DN, Baker L, Hunt R, Coetzee M. Efficacy of three insect repellents against the malaria vector *Anopheles arabiensis*. Medical and Veterinary Entomology. 2000;**14**:441-444

[42] Brenner L. Malathion. Journal of Pesticide Reform Winter. 1992;**12**(9):29

[43] Garey J, Wolff MS. Estrogenic and antiprogestagenic activities of pyrethroid insecticides. Biochemical and Biophysical Research Communications. 1998;**251**(3):855-859

[44] Rico F, Oshima A, Hinterdorfer P, Fujiyoshi Y, Scheuring S. Two-dimensional kinetics of inter-connexin interactions from single-molecule force spectroscopy. Journal of Molecular Biology. 2011;**412**:72

[45] Saler S, Saglam N. Acute toxicity of malathion on Daphnia magna Straus, 1820. Journal of Biological Sciences. 2005;**5**(3):297-299

[46] Van Wijngaarden RP, Brock TCM, Brink PJ. Threshold levels for effects of insecticides in freshwater ecosystems: A review. Ecotoxicology. 2005;**14**:355-380

[47] Akhtar Y, Yeoung YR, Isman MB. Comparative bioactivity of selected extracts from Meliaceae and some commercial botanical insecticides against two noctuid caterpillars, *Trichoplusia ni* and *Pseudaletia unipuncta*. Phytochemistry Reviews. 2008;**7**:77-88

[48] Chapagain BP, Saharan V, Wiesman Z. Larvicidal activity of saponins from *Balanites aegyptiaca* callus against *Aedes aegypti* mosquito.

Bioresource Technology.
2008;**99**:1165-1168

[49] Han Y, Li L, Hao W, Tang M, Wan S.
Larvicidal activity of lansiumamide B
from the seeds of *Clausena lansium*
against *Aedes albopictus* (Diptera:
Culicidae). Parasitology Research.
2013;**112**:511-516

[50] Perumalsamy H, Kim NJ, Ahn YJ.
Larvicidal activity of compounds
isolated from *Asarum heterotropoides*
against *Culex pipiens pallens*, *Aedes
aegypti*, and *Ochlerotatus togoi* (Diptera:
Culicidae). Journal of Medical
Entomology. 2009;**46**:1420-1423

[51] Rahuman AA, Venkatesan P,
Geetha K, Gopalakrishnan G,
Bagavan A, Kamaraj C. Mosquito
larvicidal activity of gluanol acetate, a
tetracyclic triterpenes derived from
Ficus racemosa Linn. Parasitology
Research. 2008;**103**:333-339

[52] Ansari MA, Mittal PK, Razdan RK,
Dhiman RC, Kumar A. Evaluation of
pirimiphos-methyl (50% EC) against
the immature of *Anopheles stephensi/An.
culicifacies* (malaria vectors) and *Culex
quinquefasciatus* (vector of bancroftian
filariasis). Journal of Vector Borne
Diseases. 2004;**41**:10-16

[53] Djerassi C, Shih-Coleman C,
Diekman J. Insect control of the future:
Operational and policy aspects. Science.
1974;**186**:596-607

[54] Glare TR, OCallaghan M.
Environmental and health impacts of
the insect juvenile hormone analogue,
S-methoprene. Report for the Ministry
of Health New Zealand. 1999

[55] Menn JJ, Henrick CA. Rational and
biorational design of pesticides.
Philosophical Transactions of the Royal
Society London B. 1981;**295**:57-71

[56] Cetin H, Yanikoglu A, Cilek JE.
Evaluation of the naturallyderived

insecticide spinosad against *Culex
pipiens* L. (Diptera: Culicidae) larvae in
septic tank water in Antalya, Turkey.
Journal of Vector Ecology.
2005;**30**:151-154

[57] Mulla MS, Darwazeh HA, Luna LL.
Monolayer films as mosquito control
agents and their effects on non-target
organisms. Mosquito News.
1983;**43**:489-495

[58] Stage HH. In Agricultural
Applications of Petroleum Products;
Advances in Chemistry. Washington,
DC: American Chemical Society; 1952.
DOI: 10.1021/ba-1951-0007.ch005

[59] Roberts JR, Reigart JR. Does
anything beat DEET? Pediatric Annals.
2004;**33**:443-353

[60] Raghavendra K, Subbarao SK.
Chemical insecticides in malaria vector
control in India. ICMR Bulletin.
2002;**32**:1-7

[61] Gerberich JB. Update of Annotated
Bibliography of Papers Relating to
Control of Mosquitoes by the Use of
Fish for the Years 1965. Geneva: World
Health Organization; 1985

[62] Diabate A, Tripet F. Targeting male
mosquito mating behaviour for malaria
control. Parasites & Vectors. 2015;**8**:347

[63] Charlwood JD, Pinto J, Sousa CA,
Madsen H, Ferreira C, Do Rosario VE.
The swarming and mating behaviour of
Anopheles gambiae ss (Diptera: Culicidae)
from Sao Tome Island. Journal of Vector
Ecology. 2002;**1**(27):178-183

[64] Lange L. Microbial metabolites
- an infinite source of novel chemistry.
Pure and Applied Chemistry.
1996;**68**(3):745-748

[65] Hussain AA, Mostafa SA,
Ghazal SA, Ibrahim SY. Studies on
antifungal antibiotic and bioinsecticidal
activities of some actinomycete isolates.

African Journal of Mycology and Biotechnology. 2002;**10**:63-80

[66] Sundarapandian S, Sundaram MD, Tholkappian P. Mosquitocidal properties of indigenous fungi and actinomycetes against *Culex quinquefasciatus* Say. Journal of Biological Control. 2002;**16**:89-91

[67] Newman DJ, Cragg GM, Snader KM. The influence of natural products upon drugs discovery. Natural Product Reports. 2000;**17**:215-234

[68] Kalfon A, Charles JF, Bourgouin C, de Barjac H. Sporulation of *Bacillus sphaericus* 2297: An electron microscope study of crystal like inclusion, biogenesis and toxicity to mosquito larvae. Journal of General Microbiology. 1984;**130**:893-900

[69] Wirth MC, Yang Y, Walton WE, Federici BA. Evaluation of Alternative Resistance Management Strategies for *Bacillus sphaericus*. USA: Mosquito Control Research; 2001

[70] Yousten AA, Wallis DA. Batch and continuous culture production of the mosquito larval toxin of *Bacillus sphaericus* 2362. Journal of Industrial Microbiology & Biotechnology. 1987;**2**:277-283

[71] Han B, Liu H, Hu X, Cai Y, Zheng D, Yuan Z. Molecular characterization of a glucokinase with broad hexose specificity from *Bacillus sphaericus* strain C3-41. Applied Environmental Microbiology. 2007;**73**(11):3581-3586

[72] Charles JF, Nielsen-LeRoux C, Delécluse A. *Bacillus sphaericus* toxins: Molecular biology and mode of action. Annual Review of Entomology. 1996;**41**:451-472

[73] Kovendan K, Murugan K, Vincent S, Barnard DR. Studies on larvicidal and pupicidal activity of *Leucas aspera* Willd. (Lamiaceae) and

bacterial insecticide, *Bacillus sphaericus*, against malarial vector, *Anopheles stephensi* Liston. (Diptera: Culicidae). Parasitology Research. 2012;**110**:195-203

[74] Poopathi S, Kumar KA, Kabilan L, Sekar V. Development of low-cost media for the culture of mosquito larvicides, *Bacillus sphaericus* and *Bacillus thuringiensis serovar. Israelensis*. World Journal of Microbiology and Biotechnology. 2002;**18**:209-216

[75] Kovendan K, Murugan K, Vincent S, Kamalakannan S. Larvicidal efficacy of *Jatropha curcas* and bacterial insecticide, *Bacillus thuringiensis*, against lymphatic filarial vector, *Culex quinquefasciatus* Say. (Diptera: Culicidae). Parasitology Research. 2011;**109**:1251-1257

[76] Crickmore N, Bone EJ, Williams JA, Ellar DJ. Contribution of the individual components of the D-endotoxin crystal to the mosquitocidal activity of *Bacillus thuringiensis subsp. israelensis*. FEMS Microbiology Letters. 1995;**131**:249-254

[77] Daniel T, Umarani S, Sakthivadivel M. Insecticidal action of *Ervatamia divaricata* L. and *Acalypha indica* L. against *Culex quinquefasciatus* Say. Geobias. 1995;**14**:95-98

[78] Singh CP, Singh KN, Pandey MC. Insect growth regulatory effect of neem derivative "Neemolin" on *Spilosoma obligue* Walker. Pestology. 1996;**5**:11-13

[79] Armengol G, Hernandez J, Velez JG, Orduz S. Long-lasting effects of a *Bacillus thuringiensis serovar israelensis* experimental tablet formulation for *Aedes aegypti* (Diptera: Culicidae) control. Journal of Economic Entomology. 2006;**99**:1590-1595

[80] Geetha I, Manonmani AM. Surfactin: A novel mosquitocidal biosurfactant produced by *Bacillus subtilis* sp. *subtilis* (VCRC B471) and

influence of abiotic factors on its pupicidal efficacy. Letters in Applied Microbiology. 2010;**51**(4):406-412

[81] Geetha I, Manonmani AM, Prabakaran G. *Bacillus amyloliquefaciens*: A mosquitocidal bacterium from mangrove forests of Andaman & Nicobar Islands, India. Acta Tropica. 2011;**120**(3):155-159

[82] Mittal PK. Biolarvicides in vector control: Challenges and prospects. Journal of Vector Borne Diseases. 2003;**40**(1-2):20-32

[83] Banu AN, Balasubramanian C. Myco-synthesis of silver nanoparticles using *Beauveria bassiana* against dengue vector, *Aedes aegypti* (Diptera: Culicidae). Parasitology Research. 2014;**113**:2869-2877

[84] Borase HP, Patil CD, Salunkh RB, Narkhede CP, Salunke BK. Phyto-synthesized silver nanoparticles: A potent mosquito biolarvicidal agent. Journal of Nanomedicine and Biotherapeutic Discovery. 2013;**3**(1):1-7

[85] Jeevan P, Ramya K, Rena EA. Extracellular biosynthesis of silver nanoparticles by culture supernatant of *Pseudomonas aeruginosa*. IJBT. 2012;**11**(1):72-76

[86] Luz C, Tai MH, Santos AH, Rocha LF, Albernaz DA, Silva HH. Ovicidal activity of entomopathogenic hyphomycetes on *Aedes aegypti* (Diptera: Culicidae) under laboratory conditions. Journal of Medical Entomology. 2007;**44**(5):799-804

[87] Najitha Banu A, Balasubramanian C, Vinayaga Moorthi P. Biosynthesis of silver nanoparticles using *Bacillus thuringiensis* against dengue vector, *Aedes aegypti* (Diptera: Culicidae). Parasitology Research. 2014;**113**:311-316

[88] Soni N, Prakash S. Possible mosquito control by silver nanoparticles synthesized by soil fungus (*Aspergillus niger* 2587). Adv Nanoparticles. 2013;**2**:125-132

[89] Vyas N, Dua KK, Prakash S. Bioassay of secondary metabolite of *Lagenidium giganteum* on mosquito larvae for vector control. Bulletin in Biological Science. 2006;**4**:65-69

[90] Aouiche A, Bijani C, Zitouni A, Mathieu F, Sabaou N. Antimicrobial activity of saquayamycins produced by *Streptomyces* spp. PAL114 isolated from a Saharan soil. Journal of Medical Mycology. 2014;**24**:17-23

[91] Ganesan P, Stalin A, Paulraj MG, Balakrishna K, Ignacimuthu S, Al-Dhabi NA. Biocontrol and non-target effect of fractions and compound isolated from Streptomyces rimosus on the immature stages of filarial vector Culex quinquefasciatus Say (Diptera: Culicidae) and the compound interaction with Acetylcholinesterase (AChE1). Ecotoxicology and Environmental Safety. 2018;**161**:120-128

[92] Ikeda H, Ishikawa J, Hanamoto A, Shinose M, Kikuchi H, Shiba T, et al. Complete genome sequence and comparative analysis of the industrial microorganism *Stretomyces avermitilis*. Nature Biotechnology. 2003;**21**:526-531

[93] Amador ML, Jimeno J, Paz-Ares L, Cortes-Funes H, Hidalgo M. Progress in the development and acquisition of anticancer agents from marine sources. Annals of Oncology. 2003;**14**:1607-1615

[94] Kelecom A. Secondary metabolites from marine microorganisms. Anais da Academia Brasileira de Ciências. 2002;**74**:151-170

[95] Vijayan V, Balaraman K. Metabolites of fungi and actinomycetes active against mosquito larvae. The Indian Journal of Medical Research. 1991;**93**:115-117

[96] El-Bendary MA, Rifaat HM, Keera AA. Larvicidal activity of extracellular secondary metabolites of *Streptomyces microflavus* against *Culex pipiens*. Canadian Journal of Pure & Applied Science. 2010;**4**:1021-1026

[97] Baltz RH. Antibiotic discovery from actinomycetes: Will a renaissance follow the decline and fall? SIM News. 2005;**55**:186-196

[98] Bulock JD, Kristiansen B. Basic Biotechnology. New York: Academic Press; 1997. p. 433

[99] Lam KS. Discovery of novel metabolites from marine actinomycetes. Current Opinion in Microbiology. 2006;**9**:245-251

[100] Zhao XQ, Jiao WC, Jiang B, Yuan WJ, Yang TH, Hao S. Screening and identification of actinobacteria from marine sediments: Investigation of potential producers for antimicrobial agents and type I polyketides. World Journal of Microbiology and Biotechnology. 2009;**25**:859-866

[101] Badji B, Zitouni A, Mathieu F, Lebrihi A, Sabaou N. Antimicrobial compounds produced by *Actinomadura* sp. AC104 isolated from an Algerian Saharan soil. Canadian Journal of Microbiology. 2006;**52**:373-382

[102] Chanal A, Chapon V, Benzerara K, Barakat M, Christen R, Achouak W, et al. The desert of Tataouine: An extreme environment that hosts a wide diversity of microorganisms and radio tolerant bacteria. Environmental Microbiology. 2006;**8**:514-525

[103] Maldonado LA, Stach JEM, Pathom-Aree W, Ward AC, Bull AT, Goodfellow M. Diversity of cultivable actinobacteria in geographically widespread marine sediments. Antonie Van Leeuwenhoek. 2005;**87**:11-18

[104] Thumar JT, Dhulia K, Singh SP. Isolation and partial purification of an antimicrobial agent from halotolerant alkaliphilic *Streptomyces aburaviensis* strain Kut-8. World Journal of Microbiology and Biotechnology. 2010;**26**:2081-2087

[105] Ahmed HG. Phylogenetic Diversity and Anti- MRSA Activity of Halotolerant Actinobacteria from sediments in Great Salt Plains, Oklahoma. Microbial Ecology. 2017;**1**:1-9

[106] Chaudhary HS, Yadav J, Shrivastava AR, Singh S, Singh AK, Gopalan N. Antibacterial activity of actinomycetes isolated from different soil samples of Sheopur (A city of central India). Journal of Advanced Pharmaceutical Technology & Research. 2013;**4**(2):118-123

[107] Shirling EB, Gottlieb D. Methods for characterization of *Streptomyces* species. International Journal of Systematic Bacteriology. 1966;**16**:313-340

[108] Zhang J. Improvement of an Isolation Medium for Actinomycetes. Modern Applied Science. 2011;**5**:124-127

[109] Otoguro M, Hayakawa M, Yamazaki T, Iimura Y. An integrated method for the enrichment and selective isolation of *Actinokineospora* spp. in soil and plant litter. Journal of Applied Microbiology. 2001;**91**:118-130

[110] Valan Arasu M, Duraipandiyan V, Agastian P, Ignacimuthu S. Antimicrobial activity of *Streptomyces* sp. ERI-26 recovered from Western Ghats of Tamil Nadu. Journal of Medical Mycology. 2008;**18**:147-153

[111] Oskay M. Antifungal and antibacterial compounds from *Streptomyces* strain. African Journal of Biotechnology. 2009;**8**(13):3007-3017

[112] Hayakawa M, Sadakata T, Kajiura T, Nonomura H. New methods for the

highly selective isolation of *Micromonospora* and *Microbispora* from soil. Journal of Fermentation and Bioengineering. 1991;**72**:320-326

[113] Singh G, Prakash S. Lethal effect of *Streptomyces citreofluorescens* against larvae of malaria, filaria and dengue vectors. Asian Pacific Journal of Tropical Medicine. 2012;**5**(8):594-597

[114] You KM, Park YK. A new method for the selective isolation of actinomycetes from soil. Biotechnology Techniques. 1996;**10**:541-546

[115] Duangmal K, Ward AC. Selective isolation of members of the *Streptomyces violaceoruber* clade from soil. FEMS Microbiology Letters. 2005;**245**:321-327

[116] Rehacek Z. Isolation of actinomycetes and determination of the number of their spores in soil. Microbiology USSR (English Transl.). 1959;**28**:220-225

[117] Janaki T. Larvicidal activity of *Streptomyces cacaoi* subsp. *cacaoi*-M20 against *Aedes aegypti*. International Journal of Botany Studies. 2016;**1**:47-49

[118] Bulina TI, Alferova IV, Terekhova LP. A novel approach to isolation of actinomycetes involving irradiation of soil samples with microwaves. Microbiologica. 1997;**66**:278-282

[119] Njenga WP, Mwaura FB, Wagacha JM, Gathuru EM. Methods of Isolating Actinomycetes from the Soils of Menengai Crater in Kenya. Archives of Clinical Microbiology. 2017;**8**:3

[120] Kelly LK. Central notations for the revised iscc-nbs color-name blocks. Journal of Research of the National Bureau of Standards. 1958;**61**:427-431

[121] Kuester E, Williams ST. Selection of Media for Isolation Of*Streptomycetes*. Nature. 1964;**30**:928-929

[122] Muiru WM, Mutitu EW, Mukunya DM. Identification of selected actinomycetes isolates and characterization of their antibiotic metabolites. Journal of Biological Sciences. 2008;**8**:1021-1026

[123] Slack JM. Subgroup on the taxonomy of microaerophilic actinomycetes. Report on organization, aims and procedures. International Journal of Systematic Bacteriology. 1968;**18**:253-262

[124] Slack JM, Landfried S, Gerencser MA. Morphological, Biochemical, and Serological Studies on 64 Strains of *Actinomyces israelii*. Journal of Bacteriology. 1969;**97**:873-884

[125] Farris MH, Oslon JB. Detection of actinobacteria cultivated from environmental samples reveals bias in universal primers. Letter Applied Microbiology. 2007;**45**(4):376-381

[126] Ikemoto T, Katayama T, shiraishi A, HaneishiT. Aculeximycin, a new antibiotic from *streptosporangium albidum* ii. Isolation, physicochemical and biological properties. The Journal of antibiotics. 1983;**36**(9):1097-100

[127] Govindarajan M, Jebanesan A, Reetha D. Larvicidal efficacy of extracellular metabolites of actinomycetes against dengue vector mosquito *Aedes aegypti* Linn. (Diptera: Culicidae). Research Review in BioScience. 2007;**1**:161-162

[128] Tanvir AR, Imransajida I, Hasnain S. Larvicidal potential of Asteraceae family endophytic actinomycetes against *Culex quinquefasciatus* mosquito larvae. Natural Product Research. 2014;**28**(22):2048-2052

[129] Kekuda TP, Shobha K, Onkarappa R. Potent insecticidal activity of two *Streptomyces* species isolated from the soils of the Western

Ghats of Agumbe, Karnataka. Journal of Natural Pharmacy. 2010;**1**:30-32

[130] Dhanasekaran D, Sakthi V, Thajuddin N, Panneerselvam A. Preliminary evaluation of *Anopheles* mosquito larvicidal efficacy of mangrove actinobacteria. International Journal of Applied Biology and Pharmaceutical Technology. 2010;**1**:374-381

[131] Deepika TL, Kannabiran K, Khanna VG, Rajakumar G, Jayaseelan C, Santhoshkumar T, et al. Isolation and characterization of acaricidal and larvicidal novel compound (2S,5R,6R)-2-hydroxy-3,5,6-trimethyloctan-4-one from *Streptomyces* sp. against blood-sucking parasites. Parasitology Research. 2012;**111**:1151-1163

[132] Loganathan K, Kumar G, Kirthi AV, Rao KVB, Rahuman AA. Entomopathogenic marine actinomycetes as potential and low-cost biocontrol agents against bloodsucking arthropods. Parasitology Research. 2013;**112**(11):3951-3959

[133] You J, Cao LX, Liu GF, Zhou SN, Tan HM, Lin YC. Isolation and characterization of actinomycetes antagonistic to pathogenic Vibrio spp. from near shore marine sediments. World Journal of Microbiology and Biotechnology. 2005;**21**:679-682

[134] Gomes RC, Semêedo LTAS, Soares RMA, Alviano CS, Linhares LF, Coelho RRR. Chitinolytic activity of actinomycetes from a cerrado soil and their potential in biocontrol. Letters in Applied Microbiology. 2000;**30**:146-150

[135] Dhanasekaran D, Selvamani S, Panneerselvam A, Thajuddin N. Isolation and characterization of actinomycetes in Vellar Estury, Annagkoil, Tamilnadu. African Journal of Biotechnology. 2009;**8**:4159-4162

[136] Anwar S, Ali B, Qamar F, Sajid I. Insecticidal Activity of Actinomycetes Isolated from Salt Range, Pakistan against Mosquitoes and Red Flour Beetle. Pakistan Journal of Zoology. 2014;**46**:83-92

[137] Vijayakumar R, Murugesan S, Cholarajan A, Sakthi V. Larvicidal Potentiality of Marine Actinomycetes Isolated from Muthupet Mangrove, Tamilnadu, India. International Journal of Microbiological Research. 2010;**1**(3):179-183

[138] Sanjenbam P, Kannabiran K. Antimicrobial and larvicidal activity of *Streptomyces* sp.VITPK9 isolated from a brine spring habitat of Manipur, India. Der Pharmacia Lettre. 2013;**5**(3):65-70

[139] Ando K. How to discover new antibiotics for insecticidal use. In: Takahashi T, Yoshioka H, Misato T, Matusunaka S, editors. Pesticide Chemistry: Human Welfare and the Environment. New York: Pergamon; 1983. pp. 253-259

[140] Anonymous. Biologically active KSB-1939L3 compound and its reduction—pesticide with insecticide and acaricide activity production by *Streptomyces* sp. culture. Biotechnological Abstract. 1990;**9**:58

[141] Pampiglione S, Majori G, Petrangeli G, Romi R. Avermectins, MK-933 and MK-936, for mosquito control. Transactions of the Royal Society of Tropical Medicine and Hygiene. 1985;**79**:797-799

[142] Rao KV, Chattopadhyay SK, Reddy GC. Flavonoides with mosquito larval toxicity- tangeratin, daidzein and genistein crystal production, isolation and purification; *Streptomyces* spp. culture; insecticide. Journal of Agricultural and Food Chemistry. 1990;**38**:1427-1430

[143] Zizka Z, Weiser J, Blumauerova M, Jizba J. Ultra structural effects of macroterrolides of *Sterptomyces griseus*

LKS-1 in tissues of *Culex pipiens* larvae. monactin, dinactin, triactin and nonactin preparation; insecticide activity. Cytobios. 1989;**58**:85-91

[144] Xiang WS, Wang JD, Wang XJ, Zhang J, Wang Z. Further new milbemycin antibiotics from *Streptomyces bingchenggensis*. Fermentation, isolation, structural elucidation and biological activities. The Journal of Antibiotics. 2007;**60**(10):608-613

[145] Maskey RP, Helmke E, Kayser O, Fiebig HH, Maier A, Busche A, et al. Anti-cancer and antibacterial trioxacarcins with high anti-malaria activity from a marine *streptomycete* and their absolute stereochemistry. Journal of Antibiotics (Tokyo). 2004;**57**(12):771-779

[146] Otoguro K, Ishiyama A, Kobayashi M, Sekiguchi H, Izuhara T, Sunazuka T, et al. In vitro and in vivo antimalarial activities of a carbohydrate antibiotic, prumycin, against drug-resistant strains of Plasmodia. Journal of Antibiotics (Tokyo). 2004;**57**(12):771-779

[147] Jiang Y, Mulla MS. Laboratory and field evaluation of spinosad, a biorational natural product, against larvae of *Culex* mosquitoes. Journal of the American Mosquito Control Association. 2009;**25**:456-466

[148] Isaka M, Jaturapat A, Kramyu J, Tanticharoen M, Thebtaranonth Y. Potent in vitro antimalarial activity of metacycloprodigiosin isolated from *Streptomyces spectabilis* BCC 4785. Antimicrobial Agents and Chemotherapy. 2002;**46**(4):1112-1113

[149] Prudhomme J, McDaniel E, Ponts N, Bertani S, Fenical W, Jensen P, et al. Marine actinomycetes: A new source of compounds against the human malaria parasite. PLoS One. 2008;**3**:e2335

[150] Strangman WK, Kwon HC, Broide D, Jensen PR, Fenical W. Potent inhibitors of pro-inflammatory cytokine production produced by a marine-derived bacterium. Journal of Medicinal Chemistry. 2009;**52**(8):2317-2327

[151] Fotie J, Morgan RE. Depsipeptides from microorganisms: A new class of antimalarials. Mini-Reviews in Medicinal Chemistry. 2008;**8**(11):1088-1094

[152] Molinari G, Soloneski S, Larramendy ML. New ventures in the genotoxic and cytotoxic effects of macrocyclic lactones, abamectin and ivermectin. Cytogenetic and Genome Research. 2010;**128**:37-45

[153] Saurav K, Rajakumar G, Kannabiran K, Rahuman AA, Velayutham K, Elango G, et al. Larvicidal activity of isolated compound 5-(2,4-dimethylbenzyl) pyrrolidin-2-one from marine *Streptomyces* VITSVK5 sp. against *Rhipicephalus* (Boophilus) *microplus*, *Anopheles stephensi*, and *Culex tritaeniorhynchus*. Parasitology Research. 2013;**112**:215-226

[154] Nair MG, Putnam AR, Mishra SK, Mulks MH, Taft WH, Keller JE, et al. Faeriefungin: A new broad-spectrum antibiotic from *Streptomyces griseus var. autotrophicus*. Journal of Natural Products. 1989;**52**:797-809

[155] Thenmozhi M, Gopal JV, Kannabiran K, Rajakumar G, Velayutham K, Rahuman AA. Eco-friendly approach using marine actinobacteria and its compounds to control ticks and mosquitoes. Parasitology Research. 2013;**112**(2):719-729

[156] Antonio GE, Sanchez D, Williams T, Marina CF. Paradoxical effects of sublethal exposure to the naturally derived insecticide spinosad in the dengue vector mosquito, *Aedes aegypti*. Pest Management Science. 2009;**65**:323-326

[157] Thompson GD, Dutton R, Sparks TC. Spinosad case study: An example from a natural products discovery programme. Pest Management Science. 2000;**56**:696-702

[158] Williams T, Valle J, Vinuela E. Is the naturally derived insecticide spinosad compatible with insect natural enemies? Biocontrol Science and Technology. 2003;**13**:459-475

[159] Schneider M, Smagghe C, Viñuela E. Comparative effects of several insect growth regulators and spinosad on the different developmental stages of the endoparasitoid *Hyposoter didymator* (Thunberg). Pesticides and Benefi cial Organisms. IOBC/WPRS Bulletin. 2004;**27**:13-19

[160] Darriet F, Duchon S, Hougard JM. Spinosad: A new larvicide against insecticide-resistant mosquito larvae. Journal of the American Mosquito Control Association. 2005;**21**:495-496

[161] Romi R, Proietti S, Di Luca M, Cristofaro M. Laboratory evaluation of the bioinsecticide spinosad for mosquito control. Journal of the American Mosquito Control Association. 2006;**22**:93-96

[162] Bret BL, Larson LL, Schoonover JR, et al. Biological properties of spinosad. Down to Earth. 1997;**52**:6-13

[163] Hertlein MB, Mavrotas C, Jousseaume C, et al. A review of spinosad as a natural mosquito product for larval mosquito control. Journal of the American Mosquito Control Association. 2010;**26**:67-87

[164] Allen RA, Lewis CN, Meisch MV. Residual efficacy of three spinosad formulations against *Psorophora columbiae* larvae in small rice plots. Journal of the American Mosquito Control Association. 2010;**26**:116-118

[165] Anderson JF, Ferrandino FJ, Dingman DW, Main AJ, Andreadis TG, Becnel JJ. Control of mosquitoes in catch basins in Connecticut with *Bacillus thuringiensis israelensis*, *Bacillus sphaericus*,and spinosad. Journal of the American Mosquito Control Association. 2011;**27**:45-55

[166] Bahgat IM, El Kady GA, Temerak SA, Lysandrou M. Thenatural bio-insecticide spinosad and its toxicity to combat some mosquito species in Ismailia Governorate. Egypt World Journal of Agricultural Science. 2007;**3**:396-400

[167] Darriet F, Marcombe S, Etienne M, Yébakima A, Agnew P, Yp-Tcha MM, et al. Field evaluation of pyriproxyfen and spinosad mixture for the control of insecticide resistant *Aedes aegypti* in Martinique (French West Indies). Parasites & Vectors. 2010;**3**:88

[168] Marcombe S, Darriet F, Agnew P, Etienne M, Yp-Tcha MM, Yébakima A, et al. Field efficacy of new larvicide products for control of multi-resistant *Aedes aegypti* populations in Martinique (French West Indies). The American Journal of Tropical Medicine and Hygiene. 2011;**84**:118-126

[169] Marina CF, Bond JG, Casas M, Muñoz J, Orozco A, Valle J, et al. Spinosad as an effective larvicide for control of *Aedes albopictus* and *Aedes aegypti*, vectors of dengue in southern Mexico. Pest Management Science. 2011;**67**:114-121

[170] Perez CM, Marina CF, Bond JG, Rojas JC, Valle J, Williams T. Spinosad, a naturally derived insecticide, for control of *Aedes aegypti* (Diptera: Culicidae): Efficacy, persistence, and elicited oviposition response. Journal of Medical Entomology. 2007;**44**(4):631-638

[171] Prabhu K, Murugan K, Nareshkumar A, Badeeswaran S. Larvicidal and pupicidal activity of

spinosad against the malarial vector *Anopheles stephensi*. Asian Pacific Journal of Tropical Medicine. 2011;**4**:610-613

[172] Sadanandane C, Boopathi-Doss PS, Jambulingam P, Zaim M. Efficacy of two formulations of the bioinsecticide spinosad against *Culex quinquefasciatus* in India. Journal of the American Mosquito Control Association. 2009;**25**:66-73

[173] Thavara U, Tawatsin A, Asavadachanukorn P, Mulla MS. Field evaluation in Thailand of spinosad, a larvicide derived from *Saccharopolyspora spinosa* (Actinomycetales) against *Aedes aegypti* (L.) larvae. SE Asian Journal of Tropical Medicine and Public Health. 2009;**40**:235-242

[174] Rajesh K, Dhanasekaran D, Tyagi BK. Mosquito survey and larvicidal activity of actinobacterial isolates against *Culex* larvae (Diptera: Culicidae). Journal of the Saudi Society of Agricultural Sciences. 2015;**14**(2):116-122

[175] Janardhan A, Kumar AP, Viswanath B, Gopal DS, Narasimha G. Antiviral and larvicidal properties of novel bioactive compounds produced from marine actinomycetes. Russian Journal of Marine Biology. 2018;**44**(5):424-428

[176] Hassan SED, Fouda A, Radwan AA, Salem SS, Barghoth MG, Awad MA, et al. Endophytic actinomycetes *Streptomyces* spp mediated biosynthesis of copper oxide nanoparticles as a promising tool for biotechnological applications. JBIC, Journal of Biological Inorganic Chemistry. 2019;**24**(3):377-393

[177] Ganesan P, Rajendran HAD, Appadurai DR, Gandhi MR, Michael GP, Savarimuthu I, et al. Isolation and molecular characterization of actinomycetes with antimicrobial and mosquito larvicidal properties.

Beni-Suef University Journal of Basic and Applied Sciences. 2017;**6**(2):209-217

[178] Kim JH, Choi JY, Park DH, Park DJ, Park MG, Kim SY, et al. Isolation and characterization of the insect growth regulatory substances from actinomycetes. Comparative Biochemistry and Physiology Part C: Toxicology & Pharmacology. 2020;**228**:108651

[179] Ganesan P, Jackson A, David RHA, Sivanandhan S, Gandhi MR, Paulraj MG, et al. Mosquito (Diptera: Culicidae) Larvicidal and Ovicidal Properties of Extracts from *Streptomyces vinaceusdrappus* (S12-4) Isolated from Soils1. Journal of Entomological Science. 2018;**53**(1):17-26

[180] Balakrishnan S, Santhanam P, Srinivasan M. Larvicidal potency of marine actinobacteria isolated from mangrove environment against *Aedes aegypti* and *Anopheles stephensi*. Journal of Parasitic Diseases. 2017;**41**(2):387-394

[181] Fouda A, Hassan SED, Abdo AM, El-Gamal MS. Antimicrobial, antioxidant and larvicidal activities of spherical silver nanoparticles synthesized by endophytic *Streptomyces* spp. Biological Trace Element Research. 2020;**195**(2):707-724

[182] Naine SJ, Devi CS. Larvicidal and repellent properties of *Streptomyces* sp. VITJS4 crude extract against *Anopheles stephensi, Aedes aegypti* and *Culex quinquefasciatus* (Diptera: Culicidae). Polish Journal of Microbiology. 2014;**63**(3):341-348

[183] Sivarajan A, Shanmugasundaram T, Sangeetha M, Radhakrishnan M, Balagurunathan R. Screening, production, and characterization of biologically active secondary metabolite (s) from marine *Streptomyces* sp. PA9 for antimicrobial, antioxidant, and mosquito larvicidal activity. Indian Journal of Geo Marine Sciences. 2019;**48**:1319-1326

[184] Shanmugasundaram T,
Balagurunathan R. Mosquito larvicidal
activity of silver nanoparticles
synthesised using actinobacterium,
Streptomyces sp. M25 against *Anopheles
subpictus, Culex quinquefasciatus* and
Aedes aegypti. Journal of Parasitic
Diseases. 2015;**39**(4):677-684

[185] El-Khawagh MA, Hamadah KS,
El-Sheikh TM. The insecticidal activity
of actinomycete metabolites, against the
mosquitoe *Culex pipiens*. Egypt
Academic Journal of Biology Science.
2011;**4**(1):103-113

Chapter 7

Anti-Quorum Sensing Compounds from Rare Actinobacteria

Sunita Bundale and Aashlesha Pathak

Abstract

Actinobacteria have exceptional metabolic diversity and are a rich source of several useful bioactive natural products. Most of these have been derived from Streptomyces, the dominant genus of Actinobacteria. Hence, it is necessary to explore rare actinobacteria for the production of novel bioactive compounds. Amongst the novel metabolites, anti-quorum-sensing agents, which can curb infection without killing pathogens, are gaining importance. Not many studies are targeting anti-quorum-sensing agents from rare actinobacteria and this research area is still in its infancy. This field may lead to novel bioactive compounds that can act against bacterial quorum-sensing systems. These agents can attenuate the virulence of the pathogens without challenging their growth, thereby preventing the emergence of resistant strains and facilitating the elimination of pathogens by the host's immune system. Therefore, this chapter describes the general characteristics and habitats of rare actinobacteria, isolation and cultivation methods, the methods of screening rare actinobacteria for anti-quorum sensing compounds, methods of evaluation of their properties, and future prospects in drug discovery.

Keywords: rare actinobacteria, quorum sensing, anti-quorum-sensing compounds, swarming, biofilm

1. Introduction

Actinobacteria are Gram-positive and high G + C containing bacteria with exceptional metabolic diversity. They are a rich source of several useful bioactive natural products many of which have been reported for their potential roles as antimicrobial, antibacterial, antiviral, anticancer, and antifungal compounds. More than 22,000 bioactive secondary metabolites from microorganisms have been identified and published in the scientific and patent literature, and about half of these compounds are produced by actinobacteria [1]. Currently, approximately 160 antibiotics have been used in human therapy and agriculture, and 100–120 of these compounds, including streptomycin, erythromycin, gentamicin, vancomycin, avermectin, etc., are produced by actinobacteria [2].

Most of these antibiotics in clinical use today have been developed from compounds isolated from Streptomyces, the dominant genus [3]. However, the recent search for the novel compounds from Streptomyces species has often led to the rediscovery of known compounds. Hence, the focus of screening programs has shifted to bioactive compounds from non-Streptomyces genera; also referred to as rare actinobacteria [4].

Recent evidence has demonstrated that rare actinobacteria, might represent a unique source of novel biologically active compounds, and methods designed to isolate and identify a wide variety of such actinobacteria have been developed. These methods include a variety of pre-treatment techniques in combination with appropriately supplementing selective agar media with specific antimicrobial agents.

At present, not more than 50 rare actinobacterial taxa are reported to produce 2500 bioactive compounds [5]. Thus, it is crucial that new groups of rare actinobacteria be pursued as sources of novel pharmaceutically active metabolites. Amongst the novel metabolites, anti-quorum sensing (AQS) agents, which can curb infection without a killing action, are gaining importance. Bacterial cell–cell communication, dubbed quorum sensing, is intricately related to virulence. An associated phenomenon is bacterial swarming which allows the spread of disease and virulence.

The discovery that many pathogenic bacteria employ quorum sensing (QS) to regulate their pathogenicity and virulence factor production makes the QS system an attractive target for antimicrobial therapy. Targeting the pathogenesis instead of killing the organism may provide less selective pressure for the development of resistance. Therefore, it has been suggested that inactivating the QS system in bacteria using QS inhibitors holds great promise for the treatment of infectious diseases. These compounds can attenuate the virulence of the pathogens without challenging their growth, thereby preventing the emergence of resistant strains and facilitating the elimination of pathogens by the host's immune system. Therefore, a search for anti-quorum-sensing agents as attractive alternatives to treat infection has become logical and gathered momentum [6].

Although antimicrobial properties of actinobacteria have been extensively studied, less is known about AQS activities of rare actinobacteria which may be a rich source of active compounds that can act against bacterial quorum-sensing systems.

2. Defining rare Actinobacteria

Rare Actinobacteria are defined as certain types of Actinobacteria which are abundant in various habitats but are difficult to isolate [7]. These include all non-Streptomyces actinobacterial genera like Actinomadura, Actinoplanes, Amycolatopsis, Dactylosporangium, Planomonospora, Planobispora, Salinispora, Streptosporangium, Verrucosispora and Microbiospora [8]. Even though a major percentage of antibiotics are secreted by bacteria from the genus Streptomyces [9] rare actinobacteria account for 25–30% of these. The importance of rare actinobacteria is further demonstrated by the fact that many of the successful bioactive products in the market like rifamycin, erythromycin, teicoplanin, vancomycin, and gentamycin are produced by such rare actinobacteria [10]. Consequently, rare actinobacteria are being unveiled as highly prospective sources of bioactive compounds.

Members of the genus Actinomadura have been reported to secrete about 350 different types of bioactive compounds exhibiting a wide range of mechanisms of action [11, 12]. Some strains of the genus Actinoplanes are known for the production of acarbose, a secondary metabolite that is an α-glucosidase enzyme inhibitor [13] and is currently being used in treating type 2 diabetes [14]. While numerous bioactive compounds are isolated from the strains of Amycolatopsis, vancomycin and rifamycin are the most familiar ones [15]. Research also led to the discovery of the first naturally occurring tetracycline C2 amides called dactylocyclines which are produced by members of the genus Dactylosporangium [16]. Metabolomic studies conducted on the genus Planomonospora led to the discovery of multiple compounds with biosynthetic activities such as Ureylene-containing oligopeptide antipain, the thiopeptide siomycin and sphaericin and lantibiotic 97,518 [17]. Planobispora rosea is found to secrete a novel antibiotic called GE 2770A and thiazolyl peptides which are protein synthesis inhibitors [18]. Arenimycin is another antibiotic produced by the marine actinobacterium Salinispora arenicola and is found to be effective against drug resistant S. aureus and a few other Gram-positive bacteria [19]. Molecular studies done on compounds isolated from Streptosporangium oxazolinicum, resulted in the discovery of three novel alkaloids having anti-trypanosomal activity [20]. Studies conducted using liquid chromatography mass spectrometry (LC–MS) and one strain-many compounds strategy led to the detection of another bioactive compound abyssomicin from the marine actinobacterium Verrucosispora [21]. Members of the genus Micromonospora have been a major source of Gentamicin, apart from which about 740 different antibiotics are found from other strains [22]. All the above mentioned actinobacteria and few more are a subject of intense investigations and prove to be a rich source of many novel antibiotics when modern approaches such as transcriptomics and metabolomics are employed.

Habitat	Genus	Reference
Forest Soil	Conexibacter, Actinospica,Catenulispora, Sinomonas, Longispora etc.	[25–29]
Desert Soil	Yuhushiella, Dietzia, Kineococcus, Microbacterium etc	[30–33]
Garden Soil	Ornithinimicrobium	[34]
Alkaline soil	Myceligenerans, Yonghaparkia	[35, 36]
Soil sample from Oil springs	Smaragdicoccus	[37]
Sandy Soil	Krasilnikovia, Sphaerosporangium, Micromonospora	[38–40]
Farmland Soil	Ruania, Agromyces	[41, 42]
Saline Soil	Zhihengliuella, Nocardiopsis, Nesterenkonia, Brevibacterium, Kocuria	[43–47]
Soil from tropical rainforest	Dactylosporangium, Saccharopolyspora, Pseudonocardia Planotetraspora, Sphaerisporangium	[48–52]
Soil sample from Paddy fields	Humihabitans, Humibacillus, Arthrobacter	[53–55]
Soil near wastewater Treatment facility/Activated Sludge	Flexivirga, Micrococcus, Gordonia, Nocardioides	[56–59]
Dried Seaweed	Phycicoccus, Labedella, Phycicola, Aeromicrobium	[60–63]
Marine Water	Tessaracoccus, Paraoerskovia, Marinactinospora, Demequina, Verrucosispora	[64–68]
Caves	Knoellia, Hoyosella, Jiangella	[69–71]
Roots of Plants	Plantactinospora, Phytohabitans, Actinophytocola, Flindersiella, Phytomonospora	[71–76]
Leaves of Plants	Frondicola, Rhodococcus, Nonomuraea	[77–79]
Stems of Plants	Glycomyces, Saccharopolyspora, Kineosporia, Nocardia	[80–83]
Glaciers	Cryobacterium, Leifsonia, Arthrobacter	[84–86]
Volcanic rocks	Allocatelliglobosispora, Thermoactinospora,	[87, 88]

Table 1.
Habitats of rare actinobacteria.

3. Habitats of rare Actinobacteria

Actinobacteria inhabit a wide range of habitats with diverse climatic conditions, including those of extreme temperatures and pH as well as marine waters, deserts and soil [23]. However, they are mainly found in soil. It is also observed that even though actinobacteria are found in all layers of soil, their density decreases with increasing depth [24]. Environmental factors such as pH of the soil, humus content and soil type directly influence the population density and type of rare actinobacteria present [7]. **Table 1** summarizes a few rare actinobacteria isolated from various habitats.

4. Isolation of rare Actinobacteria

4.1 Pretreatment methods

Isolation of rare actinobacteria is a strenuous task, mainly because they are slow growing. After collection of samples, the type of pretreatment method used decides the viability and isolation of the species under study. While several types of pretreatment methods are available, the requirement of each organism is different. A few common methods include suspending samples in distilled water followed by incubation in a rotary shaker, air drying the samples, or heat treatment in oven at 45°C–65°C [89]. Alferova and Terekhova have reported a modified procedure for treatment of soil samples with calcium carbonate under humid conditions [90]. This gave efficient increase in the number of isolates as well as representatives of rare genera. Chemical treatment methods such as treating the sample with 1.5% phenol or physical methods combined with other conventional or sucrose gradient centrifugation methods also proved to be highly effective [91]. Isolation of rare actinobacteria is still an area of active research and hence a particular pretreatment method does not guarantee the isolation of a specific actinobacteria. However, other prescribed pretreatment methods available in literature can be tried to achieve exclusive isolation. These methods are summarized in **Table 2**.

4.2 Use of antibiotics for selective isolation

Pretreatments do reduce a fraction of unwanted predominant fast-growing organisms. However, use of antibiotics in isolation media along with pretreatment substantially increases the chances of selective isolation as it effectively eliminates fast growing and competitive bacteria. By virtue of this property, antibiotics like gentamicin and novobiocin were successfully used to isolate members of the genus Micromonospora [24]. Similarly, members of genus Microtetraspora were isolated using nalidixic acid and trimethoprim [104]. Currently available antibiotics for selection of actinobacteria along with the type of actinobacteria isolated are summarized in **Table 2**.

4.3 Use of specific isolation media

It is observed that the growth of rare actinobacteria is highly sensitive to contamination by some known fast-growing organisms such as fungus, other bacteria and a few common streptomyces. Hence conventional methods of isolation are ineffective in isolating rare actinobacteria and there is a need to find more advanced and highly selective isolation methods [24]. Studies have shown that in abundance of nutrients, Actinobacteria prefer exponential growth over the production of secondary metabolite [105]. It is also reported that enzymes for secondary metabolite

Pretreatment method	Type of Actinobacteria	Antibiotic for selective isolation	Suggested isolation media	Reference
1. Physical methods a. Fresh samples w/o pretreatment	Thermophilic Actinobacteria	Cycloheximide, Kanamycin	Czapeck medium, glycerol asparagine medium, Oatmeal medium	[92, 93]
a. Air dried in room temperature for 2 weeks a. Samples kept in oven at 40°C-65°C	Halophilic and Alkalophilic actinobacteria	Nalidixic acid	Starch-casein agar, glycerol asparagine medium, T3 medium,	[30, 94]
a. Samples heated in oven at 110°C for 1 h a. Pre-treatment	Acidophilic Actinobacteria	Cycloheximide, Nystatin	ISP4, ISP2, Acidified oatmeal agar and modified Bennett's agar	[95, 96]
followed by calcium carbonate 2. Chemical methods a. SDS 0.05% and Yeast extract 5%	Actinobacteria from Plant origins	Pimaricin, penicillin G and polymyxin B	Sodium Propionate medium, Yeast extract medium, ISP2, ISP3, Potato dextrose agar	[97, 98]
a. Phenol 1.5% a. Chloramine-T	Actinobacteria from Soil	Actidione, Nystatin	HV Agar, Glycerol-arginine medium, Medium supplied with superoxide dismutase, PDA	[30, 99, 100]
	Marine Actinobacteria	Cycloheximide, Nystatin	Glycerol-asparagine, Glycerol-glycine, Chitin, Starch-Casein	[101–103]
	Rare Actinobacteria	Nystatin	HV, HP, Trehalose-Proline agar, ISP5, B4,	[89, 104]

Table 2.
Pretreatment method, antibiotic and isolation media used for selective isolation of rare actinobacteria.

production are inhibited in presence of glucose [106]. Factors such as the type of chemical or physical pre-treatment used [24], pH, temperature and duration of incubation as well as sources of essential nutrients used greatly affect the rate of growth of rare actinomycetes as well as metabolite production [107]. Effect of each of these factors should be considered while designing a culture medium.

A carbon source can thus either be suitable for growth or for metabolite production but not both. Generally, monosaccharides or sugars that are metabolized rapidly are found to be most suitable for growth while polysaccharides or sugars that metabolize slowly are more suitable for antibiotic production [107]. A comparison between inorganic and organic nitrogen sources used revealed that use of organic nitrogen source resulted in maximum growth as well as metabolite production [104]. Evidences have shown that antibiotic accumulation increases as soon as the nitrogen source used in medium is entirely utilized by the organism [108]. Addition of excess inorganic phosphates resulted in rapid growth since it aids the consumption of Carbon and Nitrogen sources as well as accelerates the rate of cellular respiration but it lowers the production of secondary metabolites [109].

Antibiotic production begins extensively in mid log and late log phase and is continuous in the stationary phase of bacterial growth curve [110]. Multiple studies done on various species of actinobacteria suggest that, antibiotic production is maximum at neutral pH [111] and temperature of 30°C [112, 113]. Further, requirement of other important constituents of growth medium such as trace metals and minerals vary with species and culture conditions.

Apart from all the above-mentioned common isolation media, the requirement of each bacterium is different and hence the selective media should be designed keeping in mind the nutritional requirements of target organism. In order to facilitate the process of designing selective isolation media, information from various taxonomic, phenotypic and antibiotic sensitivity databases can be used [97].

5. Anti-quorum sensing

Quorum sensing is a mechanism of cell-to-cell communication seen in bacteria which occurs by the means of certain autoinducer or chemical signal molecules. The concentration of these autoinducer molecules increases with increase in cell density. Once a certain threshold concentration is reached, these autoinducer molecules lead to alteration of gene expression in the population [114]. It is now known that quorum sensing is the underlying mechanism of a wide spectrum of bacterial physiological processes such as virulence [115, 116], bioluminescence [117], motility [118], sporulation [119], conjugation [120], development of genetic competence [121] as well as synthesis of antibiotics [122].

Considering the implications of quorum sensing in various aspects of bacterial life processes, it is evident that inhibiting quorum sensing could have potential therapeutic applications [123, 124]. There are various strategies in which the quorum sensing pathways can be inhibited. Together, they are called as quorum-quenching and the compounds or molecules used to do so are called as AQS compounds. Strategies used in quorum quenching include: Inhibition of synthesis of autoinducer molecule, designing analogues of autoinducer molecule or receptor analogues [125] and antibody or enzyme catalyzed hydrolysis of autoinducer molecule [126].

N-acyl homoserine lactones (AHLs) are an important class of signaling molecules produced by Gram negative bacteria which are known to govern the population density [127]. Lactonases are the enzymes which hydrolyse either the amide linkage between lactone and acyl side chain or affects the ester bond thereby inhibiting the signaling molecule [128]. Further, acylases [129] and oxidoreductases [130] are also found to have quorum quenching activities. Degradation of signaling molecules is also possible via antibody mediated catalysis. Lamo Marin et al. in their studies reported that hydrolysis of AHL is efficiently achieved by original squaric monoester monoamide hapten [131]. Kristina MSmith and YigongBu reported two novel compounds: N-(2-oxocyclohexyl)-3-oxododecanamide which was a moderate antagonist and N-(trans-2-hydroxycyclopentyl)-3-oxododecanamide which was a strong antagonist of autoinducer molecules of P. aeruginosa [125].

6. Methods of screening and evaluation of anti-quorum sensing compounds isolated from rare Actinobacteria

While the method of preparation of bacterial extracts remains the same, various conventional and virtual screening methods are used to screen compounds for their AQS activity.

6.1 Pigment inhibition assays

Formation of violacein, a purple pigment in the reporter strain
Chromobacterium violaceum is controlled by quorum sensing [132]. Quantifying
the inhibition of violacein synthesis, by paper disc diffusion as well as flask incubation assay is thus used as a method for screening AQS activity of a given compound.
Presence of a zone of inhibition in paper disk diffusion plates and significant reduction in violacein content in flasks, are considered as positive results. This method
was efficiently used to test the AQS activity of Vanilla extract [133], pigments
extracted from Auricularia auricular [134], essential oils [135], fruit extracts of
Passiflora edulis [136], and a few compounds isolated from plant endophytic bacteria [137]. Similarly, inhibition of pigment production in Serratia marcescens is also
a coherent test to determine the quorum quenching activity of a compound [138].

6.2 Swarming motility and biofilm formation assays

Swarming and biofilm formation are other quorum sensing controlled processes
seen in bacteria such as P. aeruginosa [139], S. marcescens [140], P. mirabilis [6]
and are known to play a major role in pathogenesis. Reduction in diameter of the
swarm zone, alteration in swarming patterns and significant reduction in biofilm
formation [141] indicate that given compound has AQS activity. While reduction
in swarm zone diameter and alteration of swarming patterns can be quantified
directly, inhibition in biofilm formation can be quantified using microtiter plate
method and observing topographical changes in biofilm by scanning electron
microscopy [125] as well as confocal laser scanning microscopy. Wijaya et al., have
reported that crude extracts from marine actinobacteria isolated from Indonesia,
show significant reduction of biofilm formation by inhibiting quorum sensing in
Gram positive and Gram negative pathogenic bacteria [142]. Similar activity is
reported in extracts from Nocardiopsis sp [143].

6.3 Molecular and physiochemical methods

Genes encoding polyketide synthase I and II (PKS-I and PKS-II), and nonribosomal peptide synthetases (NRPS), are responsible for synthesis of novel AQS
compounds in rare actinobacteria [138, 144]. PCR Screening of a given bacterium
for these genes is helpful in detecting the presence of quorum quenching activity
[145]. Purified compounds isolated from the broth cultures of such bacteria can be
further subjected to Physiochemical screening methods such as UV–Vis spectropho-
tometry, TLC to determine the nature of compound which exhibits AQS activity
[138]. Some strains of Nocardia [146] and Micromonospora [147] are subjected to
such studies and novel AQS compounds were found.

6.4 *In silico* screening methods

Biosynthetic gene clusters in a given bacteria can be identified using genome
mining studies. 20 biosynthetic gene clusters are reported in a single strain of
Pseudoalteromonads most of which were NPRS and PKS gene clusters [148].
CLUSEAN (CLUster SEquence ANalyzer) [149] and antiSMASH [150] are tools
used for analysis of bacterial secondary metabolite biosynthetic gene clusters and
can be used to predict AQS activity in a given bacterium. Using structures available
in various databases, performing molecular docking [151] studies on compounds
having AQS activities is possible and can provide remarkable information on
molecular interactions involved in inhibition of quorum sensing.

7. Conclusion

Though work on Rare actinobacteria and bioactive compounds from them is gathering momentum, not many studies are targeted towards isolation and identification of AQS compounds from rare actinobacteria and this research area is still in its infancy. These studies may lead to novel bioactive compounds that can act against bacterial quorum sensing systems. These agents can attenuate the virulence of the pathogens without challenging their growth, thereby preventing the emergence of resistant strains.

Such potent anti-quorum -sensing compounds may lead to the development of alternative therapies to address the glaring problem of antibiotic resistance.

This would be of immense medical and commercial benefit.

Author details

Sunita Bundale* and Aashlesha Pathak
Hislop School of Biotechnology, Hislop College, Nagpur, Maharashtra, India

*Address all correspondence to: sbundale@gmail.com

IntechOpen

References

[1] Berdy J. Bioactive microbial metabolites. The Journal of Antibiotics. 2005;**58**(1):1-26

[2] Sharma P, Dutta J, Thakur D. Chapter 21-Future prospects of actinobacteria in health and industry. In: New and Future Developments in Microbial Biotechnology and Bioengineering. Elsevier; 2018. pp. 305-324

[3] Barka EA, Vatsa P, Sanchez L, Gaveau-Vaillant N, Jacquard C, Klenk HP, et al. Taxonomy, physiology, and natural products of Actinobacteria. Microbiology and Molecular Biology Reviews. 2016;**80**(1):1-43

[4] Bundale S, Begde D, Pillai D, Gangwani K, Nashikkar N, Kadam T, et al. Novel aromatic polyketides from soil Streptomyces spp.: Purification, characterization and bioactivity studies. World Journal of Microbiology and Biotechnology. 2018;**34**(5):1-6

[5] Kurtböke DI. Biodiscovery from rare actinomycetes: An eco-taxonomical perspective. Applied Microbiology and Biotechnology. 2012;**93**(5):1843-1852

[6] Nashikkar N, Begde D, Bundale S, Pise M, Rudra J, Upadhyay A. Inhibition of swarming motility, biofilm formation and virulence factor expression of urinary pathogens by Euphorbia trigona latex extracts. International Journal of Pharmaceutical Sciences and Research. 2011;2(3):558

[7] Amin DH, Abdallah NA, Abolmaaty A, Tolba S, Wellington EM. Microbiological and molecular insights on rare Actinobacteria harboring bioactive prospective. Bulletin of the National Research Centre. 2020; **44**(1):1-2

[8] Arul Jose PO, Jebakumar SR. Non-streptomycete actinomycetes nourish the current microbial antibiotic drug discovery. Frontiers in Microbiology. 2013;4:240

[9] Quinn GA, Banat AM, Abdelhameed AM, Banat IM. Streptomyces from traditional medicine: Sources of new innovations in antibiotic discovery. Journal of Medical Microbiology. 2020;**69**(8):1040

[10] Lazzarini A, Cavaletti L, Toppo G, Marinelli F. Rare genera of actinomycetes as potential producers of new antibiotics. Antonie Van Leeuwenhoek. 2000;**78**(3):399-405

[11] Bhattacharjee K, Kumar S, Palepu NR, Patra PK, Rao KM, Joshi SR. Structure elucidation and in silico docking studies of a novel furopyrimidine antibiotics synthesized by endolithic bacterium Actinomadura sp. AL2. World Journal of Microbiology and Biotechnology. 2017;**33**(10):1-6

[12] Kroppenstedt RM, Stackebrandt E, Goodfellow M. Taxonomic revision of the actinomycete genera Actinomadura and Microtetraspora. Systematic and Applied Microbiology. 1990;**13**(2): 148-160

[13] Wehmeier UF. The biosynthesis and metabolism of acarbose in Actinoplanes sp. SE 50/110: A progress report. Biocatalysis and Biotransformation. 2003;**21**(4-5):279-284

[14] Acarbose LH. Clinical Drug Investigation. 2002;**22**(3):141-156

[15] Song Z, Xu T, Wang J, Hou Y, Liu C, Liu S, et al. Secondary metabolites of the genus Amycolatopsis: Structures, bioactivities and biosynthesis. Molecules. 2021;**26**(7):1884

[16] Wells JS, O'Sullivan JO, Aklonis C, HA AX, Tymiak AA, Kirsch DR, et al. Dactylocyclines, novel tetracycline derivatives produced by a

Dactylosporangium sp. I. Taxonomy, production, isolation and biological activity. The Journal of Antibiotics. 1992;**45**(12):1892-1898

[17] Zdouc MM, Iorio M, Maffioli SI, Crüsemann M, Donadio S, Sosio M. Planomonospora: A metabolomics perspective on an underexplored Actinobacteria genus. Journal of Natural Products. 2021;**84**(2):204-219

[18] Selva E, Beretta G, Montanini N, Saddler GS, Gastaldo L, Ferrari P, et al. Antibiotic GE2270 a: A novel inhibitor of bacterial protein synthesis I. isolation and characterization. The Journal of Antibiotics. 1991;**44**(7): 693-701

[19] Asolkar RN, Kirkland TN, Jensen PR, Fenical W. Arenimycin, an antibiotic effective against rifampin-and methicillin-resistant Staphylococcus aureus from the marine actinomycete Salinispora arenicola. The Journal of Antibiotics. 2010;**63**(1):37-39

[20] Inahashi Y, Iwatsuki M, Ishiyama A, Namatame M, Nishihara-Tsukashima A, Matsumoto A, et al. Spoxazomicins A–C, novel antitrypanosomal alkaloids produced by an endophytic actinomycete, Streptosporangium oxazolinicum K07-0460 T. The Journal of Antibiotics. 2011;**64**(4):303-307

[21] Zhang J, Li B, Qin Y, Karthik L, Zhu G, Hou C, et al. A new abyssomicin polyketide with anti-influenza a virus activity from a marine-derived Verrucosispora sp. MS100137. Applied Microbiology and Biotechnology. 2020;**104**(4):1533-1543

[22] Boumehira AZ, El-Enshasy HA, Hacène H, Elsayed EA, Aziz R, Park EY. Recent progress on the development of antibiotics from the genus Micromono-spora. Biotechnology and Bioprocess Engineering. 2016;**21**(2):199-223

[23] Qin S, Li WJ, Klenk HP, Hozzein WN, Ahmed I. Actinobacteria in special and extreme habitats: Diversity, function roles and environmental adaptations. Frontiers in Microbiology. 2019;**10**:944

[24] Tiwari K, Gupta RK. Diversity and isolation of rare actinomycetes: An overview. Critical Reviews in Microbiology. 2013;**39**(3):256-294

[25] Monciardini P, Cavaletti L, Schumann P, Rohde M, Donadio S. Conexibacter woesei gen. nov., sp. nov., a novel representative of a deep evolutionary line of descent within the class Actinobacteria. International Journal of Systematic and Evolutionary Microbiology. 2003;**53**(2):569-576

[26] Cavaletti L, Monciardini P, Schumann P, Rohde M, Bamonte R, Busti E, et al. Actinospica robiniae gen. nov., sp. nov. and Actinospica acidiphila sp. nov.: Proposal for Actinospicaceae fam. nov. and Catenulisporinae subord. nov. in the order Actinomycetales. International Journal of Systematic and Evolutionary Microbiology. 2006;**56**(8):1747-1753

[27] Busti E, Cavaletti L, Monciardini P, Schumann P, Rohde M, Sosio M, et al. Catenulispora acidiphila gen. nov., sp. nov., a novel, mycelium-forming actinomycete, and proposal of Catenulisporaceae fam. nov. International Journal of Systematic and Evolutionary Microbiology. 2006;**56**(8):1741-1746

[28] Zhou Y, Wei W, Wang X, Lai R. Proposal of Sinomonas flava gen. nov., sp. nov., and description of Sinomonas atrocyanea comb. nov. to accommodate Arthrobacter atrocyaneus. International Journal of Systematic and Evolutionary Microbiology. 2009;**59**(2):259-263

[29] Shiratori-Takano H, Yamada K, Beppu T, Ueda K. Longispora fulva sp.

nov., isolated from a forest soil, and emended description of the genus Longispora. International Journal of Systematic and Evolutionary Microbiology. 2011;**61**(4):804-809

[30] Mao J, Wang J, Dai HQ, Zhang ZD, Tang QY, Ren B, et al. Yuhushiella deserti gen. nov., sp. nov., a new member of the suborder Pseudonocardineae. International Journal of Systematic and Evolutionary Microbiology. 2011;**61**(3):621-630

[31] Liu M, Peng F, Wang Y, Zhang K, Chen G, Fang C. Kineococcus xinjiangensis sp. nov., isolated from desert sand. International Journal of Systematic and Evolutionary Microbiology. 2009;**59**(5):1090-1093

[32] Jiang Y, Cao Y, Zhao L, Tang S, Wang Y, Li W, et al. Large numbers of new bacterial taxa found by Yunnan Institute of Microbiology. Chinese Science Bulletin. 2011;**56**(8):709-712

[33] Zhang W, Zhu HH, Yuan M, Yao Q, Tang R, Lin M, et al. Microbacterium radiodurans sp. nov., a UV radiation-resistant bacterium isolated from soil. International Journal of Systematic and Evolutionary Microbiology. 2010;**60**(11):2665-2670

[34] Groth I, Schumann P, Weiss N, Schuetze B, Augsten K, Stackebrandt E. Ornithinimicrobium humiphilum gen. nov., sp. nov., a novel soil actinomycete with L-ornithine in the peptidoglycan. International Journal of Systematic and Evolutionary Microbiology. 2001;**51**(1):81-87

[35] Cui X, Schumann P, Stackebrandt E, Kroppenstedt RM, Pukall R, Xu L, et al. Myceligenerans xiligouense gen. nov., sp. nov., a novel hyphae-forming member of the family Promicromonosporaceae. International Journal of Systematic and Evolutionary Microbiology. 2004;**54**(4):1287-1293

[36] Yoon JH, Kang SJ, Schumann P, Oh TK. Yonghaparkia alkaliphila gen. nov., sp. nov., a novel member of the family Microbacteriaceae isolated from an alkaline soil. International journal of Systematic and Evolutionary Microbiology. 2006;**56**(10):2415-2420

[37] Adachi K, Katsuta A, Matsuda S, Peng X, Misawa N, Shizuri Y, et al. Smaragdicoccus niigatensis gen. nov., sp. nov., a novel member of the suborder Corynebacterineae. International Journal of Systematic and Evolutionary Microbiology. 2007;**57**(2):297-301

[38] Ara I, Kudo T. Krasilnikovia gen. nov., a new member of the family Micromonosporaceae and description of Krasilnikovia cinnamonea sp. nov. Actinomycetologica. 2007;**21**(1):1-0

[39] Ara I, Kudo T. Sphaerosporangium gen. nov., a new member of the family Streptosporangiaceae, with descriptions of three new species as Sphaerosporangium melleum sp. nov., Sphaerosporangium rubeum sp. nov. and Sphaerosporangium cinnabarinum sp. nov., and transfer of Streptosporangium viridialbum Nonomura and Ohara 1960 to Sphaerosporangium viridialbum comb. nov. Actinomycetologica. 2007;**21**:11-21

[40] Ara I, Kudo T. Two new species of the genus Micromonospora: Micromonospora chokoriensis sp. nov. and Micromonospora coxensis sp. nov., isolated from sandy soil. The Journal of General and Applied Microbiology. 2007;**53**(1):29-37

[41] Gu Q, Paściak M, Luo H, Gamian A, Liu Z, Huang Y. Ruania albidiflava gen. nov., sp. nov., a novel member of the suborder Micrococcineae. International Journal of Systematic and Evolutionary Microbiology. 2007;**57**(4):809-814

[42] Lee M, Ten LN, Woo SG, Park J. Agromyces soli sp. nov., isolated from

farm soil. International Journal of Systematic and Evolutionary Microbiology. 2011;**61**(6):1286-1292

[43] Zhang YQ, Schumann P, Yu LY, Liu HY, Zhang YQ, Xu LH, et al. Zhihengliuella halotolerans gen. nov., sp. nov., a novel member of the family Micrococcaceae. International Journal of Systematic and Evolutionary Microbiology. 2007;**57**(5):1018-1023

[44] Chen YG, Cui XL, Kroppenstedt RM, Stackebrandt E, Wen ML, Xu LH, et al. Nocardiopsis quinghaiensis sp. nov., isolated from saline soil in China. International Journal of Systematic and Evolutionary Microbiology. 2008;**58**(3):699-705

[45] Li WJ, Zhang YQ, Schumann P, Liu HY, Yu LY, Zhang YQ, et al. Nesterenkonia halophila sp. nov., a moderately halophilic, alkalitolerant actinobacterium isolated from a saline soil. International Journal of Systematic and Evolutionary Microbiology. 2008;**58**(6):1359-1363

[46] Tang SK, Wang Y, Schumann P, Stackebrandt E, Lou K, Jiang CL, et al. Brevibacterium album sp. nov., a novel actinobacterium isolated from a saline soil in China. International Journal of Systematic and Evolutionary Microbiology. 2008;**58**(3):574-577

[47] Tang SK, Wang Y, Lou K, Mao PH, Xu LH, Jiang CL, et al. Kocuria halotolerans sp. nov., an actinobacterium isolated from a saline soil in China. International Journal of Systematic and Evolutionary Microbiology. 2009;**59**(6):1316-1320

[48] Chiaraphongphon S, Suriyachadkun C, Tamura T, Thawai C. Dactylosporangium maewongense sp. nov., isolated from soil. International Journal of Systematic and Evolutionary Microbiology. 2010;**60**(5):1200-1205

[49] Qin S, Chen HH, Klenk HP, Kim CJ, Xu LH, Li WJ. Saccharopolyspora

gloriosae sp. nov., an endophytic actinomycete isolated from the stem of Gloriosa superba L. International Journal of Systematic and Evolutionary Microbiology. 2010;**60**(5):1147-1151

[50] Qin S, Zhu WY, Jiang JH, Klenk HP, Li J, Zhao GZ, et al. Pseudonocardia tropica sp. nov., an endophytic actinomycete isolated from the stem of Maytenus austroyunnanensis. International Journal of Systematic and Evolutionary Microbiology. 2010;**60**(11):2524-2528

[51] Suriyachadkun C, Chunhametha S, Thawai C, Tamura T, Potacharoen W, Kirtikara K, et al. Planotetraspora thailandica sp. nov., isolated from soil in Thailand. International Journal of Systematic and Evolutionary Microbiology. 2009;**59**(5):992-997

[52] Suriyachadkun C, Chunhametha S, Ngaemthao W, Tamura T, Kirtikara K, Sanglier JJ, et al. Sphaerisporangium krabiense sp. nov., isolated from soil. International Journal of Systematic and Evolutionary Microbiology. 2011;**61**(12):2890-2894

[53] Kageyama A, Takahashi Y, Ōmura S. Humihabitans oryzae gen. nov., sp. nov. International Journal of Systematic and Evolutionary Microbiology. 2007;**57**(9): 2163-2166

[54] Kageyama A, Matsumoto A, Ōmura S, Takahashi Y. Humibacillus xanthopallidus gen. nov., sp. nov. International Journal of Systematic and Evolutionary Microbiology. 2008;**58**(7):1547-1551

[55] Kageyama A, Morisaki K, Ōmura S, Takahashi Y. Arthrobacter oryzae sp. nov. and Arthrobacter humicola sp. nov. International Journal of Systematic and Evolutionary Microbiology. 2008;**58**(1): 53-56

[56] Anzai K, Sugiyama T, Sukisaki M, Sakiyama Y, Otoguro M, Ando K.

Flexivirga alba gen. nov., sp. nov., an actinobacterial taxon in the family Dermacoccaceae. The Journal of Antibiotics. 2011;**64**(9):613-616

[57] Liu XY, Wang BJ, Jiang CY, Liu SJ. Micrococcus flavus sp. nov., isolated from activated sludge in a bioreactor. International Journal of Systematic and Evolutionary Microbiology. 2007;**57**(1):66-69

[58] Yassin AF, Shen FT, Hupfer H, Arun AB, Lai WA, Rekha PD, et al. Gordonia malaquae sp. nov., isolated from sludge of a wastewater treatment plant. International Journal of Systematic and Evolutionary Microbiology. 2007;**57**(5):1065-1068

[59] Yoon JH, Kang SJ, Park S, Kim W, Oh TK. Nocardioides caeni sp. nov., isolated from wastewater. International Journal of Systematic and Evolutionary Microbiology. 2009;**59**(11):2794-2797

[60] Lee DW, Lee JM, Seo JP, Schumann P, Kim SJ, Lee SD. Phycicola gilvus gen. nov., sp. nov., an actinobacterium isolated from living seaweed. International Journal of Systematic and Evolutionary Microbiology. 2008;**58**(6):1318-1323

[61] Lee SD. Labedella gwakjiensis gen. nov., sp. nov., a novel actinomycete of the family Microbacteriaceae. International Journal of Systematic and Evolutionary Microbiology. 2007;**57**(11):2498-2502

[62] Lee SD. Phycicoccus jejuensis gen. nov., sp. nov., an actinomycete isolated from seaweed. International Journal of Systematic and Evolutionary Microbiology. 2006;**56**(10):2369-2373

[63] Lee SD, Kim SJ. Aeromicrobium tamlense sp. nov., isolated from dried seaweed. International Journal of Systematic and Evolutionary Microbiology. 2007;**57**(2):337-341

[64] Lee DW, Lee SD. Tessaracoccus flavescens sp. nov., isolated from marine sediment. International Journal of Systematic and Evolutionary Microbiology. 2008;**58**(4):785-789

[65] Khan ST, Harayama S, Tamura T, Ando K, Takagi M, Kazuo SY. Paraoerskovia marina gen. nov., sp. nov., an actinobacterium isolated from marine sediment. International Journal of Systematic and Evolutionary Microbiology. 2009;**59**(8):2094-2098

[66] Tian XP, Tang SK, Dong JD, Zhang YQ, Xu LH, Zhang S, et al. Marinactinospora thermotolerans gen. nov., sp. nov., a marine actinomycete isolated from a sediment in the northern South China Sea. International Journal of Systematic and Evolutionary Microbiology. 2009;**59**(5):948-952

[67] Ue H, Matsuo Y, Kasai H, Yokota A. Demequina globuliformis sp. nov., Demequina oxidasica sp. nov. and Demequina aurantiaca sp. nov., actinobacteria isolated from marine environments, and proposal of Demequinaceae fam. nov. International Journal of Systematic and Evolutionary Microbiology. 2011;**61**(6):1322-1329

[68] Goodfellow M, Stach JE, Brown R, Bonda AN, Jones AL, Mexson J, et al. Verrucosispora maris sp. nov., a novel deep-sea actinomycete isolated from a marine sediment which produces abyssomicins. Antonie Van Leeuwenhoek. 2012;**101**(1):185-193

[69] Groth I, Schumann P, Schütze B, Augsten K, Stackebrandt E. Knoellia sinensis gen. nov., sp. nov. and Knoellia subterranea sp. nov., two novel actinobacteria isolated from a cave. International Journal of Systematic and Evolutionary Microbiology. 2002;**52**(1):77-84

[70] Jurado V, Kroppenstedt RM, Saiz-Jimenez C, Klenk HP, Mouniee D,

Laiz L, et al. Hoyosella altamirensis gen. nov., sp. nov., a new member of the order Actinomycetales isolated from a cave biofilm. International Journal of Systematic and Evolutionary Microbiology. 2009;**59**(12):3105-3110

[71] Lee SD. Jiangella alkaliphila sp. nov., an actinobacterium isolated from a cave. International Journal of Systematic and Evolutionary Microbiology. 2008;**58**(5):1176-1179

[72] Qin S, Li J, Zhang YQ, Zhu WY, Zhao GZ, Xu LH, et al. Plantactinospora mayteni gen. nov., sp. nov., a member of the family Micromonosporaceae. International Journal of Systematic and Evolutionary Microbiology. 2009;**59**(10):2527-2533

[73] Inahashi Y, Matsumoto A, Danbara H, Ōmura S, Takahashi Y. Phytohabitans suffuscus gen. nov., sp. nov., an actinomycete of the family Micromonosporaceae isolated from plant roots. International Journal of Systematic and Evolutionary Microbiology. 2010;**60**(11):2652-2658

[74] Indananda C, Matsumoto A, Inahashi Y, Takahashi Y, Duangmal K, Thamchaipenet A. Actinophytocola oryzae gen. nov., sp. nov., isolated from the roots of Thai glutinous rice plants, a new member of the family Pseudonocardiaceae. International Journal of Systematic and Evolutionary Microbiology. 2010;**60**(5):1141-1146

[75] Kaewkla O, Franco CM. Flindersiella endophytica gen. nov., sp. nov., an endophytic actinobacterium isolated from the root of Grey box, an endemic eucalyptus tree. International Journal of Systematic and Evolutionary Microbiology. 2011;**61**(9):2135-2140

[76] Li J, Zhao GZ, Zhu WY, Huang HY, Xu LH, Zhang S, et al. Phytomonospora endophytica gen. nov., sp. nov., isolated from the roots of Artemisia annua L. International Journal of Systematic and Evolutionary Microbiology. 2011;**61**(12):2967-2973

[77] Zhang L, Xu Z, Patel BK. Frondicola australicus gen. nov., sp. nov., isolated from decaying leaf litter from a pine forest. International Journal of Systematic and Evolutionary Microbiology. 2007;**57**(6):1177-1182

[78] Li J, Zhao GZ, Chen HH, Qin S, Xu LH, Jiang CL, et al. Rhodococcus cercidiphylli sp. nov., a new endophytic actinobacterium isolated from a Cercidiphyllum japonicum leaf. Systematic and Applied Microbiology. 2008;**31**(2):108-113

[79] Qin S, Zhao GZ, Klenk HP, Li J, Zhu WY, Xu LH, et al. Nonomuraea antimicrobica sp. nov., an endophytic actinomycete isolated from a leaf of Maytenus austroyunnanensis. International Journal of Systematic and Evolutionary Microbiology. 2009;**59**(11):2747-2751

[80] Gu Q, Zheng W, Huang Y. Glycomyces sambucus sp. nov., an endophytic actinomycete isolated from the stem of Sambucus adnata wall. International Journal of Systematic and Evolutionary Microbiology. 2007;**57**(9):1995-1998

[81] Li J, Zhao GZ, Qin S, Huang HY, Zhu WY, Xu LH, et al. Saccharopolyspora tripterygii sp. nov., an endophytic actinomycete isolated from the stem of Tripterygium hypoglaucum. International Journal of Systematic and Evolutionary Microbiology. 2009;**59**(12):3040-3044

[82] Li J, Zhao GZ, Huang HY, Qin S, Zhu WY, Xu LH, et al. Kineosporia mesophila sp. nov., isolated from surface-sterilized stems of Tripterygium wilfordii. International Journal of Systematic and Evolutionary Microbiology. 2009;**59**(12):3150-3154

[83] Zhao GZ, Li J, Zhu WY, Klenk HP, Xu LH, Li WJ. Nocardia artemisiae sp.

nov., an endophytic actinobacterium isolated from a surface-sterilized stem of Artemisia annua L. International Journal of Systematic and Evolutionary Microbiology. 2011;**61**(12):2933-2937

[84] Zhang DC, Wang HX, Cui HL, Yang Y, Liu HC, Dong XZ, et al. Cryobacterium psychrotolerans sp. nov., a novel psychrotolerant bacterium isolated from the China No. 1 glacier. International Journal of Systematic and Evolutionary Microbiology. 2007;**57**(4): 866-869

[85] Reddy GS, Prabagaran SR, Shivaji S. Leifsonia pindariensis sp. nov., isolated from the Pindari glacier of the Indian Himalayas, and emended description of the genus Leifsonia. International Journal of Systematic and Evolutionary Microbiology. 2008;**58**(9):2229-2234

[86] Margesin R, Schumann P, Zhang DC, Redzic M, Zhou YG, Liu HC, et al. Arthrobacter cryoconiti sp. nov., a psychrophilic bacterium isolated from alpine glacier cryoconite. International Journal of Systematic and Evolutionary Microbiology. 2012;**62**(2):397-402

[87] Lee DW, Lee SD. Allocate-lliglobosispora scoriae gen. nov., sp. nov., isolated from volcanic ash. International Journal of Systematic and Evolutionary Microbiology. 2011;**61**(2): 264-270

[88] Zhou EM, Tang SK, Sjøholm C, Song ZQ, Yu TT, Yang LL, et al. Thermoactinospora rubra gen. nov., sp. nov., a thermophilic actinomycete isolated from Tengchong, Yunnan province, south–West China. Antonie Van Leeuwenhoek. 2012;**102**(1):177-185

[89] Fang BZ, Salam N, Han MX, Jiao JY, Cheng J, Wei DQ, et al. Insights on the effects of heat pretreatment, pH, and calcium salts on isolation of rare Actinobacteria from karstic caves. Frontiers in Microbiology. 2017;**8**:1535

[90] Alferova IV, Terekhova LP. Use of the method of enriching of soil samples with calcium carbonate for isolation of Actinomyces. Antibiotiki i khimioterapiia= Antibiotics and Chemoterapy [sic]. 1988;**33**(12):888-890

[91] Hayakawa M. Studies on the isolation and distribution of rare actinomycetes in soil. Actinomycetologica. 2008;**22**(1): 12-19

[92] McCarthy AJ, Cross T. A note on a selective isolation medium for the thermophilic actinomycete Thermomonospora chromogena. Journal of Applied Bacteriology. 1981;**51**(2):299-302

[93] Shirling ET, Gottlieb D. Methods for characterization of Streptomyces species1. International Journal of Systematic and Evolutionary Microbiology. 1966;**16**(3):313-340

[94] Meena B, Anburajan L, Vinithkumar NV, Kirubagaran R, Dharani G. Biodiversity and antibacterial potential of cultivable halophilic actinobacteria from the deep sea sediments of active volcanic Barren Island. Microbial Pathogenesis. 2019;**132**:129-136

[95] Cho SH, Han JH, Ko HY, Kim SB. Streptacidiphilus anmyonensis sp. nov., Streptacidiphilus rugosus sp. nov. and Streptacidiphilus melanogenes sp. nov., acidophilic actinobacteria isolated from Pinus soils. International Journal of Systematic and Evolutionary Microbiology. 2008;**58**(7):1566-1570

[96] Hallberg KB, Johnson DB. Biodiversity of acidophilic microorganisms. Advances in Applied Microbiology. 2001;**49**:37-84

[97] Goodfellow M. Selective isolation of Actinobacteria. In: Manual of Industrial Microbiology and Biotechnology. Wiley Online Library; 2010. pp. 13-27

[98] Sengupta S, Pramanik A, Ghosh A, Bhattacharyya M. Antimicrobial activities of actinomycetes isolated from unexplored regions of Sundarbans mangrove ecosystem. BMC Microbiology. 2015;**15**(1):1-6

[99] Hayakawa M, Nonomura H. A new method for the intensive isolation of actinomycetes from soil. Actinomycetologica. 1989;**3**(2):95-104

[100] Takahashi Y, Katoh S, Shikura N, Tomoda H, Omura S. Superoxide dismutase produced by soil bacteria increases bacterial colony growth from soil samples. The Journal of General and Applied Microbiology. 2003;**49**(4): 263-266

[101] Porter JN, Wilhelm JJ, Tresner HD. Method for the preferential isolation of actinomycetes from soils. Applied Microbiology. 1960;**8**(3):174-178

[102] Barcina I, Iriberri J, Egea L. Enumeration, isolation and some physiological properties of actinomycetes from sea water and sediment. Systematic and Applied Microbiology. 1987;**10**(1):85-91

[103] Ghanem NB, Sabry SA, El-Sherif ZM, El-Ela GA. Isolation and enumeration of marine actinomycetes from seawater and sediments in Alexandria. The Journal of General and Applied Microbiology. 2000;**46**(3): 105-111

[104] Bundale S, Begde D, Nashikkar N, Kadam T, Upadhyay A. Optimization of culture conditions for production of bioactive metabolites by Streptomyces spp. isolated from soil. Advances in Microbiology. 2015;**5**(06):441-451

[105] Gallo M, Katz E. Regulation of secondary metabolite biosynthesis: Catabolite repression of phenoxazinone synthase and actinomycin formation by glucose. Journal of Bacteriology. 1972;**109**(2):659-667

[106] Reddy NG, Ramakrishna DP, Rajagopal SV. Optimization of culture conditions of Streptomyces rochei (MTCC 10109) for the production of antimicrobial metabolites. Egyptian Journal of Biology. 2011;**13**:21-29

[107] Iwai Y, Omura S. Culture conditions for screening of new antibiotics. The Journal of Antibiotics. 1982;**35**(2):123-141

[108] Aharonowitz Y. Nitrogen metabolite regulation of antibiotic biosynthesis. Annual Reviews in Microbiology. 1980;**34**(1):209-233

[109] Demain AL, Fang A. Emerging concepts of secondary metabolism in actinomycetes. Actinomycetologica. 1995;**9**(2):98-117

[110] Sejiny M. Growth phases of some antibiotics producing Streptomyces and their identification. Journal of King Abdulaziz University. 1991;**3**:21-29

[111] Singh V, Khan M, Khan S, Tripathi CK. Optimization of actinomycin V production by Streptomyces triostinicus using artificial neural network and genetic algorithm. Applied Microbiology and Biotechnology. 2009;**82**(2):379-385

[112] Hassan MA, El-Naggar MY, Said WY. Physiological factors affecting the production of an antimicrobial substance by Streptomyces violatus in batch cultures. Egyptian Journal of Biology 2001;**3**(1):1-0

[113] Oskay M. Effects of some environmental conditions on biomass and antimicrobial metabolite production by Streptomyces Sp., KGG32. International Journal of Agriculture & Biology. 2011;**13**(3): 317-324

[114] Miller MB, Bassler BL. Quorum sensing in bacteria. Annual Reviews in Microbiology. 2001;**55**(1):165-199

[115] Castillo-Juárez I, Maeda T, Mandujano-Tinoco EA, Tomás M, Pérez-Eretza B, García-Contreras SJ, et al. Role of quorum sensing in bacterial infections. World Journal of Clinical Cases: WJCC. 2015;**3**(7):575

[116] Deep A, Chaudhary U, Gupta V. Quorum sensing and bacterial pathogenicity: From molecules to disease. Journal of Laboratory Physicians. 2011;**3**(01):004-011

[117] Popham DL, Stevens AM. Bacterial quorum sensing and bioluminescence. In: 2005 Proc. Assoc. Biol. Lab. Educ. (ABLE). Vol. 27. Association for Biology Laboratory Education (ABLE); 2005. pp. 201-215

[118] Asfour HZ. Anti-quorum sensing natural compounds. Journal of Microscopy and Ultrastructure. 2018;**6**(1):1

[119] Steiner E, Scott J, Minton NP, Winzer K. An agr quorum sensing system that regulates granulose formation and sporulation in clostridium acetobutylicum. Applied and Environmental Microbiology. 2012;**78**(4):1113-1122

[120] Koraimann G, Wagner MA. Social behavior and decision making in bacterial conjugation. Frontiers in Cellular and Infection Microbiology. 2014;**4**:54

[121] van der Ploeg JR. Regulation of bacteriocin production in Streptococcus mutans by the quorum-sensing system required for development of genetic competence. Journal of Bacteriology. 2005;**187**(12):3980-3989

[122] Duerkop BA, Varga J, Chandler JR, Peterson SB, Herman JP, Churchill ME, et al. Quorum-sensing control of antibiotic synthesis in Burkholderia thailandensis. Journal of Bacteriology. 2009;**191**(12):3909-3918

[123] Defoirdt T. Quorum-sensing systems as targets for antivirulence therapy. Trends in Microbiology. 2018;**26**(4):313-328

[124] Rasmussen TB, Givskov M. Quorum-sensing inhibitors as anti-pathogenic drugs. International Journal of Medical Microbiology. 2006;**296**(2-3):149-161

[125] Smith KM, Bu Y, Suga H. Induction and inhibition of Pseudomonas aeruginosa quorum sensing by synthetic autoinducer analogs. Chemistry & Biology. 2003;**10**(1):81-89

[126] Vadakkan K, Choudhury AA, Gunasekaran R, Hemapriya J, Vijayanand S. Quorum sensing intervened bacterial signaling: Pursuit of its cognizance and repression. Journal of Genetic Engineering and Biotechnology. 2018;**16**(2):239-252

[127] Papenfort K, Bassler BL. Quorum sensing signal-response systems in gram-negative bacteria. Nature Reviews. Microbiology. 2016;**14**(9):576-588. DOI: 10.1038/nrmicro.2016.89 PMID: 27510864; PMCID: PMC5056591

[128] Dong YH, Wang LH, Xu JL, Zhang HB, Zhang XF, Zhang LH. Quenching quorum-sensing-dependent bacterial infection by an N-acyl homoserine lactonase. Nature. 2001;**411**(6839):813-817

[129] Czajkowski R, Krzyżanowska D, Karczewska J, Atkinson S, Przysowa J, Lojkowska E, et al. Inactivation of AHLs by Ochrobactrum sp. A44 depends on the activity of a novel class of AHL acylase. Environmental Microbiology Reports. 2011;**3**(1):59-68

[130] Bijtenhoorn P, Mayerhofer H, Müller-Dieckmann J, Utpatel C, Schipper C, Hornung C, et al. A novel metagenomic short-chain dehydrogenase/reductase attenuates Pseudomonas aeruginosa biofilm formation and

virulence on Caenorhabditis elegans. PLoS One. 2011;**6**(10):e26278

[131] Marin SD, Xu Y, Meijler MM, Janda KD. Antibody catalyzed hydrolysis of a quorum sensing signal found in gram-negative bacteria. Bioorganic & Medicinal Chemistry Letters. 2007;**17**(6):1549-1552

[132] Lichstein HC, Van De Sand VF. Violacein, an antibiotic pigment produced by Chromobacterium violaceum. The Journal of Infectious Diseases. 1945;**76**(1):47-51

[133] Choo JH, Rukayadi Y, Hwang JK. Inhibition of bacterial quorum sensing by vanilla extract. Letters in Applied Microbiology. 2006;**42**(6):637-641

[134] Zhu H, He CC, Chu QH. Inhibition of quorum sensing in Chromobacterium violaceum by pigments extracted from Auricularia auricular. Letters in Applied Microbiology. 2011;**52**(3):269-274

[135] Poli JP, Guinoiseau E, de Rocca SD, Sutour S, Paoli M, Tomi F, et al. Anti-quorum sensing activity of 12 essential oils on chromobacterium violaceum and specific action of cis-cis-p-Menthenolide from Corsican Mentha suaveolens ssp. Insularis. Molecules. 2018;**23**(9):2125

[136] Venkatramanan M, Sankar Ganesh P, Senthil R, Akshay J, Veera Ravi A, Langeswaran K, et al. Inhibition of quorum sensing and biofilm formation in Chromobacterium violaceum by fruit extracts of Passiflora edulis. ACS Omega. 2020;**5**(40):25605-25616

[137] Koh KH, Tham FY. Screening of traditional Chinese medicinal plants for quorum-sensing inhibitors activity. Journal of Microbiology, Immunology and Infection. 2011;**44**(2):144-148

[138] Bundale S, Singh J, Begde D, Nashikkar N, Upadhyay A. Rare actinobacteria: A potential source of

bioactive polyketides and peptides. World Journal of Microbiology and Biotechnology. 2019;**35**(6):1-1

[139] Kalia M, Yadav VK, Singh PK, Sharma D, Pandey H, Narvi SS, et al. Effect of cinnamon oil on quorum sensing-controlled virulence factors and biofilm formation in Pseudomonas aeruginosa. PLoS One. 2015;**10**(8): e0135495

[140] Bakkiyaraj D, Sivasankar C, Pandian SK. Inhibition of quorum sensing regulated biofilm formation in Serratia marcescens causing nosocomial infections. Bioorganic & Medicinal Chemistry Letters. 2012;**22**(9):3089-3094

[141] Brackman G, Coenye T. Quorum sensing inhibitors as anti-biofilm agents. Current Pharmaceutical Design. 2015;**21**(1):5-11

[142] Wijaya M, Delicia D, Waturangi DE. Screening and quantification of antiquorum-sensing activity of Actinobacteria isolates against gram-positive and gram-negative biofilm associated bacteria. Research Square. 2020 [Preprint]

[143] Rajivgandhi G, Vijayan R, Maruthupandy M, Vaseeharan B, Manoharan N. Antibiofilm effect of Nocardiopsis sp. GRG 1 (KT235640) compound against biofilm forming gram negative bacteria on UTIs. Microbial Pathogenesis. 2018;**118**:190-198

[144] Bundale SB, Begde DN, Nashikkar NN, Kadam TA, Upadhyay AA. Isolation of aromatic polyketide producing soil Streptomyces using combinatorial screening strategies. Open Access Library Journal. 2014;**1**:1-6

[145] Hamedi J, Imanparast S, Mohammadipanah F. Molecular, chemical and biological screening of soil actinomycete isolates in seeking bioactive peptide metabolites. Iranian Journal of Microbiology. 2015;**7**(1):23

[146] El-Kurdi N, Abdulla H, Hanora A. Anti-quorum sensing activity of some marine bacteria isolated from different marine resources in Egypt. Biotechnology Letters. 2021;**43**(2):455-468

[147] Kavimalar DN. Quorum Quenching Potential of Actinobacteria Isolated from Selected Habitats in Peninsular Malaysia/Kavimalar Devaraj Naidu [Doctoral dissertation]. University of Malaya; 2017

[148] Ong JF, Goh HC, Lim SC, Pang LM, Chin JS, Tan KS, et al. Integrated genomic and metabolomic approach to the discovery of potential anti-quorum sensing natural products from microbes associated with marine samples from Singapore. Marine Drugs. 2019;**17**(1):72

[149] Weber T, Rausch C, Lopez P, Hoof I, Gaykova V, Huson DH, et al. CLUSEAN: A computer-based framework for the automated analysis of bacterial secondary metabolite biosynthetic gene clusters. Journal of Biotechnology. 2009;**140**(1-2):13-17

[150] Medema MH, Blin K, Cimermancic P, De Jager V, Zakrzewski P, Fischbach MA, et al. antiSMASH: Rapid identification, annotation and analysis of secondary metabolite biosynthesis gene clusters in bacterial and fungal genome sequences. Nucleic Acids Research. 2011;**39**(suppl_2):W339-W346

[151] Priyanka S, Jayashree M, Shivani R, Anwesha S, Rao KB. Characterisation and identification of antibacterial compound from marine actinobacteria: In vitro and in silico analysis. Journal of Infection and Public Health. 2019;**12**(1):83-89

Section 3

Applications of Actinobacteria

Application of Actinobacteria in Agriculture, Nanotechnology, and Bioremediation

Saloni Jain, Ishita Gupta, Priyanshu Walia and Shalini Swami

Abstract

"Actinobacteria" are of significant economic value to mankind since agriculture and forestry depend on their soil system contribution. The organic stuff of deceased creatures is broken down into soil, and plants are able to take the molecule up again. Actinobacteria can be used for sustainable agriculture as biofertilizers for the improvement of plant growth or soil health by promoting different plant growth attributes, such as phosphorus and potassium solubilization, production of iron-chelating compounds, phytohormones, and biological nitrogen attachment even under the circumstances of natural and abiotic stress. Nanotechnology has received considerable interest in recent years due to its predicted impacts on several key fields such as health, energy, electronics, and the space industries. Actinobacterial biosynthesis of nanoparticles is a dependable, environmentally benign, and significant element toward green chemistry, which links together microbial biotechnology and nanobiology. Actinobacterial-produced antibiotics are common in nearly all of the medical treatments, and they are also recognized to aid in the biosynthesis of excellent surface and size properties of nanoparticles. Bioremediation using microorganisms is relatively safe and more efficient. Actinobacteria use carbon toxins to synthesize economically viable antibiotics, enzymes, and proteins as well. These bacteria are the leading microbial phyla that are beneficial for deterioration and transformation of organic and metal substrates.

Keywords: Actinobacteria, *Streptomyces*, PGPR, agriculture, nanoparticles, bioremediation

1. Introduction

One of the largest taxonomic groups of bacteria, Actinobacteria, are generally gram-positive with high Guanine + Cytosine (G + C) content (usually around 70%), a common marker in bacterial systematics [1, 2]. They are unicellular, filamentous, spore-forming, motile, or nonmotile and can be aerobic or anaerobic in nature [3]. The morphological structure ranges from coccoid (*Micrococcus*) and rod-coccoid (*Arthrobacter*) to fragmenting hyphal forms

(*Nocardia*) and branched mycelium (*Streptomyces*) [4]. In culture media, actinobacterial colonies have a powdery consistency and adhere tightly to the agar surface, forming hyphae and conidia/sporangia-like fungi on (aerial mycelium) or under (substrate mycelium) the agar surface [2, 5]. Found in a plethora of environment, including terrestrial and aquatic (both marine and freshwater), they share features of both the bacteria (chromosomes organized in a nucleoid and cell wall made of peptidoglycan) as well as fungi (presence of mycelium) [1, 2, 6]. Actinobacteria possess high ecological significance with an immense ability to produce organic acids, fix nitrogen from the atmosphere, and impart an essential role in the decomposition of organic compounds including cellulose and chitin, thus contributing to organic matter turnover and the carbon cycle. This further renews the supply of nutrients in the soil and forms humus [2, 3]. They play a crucial role not only in agriculture but also in the clinical and pharmaceutical industry [7]. Antibiotics, antifungals, enzymes, enzyme inhibitors, antivirals, antioxidants, anticholesterol, antiprotozoal, anticancer, and immunosuppressant are few of the beneficial secondary metabolites with therapeutic implications produced by Actinobacteria [1, 6, 7]. Some of the important genera of Actinobacteria found in soil are *Actinoplanes*, *Micromonospora*, *Nocardia*, *Streptomyces*, and *Streptosporangium* [2], and found as plant or animal pathogens are *Corynebacterium*, *Mycobacterium*, or *Nocardia* [1].

2. Types of Actinobacteria

Actinobacteria, along with normal environments, can thrive in extreme environments like acidic/alkaline pH, low/high temperatures, high salt concentration, high level of radiation, low moisture content, and nutrients [2]. Based on the above environmental conditions, the different types of Actinobacteria along with their ecological significance have been summarized in **Table 1**.

Considering capabilities of Actinobacteria (**Table 1**), they can potentially be exploited as a candidate for agriculture and environmental biotechnology.

Types of Actinobacteria	Ecological significance	Important genera	References
Thermophilic	• Used in composting, antimicrobial activity, and plant growth promotion and in the production of polyester-hydrolyzing enzymes. • Responsible for causing severe respiratory diseases such as Farmer's lung and bagassosis.	• *Amycolatopsis, Cellulosimicrobium, Micrococcus, Micromonospora, Planomonospora, Saccharopolyspora, Streptomyces, Thermobifida,* and *Thermomonospora.* • *Saccharopolyspora rectivirgula, S. viridis, Thermoactinomyces viridis,* and *T. vulgaris.*	[4, 5]
Acidophilic	• Exhibit strong antagonistic effect toward multiple fungal root pathogens (for example, inhibit the rice pathogenic fungi *Fusarium moniliforme* and *Rhizoctonia solani*), phosphate solubilization activity, and produce siderophores.	• *Actinospica, Catenulispora,* and *Streptomyces acidiphilus.*	[8]

Types of Actinobacteria	Ecological significance	Important genera	References
Halophilic	• Produce vital metabolites and enzymes (amylase, cellulase, lipase, and protease) with respect to stress response.	• *Actinomycete, Actinokineospora, Actinopolyspora, Dactylosporangium, Halothermothrix orenii, Marinophilus, Microbacterium, Micrococcus, Microtetraspora, Mycobacterium, Nocardiopsis, Rhodococcus, Saccharopolyspora, Salinispora, Streptomyces,* and *Streptoverticillium.*	[9]
Endophytic	• Protect and guard the host plants against insects and diseases. • Produce secondary metabolites such as alkaloids, polyketides, terpenes, and terpenoids benzopyrones, quinones, peptides and fatty acids derivatives, which are of therapeutic importance.	• *Actinomadura, Actinopolyspora, Brevibacterium, Kibdelosporangium, Nocardioides,* and *Streptomyces.* • *Aeromicrobium, Kitasatospora, Microbispora, Micromonospora, Nocardia caishijiensis,* and *Pseudonocardia, carboxydivorans, Streptomyces,* and *Verrucosispora maris.*	[2, 6]
Symbiotic	• Form nitrogen-reducing (NIR) vesicles in actinorhizal plants, aid in nitrogen fixation, and facilitate early colonization of plants during primary succession. • Inhibit higher plants and cause diseases such as potato scab and responsible for ratoon stunting disease in sugarcane. Infect xylem and responsible for plant wilting in alfalfa, corn, tomato, and potato. Cause leafy gall syndrome in dicotyledonous, herbaceous plants.	• *Frankia* • *Streptomyces scabies* • *Leifsonia xyli.* • *Clavibacter michiganensis.* • *Rhodococcus fascians.*	[2, 10]
Endosymbiotic	• Produce bioactive compounds or plant growth regulators (PGRs) and protect crops from fungal infection. • Associate with marine sponges and serve as a promising source of novel antibiotic leads.	• *Streptomyces griseoviridis.* • *Arthrobacter, Brachybacterium, Brevibacterium, Corynebacterium, Dietzia, Microbacterium, Micrococcus, Micromonospora, Mycobacterium, Nocardiopsis, Rhodococcus, Rubrobacter, Salinispora,* and *Streptomyces.*	[2, 11, 12]
Gut	• Detoxify certain compounds, supply nutrients and vitamins, enhance growth performance, digest complex food sources, and provide protection against pathogenic	• *Rhodococcus rhodnii, Coriobacteriaceae, Bifidobacterium, Streptomyces,* and *Micromonospora.*	[2, 10]

Table 1.
Types of Actinobacteria and their relevant functions.

3. Applications of Actinobacteria

3.1 Applications of Actinobacteria in agriculture

Overuse of agrochemicals has led to significant deterioration in soil fertility and threatens to deprive a major population of essential food sources. This necessitates the need to implement natural methods for sustaining as well as developing our precious agricultural areas. Actinobacteria, a naturally occurring microorganism in the bulk soil or rhizospheric soil has caught the attention of almost all the researchers. Due to their extraordinary properties compared to other microbes, they are beneficial for improving the soil quality, enhancing plant growth, and thereby contributing toward the "Green Revolution" [2].

3.1.1 Actinobacteria as plant growth-promoting Rhizobacteria (PGPR)

Actinobacteria are ubiquitously present in soil with an average count of 5 × 10^{10}–6 × 10^{10} CFU/gm of soil [13]. They are usually found as dormant spores and develop into mycelial forms only in favorable environmental conditions [13]. As the soil depth increases, their population expands but only up to horizon C (regolith) [14]. Some of the important genera of Actinobacteria found in soil are *Streptomyces*, *Nocardia*, *Micromonospora*, *Actinoplanes*, and *Streptosporangium*, wherein *Streptomyces* alone can contribute to nearly 70% of the population [2]. Actinobacteria like other plant growth-promoting rhizobacteria (PGPR) can enhance plant growth either directly or indirectly (**Figure 1**) [14].

3.1.1.1 Role as biofertilizer

The three essential nutrients required by the plants for their proper growth are nitrogen, phosphorus, and potassium (NPK). These requirements are fulfilled by different soil microbes—Actinobacteria being the chief contender. NPK is required by plants for the synthesis of several macromolecules, biosynthesis of ATP, photosynthesis, and other cellular processes.

3.1.1.1.1 Nitrogen fixation

Nitrogen is a highly inert gas and has to be converted into readily bioavailable forms like ammonia, nitrates, or nitrites. This is attained through the process known as nitrogen fixation. Actinobacteria have been recognized to fix atmospheric nitrogen either symbiotically or under free-living conditions (**Table 2**). Two important genes required for this process are *nif* and *nod* genes. The *nif* gene encodes nitrogenase enzyme which is required for nitrogen-fixing (N-fixing) and the *nod* gene encodes Nod factors which are responsible for nodule formation [16]. Chemoattractant signals elicited by hosts lead to sequential events – attachment of bacteria to the root hair of host plants, curling of root hair, formation of infection thread, and bacterial establishment into the nodules [17].

Some endophytic Actinobacteria like *Arthrobacter*, *Agromyces* sp. ORS 1437, *Microbacterium FS-01*, *Mycobacteria*, and *Propionibacteria* can also fix nitrogen [14, 18]. With the advancement of molecular studies, several nifH-containing Actinobacteria (other than *Frankia* sp) as well as non-*Frankia* Actinomycetes like *Gordonibacter pamelaecae*, *Rothia mucilaginosa*, and *Slackia exigua* have been discovered leaving behind questions about diazotrophic origin and emergence

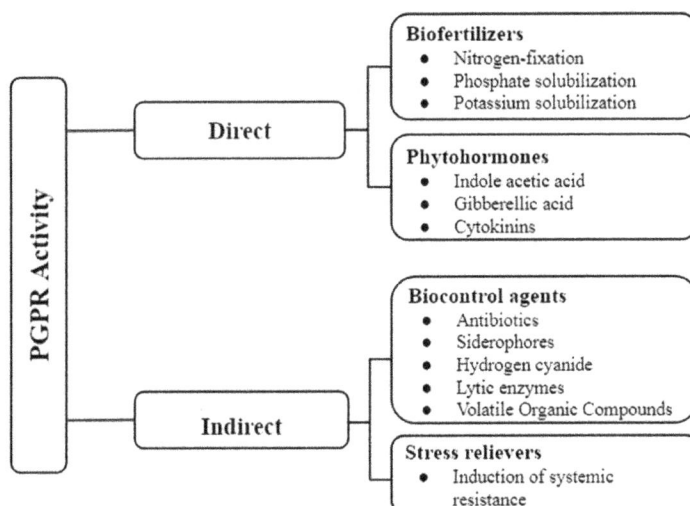

Figure 1.
Flow chart representing PGPR activity of Actinobacteria through direct and indirect methods.

N-fixing Actinobacteria	Association with plants	References
Corynebacterium sp. AN1	Promotes the growth of maize crop by relegating the level of acetylene.	[1]
Pseudonocardia dioxanivorans CB1190	Fixes dinitrogen symbionts. The only source of carbon and energy is 1,4-dioxane.	[1]
Streptomyces sp.	Provides nitrogen source as well as protects the leguminous plants like pea from pathogens.	[1, 15]
Frankia	Fixes nitrogen either under free-living conditions or symbiotically via actinonodules in higher non-leguminous plants like *Alnus*, *Casuarina*, and *Hippophae*.	[1, 14]
Micromonospora	Fixes nitrogen symbiotically via actinonodules in trees and shrubs.	[1]

Table 2.
N-fixing Actinobacteria and associated plants.

among Actinobacteria [1, 15]. Apart from this, Actinobacteria also forms symbiotic association with mycorrhiza by promoting hyphal elongation of symbiotic fungi. An example of such a symbiosis is found on the roots of sorghum and clover associated with *Streptomyces coelicolor* and *Streptomyces* sp. MCR9 and MCR24, respectively [14].

3.1.1.1.2 Phosphate solubilization

Phosphorus (P) is generally present in the soil in insoluble form and hence cannot be taken up by the plants for their nourishment. Not all the P provided by agrochemicals are utilized by plants. Unused soluble forms of P are fixed in the process with the aid of large quantities of cations (Zn^{2+}, Ca^{2+}, Al^{3+}, and Fe^{3+}). This in turn may result in eutrophication and depleted soil fertility [19]. Phosphorus-solubilizing microbes like Actinobacteria is an eco-friendly substitute to this, since they provide soluble P constantly due to their steady degrading activities. Two known mechanisms used by them are as follows:i) they secrete extracellular enzyme phytases which degrade phytate and ii) they create acidic environment near the rhizosphere by releasing various acids such as citric, gluconic, malic, oxalic, propionic, and succinic acids which solubilize the insoluble forms. Characteristic examples include *Arthrobacter*, *Gordonia*, *Kitasatospora*, *Kocuria kristinae* IARI-HHS2–64, *Micrococcus*, *Micromonospora* sp., *Micromonospora aurantiaca*, *Micromonospora endolithica*, *Rhodococcus*, *Streptomyces* sp., *Streptomyces griseus*, and *Thermobifida* [14, 15, 18, 20].

3.1.1.1.3 Potassium solubilization

Just like phosphorus, potassium (K) is also present in insoluble form in the soil and can be solubilized with the help of potassium-solubilizing microbes like Actinobacteria. The mechanisms implemented by them areas follows: (i) exchange and complexation reaction; (ii) production of organic acid which is subsequently followed by acidolysis; and (iii) chelation. Characteristic examples include *Arthrobacter* sp., *Microbacterium* FS-01, *Streptomyces* sp. KNC-2, and *Streptomyces* sp. TNC-1 [15, 18].

3.1.1.2 Production of phytohormones

Phytohormones like auxins (indole-acetic acid, IAA), gibberellins (GA3), and cytokinins are responsible for increasing the branching of root hair and widening the surface area, allowing the plants to take up more nutrients for their growth. Several Actinobacteria are responsible for the production of such phytohormones which have been listed in **Table 3**.

3.1.1.3 Role as biocontrol agents (BCAs)

Biological control simply means suppression of plant pathogens by other living organisms and controlling a variety of diseases. The microbial biocontrol agents (MBCAs) are target-specific with minimal impact on the rest of the plant population. They can sustain their effect for a longer duration and promote plant growth in an eco-friendly manner. MBCAs like Actinobacteria produce multifarious substances such as antibiotics, siderophores, hydrolytic enzymes, hydrogen cyanide (HCN), and other volatile organic compounds (VOCs) and guard the plants from the attacking phytopathogens via antagonistic effect [2, 22, 23].

3.1.1.3.1 Production of antibiotics

Streptothricin became the first antibiotic obtained from *Streptomyces* in the year 1942, and in 1944, Streptomycin was discovered. Since then, this microbe has been exploited for the discovery of many novel antibiotics [20]. Today *Streptomyces* alone contribute to two-third of the world's antibiotic production due to its extra-large DNA complement [2]. Antibiosis is enabled by the production of several groups of antibiotics ranging from aminoglycosides (streptomycin and kanamycin), ansamycins (rifampin), anthracyclines (doxorubicin), β-lactams (cephalosporins), macrolides (erythromycin and oleandomycin), and polyene (nystatin and levorin) to tetracycline [2, 15]. Some of them have been listed in **Table 4**.

Some Actinobacteria can produce a combination of antibiotics. For example, *Streptomyces violaceusniger* YCED9 produces three antifungal compounds—nigrecine, geldanamycin, and guanidyl fingine to keep a stringent

Actinobacteria	Phytohormones	Role
Nocardiopsis, Streptomyces atrovirens, S. olivaceoviridis, S. rimosus, S. rochei, and *S. viridis.*	IAA	Enhances the germination of seed, elongation of root as well as growth.
S. atroolivaceus	IAA	Promotes cell differentiation, hyphal elongation, and sporulation.
S. purpurascens NBRC 13077	IAA (low level)	Regulates the expression of biosynthetic gene, rhodomycin.
S. hygroscopicus TP_A045	Pteridic acids A and B (metabolites that exhibit auxin- like activity)	Promotes root elongation in common beans.
Actinomyces sp., *Arthrobacter, Micrococcus, Nocardia* sp., and *Streptomyces* sp.	GA3	Extends the tissues of stem leading to alteration in plant morphology and raises the height of the plant, overall biomass, and grain in common beans.
Arthrobacter, Frankia sp., *Leifsonia soli, Rhodococcus fascians,* and *S. turgidiscabies*	Cytokinins	Promotes cell division and enlargement and transfers signals from roots to shoots under environmental stresses, for example, in soybean.

Table 3.
Actinobacteria-producing phytohormones [14, 21].

check on the attacking pathogen [1]. Other antibiotic producers belonging to Actinobacteria are *Actinoplanes* (purpuromycins), *Microbispora* (microbiaeratin), *Micromonospora* (clostomicins), *Nocardia* (nocathiacins), and *Nocardiopsis* (thiopeptide antibiotic) [21].

3.1.1.3.2 Production of siderophores

Iron (Fe) is present in their insoluble forms, hydroxides, and oxyhydroxides in the soil which is unavailable to both the plants and the microbes. In order to cope with Fe deficiency, microbes started producing small-molecular-weight compounds called siderophores which are a specific carrier of ferric ions (Fe^{3+}). In addition to fulfilling the nutrient requirement for plant growth, the siderophores also act as BCA. They sequester (chelate) iron, form complexes with iron in a 1:1 ratio, create a competitive surrounding for pathogenic microorganisms, and remove the low-affinity siderophores of the pathogens. The process involves conversion of Fe^{3+} ions (insoluble form) to ferrous (Fe^{2+}) ions (soluble form) with the assistance of esterase enzymes. The Fe^{2+} ions are then released into the cells with the help of ATPase activity/proton motive force (PMF). For instance, *Streptomyces* protect against *Fusarium oxysporum* f. sp. *ciceri* under wilt sick field conditions on chickpea [14, 21]. *Streptomyces* sp. CMU.MH021 produces hydroxamate siderophores as well as IAA and slows down the hatching rate of eggs of nematode pathogens like *Meloidogyne incognita* [15]. Heterobactin siderophore of *Rhodococcus* and *Nocardia;* coelichelin and coelibactin peptide siderophores of *Streptomyces coelicolor*; enterobactin of *Streptomyces tendae*; oxachelin of *Streptomyces* sp. GW9/1258; erythrobactin, a hydroxamate-type siderophore of *Saccharopolyspora erythraea* SGT2; nocardamine, a cyclic siderophore of *Citricoccus* sp. KMM3890; desferrioxamine (DFO) B and E of *Salinispora*; tsukubachelin, a siderophore of *Streptomyces* sp. TM-34; foroxymithine of *Streptomyces* sp.; and amychelin, an uncommon mixed-ligand siderophore of *Amycolatopsis* sp. AA4 that modifies the developmental processes of *Streptomycetes*

Antibiotics	Actinobacteria	Role	References
Antibacterial and antifungal elements	*Streptomyces* sp.	Inhibits the growth of *Rhizoctonia solani*, a fungal pathogen of tomato.	[15]
Antifungal metabolite polyoxin B	*S. cacaoi* var. *asoensis*	Inhibits fungal pathogens in fruit, vegetables, and ornamental plants by interfering with fungal cell wall formation and inhibition of chitin synthase enzyme.	[20]
Antifungal metabolite polyoxin D	*S. cacaoi* var. *asoensis*	Inhibits rice sheath blight caused by *R. solani*.	[20]
Avermectins (a class of macrocyclic lactones)	*S. avermitilis*	Protects the host plant from nematode pathogens like *Meloidogyne incognita* and *Caenorhabditis elegans*.	[15]
Germicidin and hypnosin	*S. alboniger*	Inhibits spore germination.	[21]
Geldanamycin and elixophyllin	*S. hygroscopicus*	Suppresses *Rhizoctonia* root rot of pea.	[21]
Kasugamycin	*S. kasugaensis*	Exhibits antagonistic effect against fungal pathogen *Magnaporthe oryzae* and inhibits rice blast.	[24]
Mildiomycin	*Streptoverticillium rimofaciens*	Inhibits powdery mildews on various crops.	[5]
Naphthoquinone	*Streptomyces* sp.	Protects the host plant *Alnus glutinosa* from many bacterial and fungal pathogens.	[15]
Polyketides	*Streptomyces* sp. AP-123	Exhibits toxic effects on *Helicoverpa armigera* and *Spodoptera litura* larvae.	[20]
Polyene-like compounds related to guanidyl-containing macrocyclic lactones	*Streptomyces* sp.	Exhibits anti-*Fusarium* activity (AFA) against *Fusarium oxysporum*.	[21]
Streptomycin	*Streptomyces* sp.	Inhibits fire blight of pear caused by *Erwinia amylovora* (a pome fruit pathogen).	[25]

Table 4.
Antibiotic-producing Streptomyces along with their inhibitory role.

surrounding them are some of the few examples of siderophores produced by Actinobacteria [14, 21, 26].

3.1.1.3.3 Production of hydrogen cyanide (HCN)

Hydrogen cyanide (HCN) acts as another BCA and inhibits the phytopathogens by hampering the respiratory electron transport chain system. Moreover, the production of HCN also boosts up other mineral solubilization like phosphorus, improving the quality of the soil and hence crop production. *Arthrobacter* and *Streptomyces* are capable of producing HCN. *Streptomyces* sp. from roots of *Solanum nigrum* inhibit fungal disease—root rot and damping-off of tomato caused by *Fusarium oxysporum* f. sp. *radicis lycopersici* [15, 27].

3.1.1.3.4 Production of lytic enzymes

The cell walls of most of the phytopathogens are composed of chitin, glucan, cellulose, hemicellulose, lignins, pectins, proteins, keratins, xylans, dextrans, and lipids. The soil microbes can target the cell wall through the specific enzymes produced by them and thus inhibit the growth of these pathogens. Several enzymes produced by Actinobacteria are amylases, cellulases, chitinases, dextranases, glucanases, hemicellulases, keratinases, ligninases, lipases, nucleases, pectinases, peptidases, peroxidases, proteinases, and xylanases [14]. Some of them have been listed in **Table 5**.

The extracellular enzymes show an enhanced effect when used synergistically with the antibiotics. For example, antibiotics along with enzyme chitinase produced by *S. lydicus* WYEC108 works synergistically against pathogen *Pythium ultimum* which is responsible for causing fungal root and seed diseases [20].

3.1.1.3.5 Production of volatile organic compounds (VOCs)

Actinomycetes are known to produce geosmin. These volatile organic compounds result in the characteristic odor of the soil and at times also translate into an earthly taste of potable water. Besides imparting odor and taste, these actinomycetes-derived VOCs are also known to have biocontrol attributes [5]. The very ability to diffuse comfortably through soil particles and damage pathogens makes it a potent and sustainable alternative for agrochemicals. For instance, germination of *Botrytis cinerea* and *Penicillium chrysogenum* spores are inhibited by *Streptomyces coelicolor*. Moreover, VOCs from *S. globisporus* and *S. philanthi* have shown activity against *Botrytis cinerea* and *Fusarium moniliforme*, respectively. Pathogen-causing downy blight in litchi—*Peronophythora litchii*, can also be actively targeted by VOCs from *S. fimicarius* [15]. Another VOC, methyl vinyl ketone from *S. griseoruber* has been reported to inhibit *Cladosporium cladosporioides* spore germination [20].

3.1.1.4 Role as stress reliever

It is the genetic makeup of the plant which decides the productivity and their ability to adapt resistance against various abiotic stresses and phytopathogens [15]. Plants have adapted certain mechanisms like the induced systemic resistance (ISR) and systemic acquired resistance (SAR). Upon arrival of stressful conditions, plants start synthesizing elevated levels of stress-responsive hormone—ethylene (ET) that causes premature death of plants. 1-aminocyclopropane-1-carboxylate (ACC) is the precursor of ethylene hormone. Actinobacteria have the capability to survive in different types of biotic and abiotic stress factors, such as drought, extreme temperatures, floods, and salinity, but the plants might get affected, resulting in the low production of crops [14]. To enhance the plant growth, tolerant strains like Actinobacteria are inoculated. *Amycolatopsis*, *Mycobacterium*, *Nocardia*, *Rhodococcus*, and *Streptomyces* produce a specific enzyme 1-aminocyclopropane-1-carboxylate (ACC) deaminase to target ACC and convert it into ammonia and α-ketobutyrate. Some of the strategies adopted by *Streptomyces padanus* for drought tolerance involve accumulation of callose, cell wall lignification, and stimulation of high levels of osmotic pressure of plant cells [14].

For instance, under the onset of saline conditions, *Streptomyces* sp. enhances the growth of maize and wheat. It has been found that under *in vitro* conditions of high concentration of NaCl, *Arabidopsis* seedlings showed enhanced growth of biomass and lateral roots when inoculated with *Streptomyces* sp. [14]. It has also

Enzymes	Actinobacteria	Comments	References
L-asparaginase	*Nocardia* sp, *Streptomyces karnatakensis, S. albidoflavus,* and *S. griseus*	—	[2]
β-1,3; β-1,4; and β-1,6glucanases	*Actinoplanes philippinensis, A. campanulatus, Microbispora rosea, Micromonospora chalcea, Streptomyces griseoloalbus,* and S. spiralis.	Inhibit *Pythium aphanidermatum* and *Phytophthora fragariae,* causal agent of damping-off disease in seedlings of cucumber (*Cucumis sativus* L.) and raspberry.	[15, 20]
Proteases	*Streptomyces* sp. strain A6	Manages anthracnose on tomato fruits and inhibits diseases associated with *Fusarium udum*.	[20]
Keratinase	*Nocardiopsis* sp. SD5	Degrades the poultry chicken feather.	[2]
Chitinase and glucanase	*S. cavourensis* SY224	Manages anthracnose in pepper.	[20]
Chitinase	*S. plymuthica* C48 *S. violaceusniger* XL-2	• Inhibits spore germination of *Botrytis cinerea*. • Suppresses wood-rotting fungi.	[20]
Chitinases, glucanases, cellulases, lipases, and proteases	*Streptomyces* sp. 9P	• Inhibits *Alternaria brassicae* infecting plants of *Brassica* species. • Inhibits *Colletotrichum gloeo-sporioides*, infecting perennial plants. • Inhibits *Rhizoctonia solani*, a phytopathogen with a wide host range. • Inhibits *Phytophthora capsici*, infecting commercial crops like peppers.	[15]

Table 5.
Lytic enzymes produced by Actinobacteria and their inhibitory effect.

been revealed that *Streptomyces* sp. produces the enzyme ACC deaminase which in turn resulted in an increase in the level of calcium and potassium and allows the plant *Oryza sativa* to survive under the saline conditions. In addition, siderophore production and other PGP traits enable them to resist heavy metal toxicity [15]. *S. coelicolor* and *S. olivaceus* are examples of drought-tolerant species and have a tremendous plant growth-enhancing capacity. *Citricoccus zhacaiensis* promotes germination rate and plant growth as well as produces different enzymes and hormones like phosphate-solubilizing enzymes, ACC deaminase, IAA, and GA3 to cope up with the high osmotic pressure conditions [15].

3.2 Applications of Actinobacteria in nanotechnology

Nanotechnology research is among the most rapidly developing scientific and technological fields [28]. It is a transdisciplinary field which has an impact in the domains of agriculture, medicine, and industry [29]. Nanotechnology allows us to produce nanoparticles with specific properties for use in a wide range of applications [30]. Integration of nanotechnology with biotechnology has evolved as a new biosynthetic and environment-friendly approach for the production of nanomaterials [31]. Nanoparticles have received a lot of attention recently because of their unique qualities, and they are being employed in a lot of different fields like pharmaceuticals, nanoengineering, drug delivery, nanoantibiotics, catalysis, electronics, sensor creation, and other areas [30, 32]. There are two techniques which are used for the synthesis of nanoparticles: (1) the top-down technique, which involves breaking down bulk materials into nanosized materials and (2) the bottom-up technique, which involves assembling the atoms and molecules into molecular structures in the nanoscale range [33, 34]. The top-down technique is quite expensive, and it also produces exceedingly poisonous substances as by-products and consumes a lot of energy. As a result, a biological, ecological-friendly strategy for pollution-free, nontoxic, biodegradable synthesis of technologically relevant nanomaterials becomes critical [34].

3.2.1 Biological synthesis of nanoparticles by Actinobacteria

Synthesis of nanoparticles using a biological system is a rapid, efficient, economical, nontoxic, and environmental-friendly method. Many researchers have investigated the production of desired nanoparticles using Actinobacteria, bacteria, microalgae, yeast, viruses, and fungi [30, 35]. The use of microorganisms, enzymes, and plant extracts to produce nanoparticles has also been proposed as a feasible biological technique [36]. Microorganisms such as Actinobacteria are capable of producing nanoparticles which are widely used as novel therapeutics such as antimicrobial, anticancer agents, anti-biofouling agents, antifungals, and antiparasitic (**Figure 2**) [37]. Inorganic compounds are produced by Actinobacteria either intracellularly or extracellularly, and they are often nanoscale in size and morphology. Most harmful heavy metals are resistant to Actinobacteria due to chemical detoxification as well as energy-dependent ion efflux from the cell through membrane proteins that operate as ATPase, chemiosmotic cation, or proton anti-transporters [38, 39]. The main principle behind the synthesis of nanoparticles is that actinobacterial enzymes reduce metal ions to stable nanoparticles when provided with metal ions as substrates. For example, the synthesis of silver nanoparticles (AgNPs) usually uses silver nitrate solution ($AgNO_3$) as a substrate for the secreting enzymes, and the substrates used for the production of gold nanoparticles (AuNPs) are chloroauric acid solutions ($AuCl_4$). Nanoparticles can also be produced with other metals like zinc, copper, and manganese [34]. Actinobacterial detoxification can occur via extracellular biosorption, precipitation biomineralization, or complexation, or through intracellular bioaccumulation [28]. Studies of the *Arthrobacter* and *Streptomyces* genera as potential nanofactories have been conducted in an effort to discover safe and clean techniques for synthesizing gold and silver nanoparticles [2]. There was a wide variety of silver nanoparticle synthesizing Actinobacteria, found in the marine environment, with 25 isolates out of 49 generating silver nanoparticles. The genera of bacteria synthesizing silver nanoparticles are *Actinopolyspora* sp., *Kibdelosporangium* sp., *Nocardiopsis* sp., *Saccharopolyspora* sp., *Streptomyces* sp., *Thermoactinomyces* sp., and *Thermomonospora* sp. [2].

3.2.1.1 Intracellular synthesis of actinobacterial nanoparticles

Additional processing procedures, such as ultrasonic treatment or interaction with appropriate detergents, are necessary to liberate the intracellularly produced nanoparticles. This can be used to recover valuable metals from mine wastes and metal leachates. Metal nanoparticles that have been biomatrixed could be employed as catalysts in a variety of chemical processes [28, 40]. Gold nanoparticles

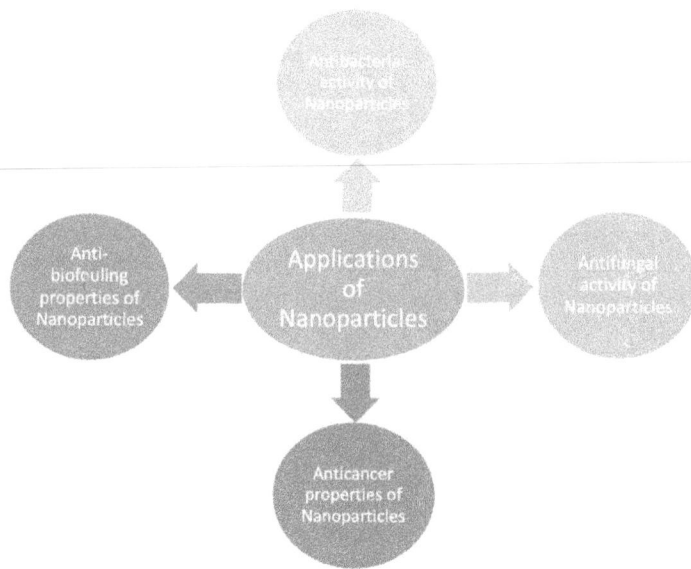

Figure 2.
Applications of biologically synthesized nanoparticle.

synthesized by alkalotolerant *Actinomycetes*, *Rhodococcus* sp., were characterized for the first time by Ahmad et al. [41]. The reduction of zinc sulfate ($ZnSO_4$) and manganese sulfate ($MnSO_4$) using *Streptomyces* sp. HBUM171191 proved to be a suitable intracellular method of producing zinc and manganese nanoparticles (10–20 nm) [34]. The intracellular synthesis of silver nanoparticles (AgNPs) from *Aspergillus fumigatus* and *Streptomyces* sp. was compared by Alani et al. The change in color from colorless to a light brownish to dark brownish was used by them to identify nanoparticle production [42]. *Streptomyces* sp. (strains: D10, ANS2, HM10 and MSU) isolated from the Himalayan Mountain ranges were capable of producing spherical and rod-shaped intracellular gold nanoparticles (AuNPs) while also exhibiting antibacterial activity [43].

3.2.1.2 Extracellular synthesis of actinobacterial nanoparticles

The ability of Actinobacteria to synthesize extracellular metal nanoparticles is dependent on the location of reductive elements within the cell. It entails the use of soluble secretory enzymes or cell wall reductive enzymes that can recognize metal ions and reduce them to nanoparticles [44]. A study focused on different actinobacterial strains for gold and AgNP production with diverse morphologies and size distributions. They discovered that when an alkali thermophilic *Thermomonospora* sp. is exposed to gold chloride, it produces spherical AuNPs with a limited size distribution with a diameter of 8 nm [45]. Extracellular AgNPs with a diameter of 68.33 nm were generated by a soil isolate, *Streptomyces* sp. JAR, and showed antibacterial efficacy against a wide range of fungal and bacterial diseases [46]. The antidermatophytic characteristics of biologically synthesized, cubical shaped AuNPs (90 nm size) obtained from the culture extract of *Streptomyces* sp. VITDDK3 were documented, as well as their antifungal activities against *Microphyton gypseum* and *Trichophyton rubrum* [47]. Other extracellular producers of AgNPs have been identified as *Rhodococcus* sp., a metabolically flexible Actinobacteria, and *Streptomyces glaucus* 71MD, a novel actinobacterial strain [48, 49].

3.2.2 Antibacterial activity of nanoparticles

Nanoparticles produced using a variety of technologies have been applied in a variety of *in vitro* diagnostic procedures [50, 51]. The antibacterial activity of gold and silver nanoparticles against human and animal diseases has been widely reported [28, 43, 47]. For achieving synergistic effects with biomolecules, the antibacterial mechanism of developed nanoparticles is crucial. Actinobacteria, primarily the species *Streptomyces* and *Micromonospora*, are known to be the source of about 80% of the world's antibiotics [2]. *Streptomyces* are the primary producers of antibiotics in the pharmaceutical industry since they produce around 7600 compounds, many of which are secondary metabolic products that are potent antibiotics [52, 53].

The production of reactive oxygen species (ROS) by metal nanoparticles is the most common mechanism of cellular toxicity [54]. Gold and silver nanoparticles have antibacterial capabilities due to their slow oxidation and release of Ag^+ and Au^{3+} ions into the environment, making them suitable biocidal agents [34]. Nanoparticles having a large surface area possess high antibacterial capabilities, allowing them to make the most contact with the environment possible [55]. By disrupting cellular permeability and respiration, metal nanoparticles have proved to have good antibacterial capabilities. The positively charged metal ions breach the bacterial cell wall by adhering to and breaking the negatively charged bacterial cell wall, causing protein denaturation, DNA replication interference, and eventually the organism's death [56, 57]. Silver nanoparticles induce cell death by breaking the plasma membrane or inhibiting respiration by converting the cell wall oxygen and sulfhydryl (–SH) groups to RS-SR groups [58, 59]. *Streptomyces viridogens*-derived gold nanoparticles (AuNPs) exhibited remarkable antibacterial efficacy against *Staphylococcus aureus* and *Escherichia coli* [43]. The potential antibacterial impact of silver nanoparticles synthesized from *Streptomyces albidoflavus* using an environmentally benign approach was revealed against several gram-positive and gram-negative species. *Streptomyces* sp. [60]. AgNPs were found to be active against the anti-extensive spectrum beta-lactamase-producing strain Klebsiella pneumoniae (ATCC 700603), as well as other therapeutically relevant pathogens like *E. coli* and *Citrobacter* species [34]. Silver nanoparticles (AgNPs) from a new *Streptomyces* sp. BDUKAS10 strain also showed improved bactericidal action against some bacteria [61]. Some food microbe pathogens, such as *Bacillus cereus*, *E. coli*, and *S. aureus*, were eliminated using AgNPs from *Streptomyces albogriseolus* [34, 62].

3.2.3 Antifungal activity of nanoparticles

In recent years, fungal infections have become increasingly widespread, and silver nanoparticles have evolved as prospective antifungal medicines. Due to cancer chemotherapy or human immunodeficiency virus infections, fungal infections are more typically encountered in immune-deficient patients [28, 63]. Gold nanoparticles (AuNPs) produced utilizing a sustainable technique with *Streptomyces* sp. VITDDK3 have good antifungal action against *Microsporum gypseum* and *Trichophyton rubrum* by causing membrane potential to fluctuate and by inhibiting the ATP synthase activity, which causes a general decline in the metabolic activities. The vulnerability of the pathogen's cell wall and the toxicity of metallic gold could explain the antidermatophytic activity of the produced AuNPs [47]. *Fusarium* sp. and *Aspergillus terreus* JAS1 were suppressed by biologically produced silver nanoparticles made with *Streptomyces* sp. JAR1 [46]. Silver nanoparticles produced from *Streptomyces* sp. VITBT7 showed inhibitory action against *Aspergillus niger* and *Aspergillus fumigatus* (MTCC 3002), whereas silver nanoparticles produced from *Streptomyces* sp. VITPK1 demonstrated promising antifungal activity against *Candida krusei*, *Candida tropicalis*, and *Candida albicans* [30, 64].

3.2.4 Anticancer properties of nanoparticles

Cancer is one of the most common causes of death, accounting for one out of every six fatalities in 2018. However, 70% of cancer deaths occur in middle- and low-income nations [65]. The most frequent cancer treatment and management methods include surgery, chemotherapy, hormone therapy, and radiation therapy. However, in recent times, nanotechnology-based therapeutic and diagnostic techniques have demonstrated potential for improving cancer treatment [66]. The production of nanoparticles was reported by utilizing a novel Nocardiopsis sp. MBRC-1 isolated from marine sediment samples off the coast of Busan, South Korea [67]. In vitro cytotoxicity of the biosynthesized AgNPs against the human cervical cancer cell line (HeLa) was observed, along with high antimicrobial activity against bacteria and fungi [67]. Silver nanoparticles synthesized with *S. naganishii* (MA7) from the Salem area of Tamil Nadu, India, were also found to have cytotoxic properties against HeLa cancer cell lines [68].

3.2.5 Anti-biofouling properties of nanoparticles

Anti-biofouling is a method of removing biofouling, which occurs when bacteria cluster on wetted surfaces, forming biofilms and emitting a foul odor. In industries such as medicine, treatment plants, sensor sensitivity, and transportation, biofilms cause operational issues. Biofilm accumulations can be efficiently prevented or eliminated by utilizing the anti-biofouling characteristics of biosynthesized nanoparticles [69]. According to Shanmugasundaram et al., *Streptomyces naganishii* MA7 biosynthesized spherical, 5–50-nm-sized silver nanoparticles that were efficient against 10 different biofouling microorganisms *in vitro* [34, 68].

3.3 Application of Actinobacteria in bioremediation

Heavy metals are natural components of soil, and they work as cofactors in a variety of enzymes. Heavy metal pollution of the biosphere has increased as a result of industrial evolution, which often becomes hazardous at high concentrations. The discharge of heavy metals from the electroplating industry is one of the most significant sources of heavy metal toxicity around the globe [70]. Heavy metals such as copper, mercury, chromium, lead, zinc, and cadmium are commonly found in the effluents/wastewater generated by the industry. Continuous exposure to heavy metals has been linked to infant growth retardation, the onset of numerous cancers, and liver and kidney damage. Bioremediation is an efficient and sustainable process of reverting a contaminated environment to its original state using microbes or their enzymes [71]. In soils, *Actinomycetes* comprise a significant microbial population, and they are also extensively distributed in nature [72]. Heavy metal tolerance, as well as metabolic diversity and unique growth properties of *Actinomycetes*, such as mycelium development and relatively quick colonization of selected substrates, vindicate their capabilities as excellent bioremediation agents [73].

3.3.1 Bioremediation of toxic heavy metals

3.3.1.1 Copper bioremediation

Copper (Cu) is a vital heavy metal with numerous functions in biological systems, such as cellular respiration, pigment formation, connective tissue growth, and neurotransmitter generation [74]. Copper becomes hazardous at high concentrations [75], causing behavioral and mental problems, renal damage, sickle cell anemia, dermatitis, schizophrenia, and nervous system disorders like Parkinson's and Alzheimer's [76]. Copper has been widely distributed in soil, silt, trash, and wastewater as a result of industrial use and discharge, posing significant environmental concerns. *Streptomyces* AB5A [77], *Amycolatopsis* [78], and *Kineococcus radiotolerans* [79] are some of the Actinobacteria involved in copper bioremediation. Extracellular cupric reductase activity was found in *Streptomyces* sp. AB2A. In both copper-adapted and non-adapted cells, *Amycolatopsis tucumanensis* DSM 45259 displayed effective cupric reductase activity. The copper-specific biosorption capacity of *A. tucumanensis* DSM 45259 was validated by subcellular fractionation experiments, which revealed that the retained copper was connected with the extracellular fraction (exopolymer, 40%), but mostly within the cells [80]. *Streptomyces* sp. WW1 identified from the wastewater treatment plant in Saudi Arabia has been found to successfully remove copper.

3.3.1.2 Chromium bioremediation

Chromium is most commonly found as chromite ($FeCr_2O_4$) in nature. Cr (VI), an oxidized form of chromium, is potentially poisonous, induces allergic dermatitis, and has carcinogenic, mutagenic, and teratogenic effects on biological organisms [81]. Trivalent chromium is 100 times more hazardous and 1000 times more mutagenic than hexavalent chromium compounds [73]. Das and Chandra [82] were the first to document the reduction of Cr (VI) by *Streptomyces*, while Laxman and More [83] were the first to report the reduction of Cr (VI) by *Streptomyces griseus* [82, 83]. *Microbacterium*, *Arthrobacter*, and *Streptomyces* have all been found to reduce Cr (VI) [83, 84]. The reduction of Cr (VI) by *Streptomyces* sp. MC1 bioemulsifiers were utilized as a washing agent to improve soil-bound metal desorption [85]. When glycerol and urea were employed as sources of carbon and nitrogen, *A. tucumanensis* DSM 45259 generated an emulsifier. Under harsh conditions of pH, temperature, and salt content, the bioemulsifiers demonstrated remarkable levels of stability. For the remediation of hexavalent chromium compounds, microbial emulsifiers based on remediation technologies appear to be more promising. *Arthrobacter* and *Amycolatopsis* are two actinobacterial genera active in chromium bioremediation [86, 87]. Other species involved in chromium bioremediation are *Halomonas* sp. [88], *Flexivirga alba* [89], *Friedmanniella antarctica* [90], and *Intrasporangium chromatireducens* [91].

3.3.1.3 Mercury bioremediation

Mercury is a highly hazardous heavy metal that has been associated with kidney damage and cardiovascular problems. Mercury pollution is mostly caused by discharges from refineries and industries, as well as human activities such as the burning of coal and petroleum, the use of mercurial fungicides in agriculture, and the use of mercury as a catalyst in industry [92]. Mercury resistance has been demonstrated in two *Actinomycete* strains, CHR3 and CHR28, obtained from metal-contaminated areas in Baltimore's Inner Harbor, USA [93]. The biomass of *Streptomyces* VITSVK9 was employed for mercury biosorption, and it showed a high metal tolerance capacity [94]. TY046-017, a *Streptomyces* isolated from tin tailings, also demonstrated possible tolerance to mercury.

3.3.1.4 Lead bioremediation

Lead is a neurotoxic substance that can build up in both soft and hard tissues, causing neurological problems and affecting physical development. Corrosion of household plumbing, brass and bronze fittings, and lead-based solders are prominent sources of these contaminants [95]. Metal tolerance was shown in *Streptomyces* VITSVK9 biomass and *Streptomyces* sp. BN3 which was discovered in Moroccan mine waste [96]. The biosorption of heavy metal Pb (II) by *Streptomyces* VITSVK5 spp. biomass was concentration- and pH-dependent. Heavy metal tolerance and lead buildup were observed in *Streptomyces* isolated from abandoned Moroccan mines [94, 96].

3.3.1.5 Zinc bioremediation

The free zinc ion in solution is extremely harmful to bacteria, plants, invertebrates, and even vertebrate fish, although it is less dangerous to humans. In humans, zinc toxicity is caused by zinc overload and hazardous overexposure [70]. Three strains of *Streptomyces* NGP (JX843532), *Streptomyces albogriseolus* (JX843531), and *Streptomyces variabilis* (JX43530) were recovered from a coastal marine soil sample in Tamil Nadu, India, and showed high levels of zinc biosorption. Strain WW1 of *Streptomyces* sp. obtained from the wastewater treatment plant in Saudi Arabia exhibited biosorption of zinc.

3.3.2 Bioremediation of pesticides

Agricultural production is one of the greatest and also most important economic activities on earth; thus, protecting it against pest infestations is a must. Agricultural runoff contaminates aquatic habitats with numerous residues of pollutants such as insecticides. Pesticides and fertilizers pollute local water bodies, causing detrimental effects in humans through food and drinking water. Pesticide residues have been found in groundwater and drinking water in India and around the world, according to several researchers [71, 97]. Yadav et al. [98] published a comprehensive review that found long-lasting pesticides in multi-component settings. Pesticides such as dichlorodiphenyltrichloroethane (DDT), endosulfan, hexachlorocyclohexane (HCH), and parathion methyl were detected in freshwater bodies, and several of them were classified as persistent organic pollutants (POPs). Since soil provides varied binding sites for these hydrophobic contaminants, the preservation of HCH isomers in various soil types inhibits the breakdown process [98]. The solubility of pesticides in water, their adsorption by soil particles, and their persistence all play a role in their mobility in soil compartments. Since it provides a multitude of binding sites for organic contaminants, especially hydrophobic chemicals, organic matter concentration is a characteristic that defines pesticide retention in soil and sediments [99].

A class of synthetic organic compounds known as organochlorine pesticides (POs) is composed of chlorine-containing hydrocarbons that have had one or more hydrogen atoms exchanged for chlorine atoms. These compounds may also contain other elements like oxygen or sulfur. Due to their toxicity, prolonged persistence, low biodegradability, widespread availability in the environment, and long-term consequences on wildlife and humans, many insecticides have been phased out of usage. Furthermore, their physicochemical qualities combine to allow them to traverse great distances [71].

3.3.2.1 Harmful effects of pesticides on human health

Acute intoxication is caused by high dosage of organophosphorus (OP) insecticides. Gastrointestinal discomfort, perspiration, muscle spasms, muscle weakness, bronchospasm, high blood pressure, bradycardia, central nervous system depression, and coma are all indications of this type of poisoning. People who have been exposed to OP for a long time develop a pesticide-related sickness, which can include headaches, dizziness, abdominal discomfort, nausea, blurred vision, vomiting, and chest tightness [100]. Insecticides containing carbamates have been implicated with the development of respiratory illnesses. Organochlorine (OC)-based insecticides pose the most serious and dangerous harm. They are difficult to degrade because they are lipid-soluble and have a high rate of persistence. Infertility, genital malignancies, tumors of the reproductive organs, neurotoxicity, and immunotoxicity are among the harmful effects of these chemicals [101].

3.3.2.2 Degradation of pesticides by Actinobacteria

Potential candidates for the degradation of resistant inorganic and organic contaminants are Actinobacteria. The most common pesticide-degrading Actinobacteria are *Arthrobacter*, *Streptomyces*, *Janibacter*, *Kokuria*, *Rhodococcus*, *Mycobacterium*, *Nocardia*, *Frankia*, *Pseudonocardia*, and *Mycobacterium* (**Table 6**). These bacteria are capable of growing and degrading a variety of pesticide chemical families, including carbamate (CB), organophosphorus (OP), organochlorine (OC), ureas, pyrethroids, and chloroacetanilide, among others [71]. Since members of the *Arthrobacter* genus exhibit diverse catabolic pathways for the detoxification of these substances, the majority of which are plasmid-encoded, the genus has been recognized as a degrader of several xenobiotics. Because of their dietary flexibility and resistance to environmental stress, this genus of microorganisms is found all over the world. *Arthrobacter* sp. AK-YN10 has been found to use plasmid-encoded information to degrade atrazine to cyanuric acid [105]. Endosulfan, which is based on organochlorines, is detoxified to endosulfan sulfate, which is then eliminated metabolically [102]. Atrazine is an effective nitrogen and carbon source for *Rhodococcus* sp. BCH2 [106].

Streptomyces aureus HP-S 01 was found to detoxify deltamethrin to 2-hydroxy-4-methoxy benzophenone and several other pyrethroids [104]. *Streptomyces* sp. M-7 has been discovered to have multi-pesticide resistance and can detoxify a variety of organochlorine pesticides such as aldrin, DDT, chlordecone (CLD), heptachlor, and dieldrin [107]. In soil microcosm assays, it may remove up to 78% gamma-HCH. *Streptomyces* sp. AC1-6 and ISP-4 can remove diazinon by up to 90%. Immobilized cells have various advantages over free suspended cells, including increased microbe retention in the reactor, improved cellular viability, and cell toxicity prevention, among other things [108]. Microbial and enzyme immobilization-mediated bioremediation procedures are more efficient, with a higher biodegradation rate [109]. Four distinct matrices were used to immobilize *Streptomyces* strains, either as pure cultures or as part of a consortium (polyvinyl alcohol, cloth sachets, silicone tubes, and agar). Immobilized microorganisms removed considerably more lindane than free cells. Additionally, the cells might be reused twice more before being discarded, lowering the overall cost of the biotechnology process [110].

Microorganism	Pesticide	Isolation sample	References
Arthrobacter sp.	Organochlorines (α-endosulfan)	Soil from different agricultural fields, contaminated by pesticides, India	[102]
Arthrobacter sp. BS1, BS2 and SED1	Urea (diuron)	Soil from the interface between a vineyard and the Morcille river, France	[103]
Streptomyces aureus HP-S-01	Pyrethroid (deltamethrin)	Activated sludge samples from an aerobic pyrethroid-manufacturing wastewater	[104]

Table 6.
General characteristics of main genera of pesticide-degrading Actinobacteria.

3.3.3 Bioremediation of petroleum refinery effluent

Petroleum is a heterogenous mixture of hydrocarbons and resins which contains toluene, benzopyrene, benzene, and naphthalene. The majority of them are stable and poisonous and can cause cancer [111]. Bacteria and Actinobacteria (**Table 7**) are both excellent options for microbial oil recovery. Natural attenuation processes and biodegradation are being used to bioremediate petroleum-contaminated soil. For petroleum refinery effluent, bioaugmentation and compositing are effective remediation strategies [121]. However, because of the negative effects of the environment on microbial life, such as disintegration of cell membranes, denaturation of enzymes, poor solubility of oxygen, low solubility of hydrocarbons, and desication, employing Actinobacteria is limited [122]. *Pseudomonas* sp. and *Azotobacter vinelandii* are known to decompose petroleum. *Burkholderia cepacia* is capable of degrading hundreds of organic compounds. Microbial growth and activity are aided by the conversion of hydrocarbons into carbon dioxide and water, which releases energy [123]. Diesel was degraded by *Pseudomonas* sp., which removed long- and medium-chain alkanes [124]. In several treatment techniques devised by Wang et al., a microbial consortium consisting of *Actinomadura* sp., *Brevibacillus* sp., and an uncultured bacterial clone improved oil recovery for biopolymer manufacture [125].

By introducing bioemulsifiers and biosurfactants into the environment, the rate of bioremediation/biodegradation of organic contaminants improves [126]. It is dependent on the mechanism that is engaged in the interactions between microbial cells and insoluble hydrocarbons in surface-active compounds (SACs): (i) emulsification; (ii) micellarization; (iii) adhesion-deadhesion of microorganisms to and from hydrocarbons; and (iv) desorption of contaminants [126, 127]. The use of

Compounds	Microorganisms	References
Alkanes	*Acinetobacter calcoaceticus*	[112]
	Nocardia erythropolis	[113]
	Pseudomonas sp.	[114]
Mono-aromatic hydrocarbons	*Pseudomonas sp.*	[115]
	Sphingomonas paucimobilis	[116]
Poly-aromatic hydrocarbons	*Achromobacter sp., Mycobacterium sp., Pseudomonas sp., Mycobacterium flavescens, Rhodococcus sp.*	[115, 117–120]

Table 7.
Actinobacteria capable of degrading petroleum hydrocarbon.

surfactants aids in the solubility of petroleum components because diesel oil biosurfactants increase oil mobility and bioavailability, hence improving biodegradation rates [128]. As a possible biosurfactant producer, *Nocardiopsis* B4 was discovered; this strain is important in the breakdown of poly-aromatic hydrocarbons (PAHs) in soils. A wide range of temperature, pH, and salt concentrations did not affect the biosurfactant activity, demonstrating its suitability for bioremediation [129].

4. Conclusion

Actinobacteria through its unique capabilities have gained importance in the field of agriculture, pharmaceuticals, industry, nanotechnology, and many more. *Streptomyces* is the most common genera among them. They possess several PGP traits such as biofertilizers, phosphorus and potassium solubilization, production of siderophores, antibiotics, phytohormones, and biological nitrogen fixation. Furthermore, nanotechnology research being the most rapidly developing fields is using actinobacterial biosynthesis of nanoparticles which is both environment-friendly and cost-effective, and the nanoparticles which are produced as a result show potential biological properties such as antibacterial activities, antifungal activities, anticancer properties, anti-biofouling properties. The combination of synthesizing methodologies with biological processes has resulted in the development of nanoparticles which are used in a number of *in vitro* diagnostic methods. Apart from this, toxic heavy metals like chromium, zinc, lead, and copper and pesticides can be sustainably removed using this microbe. The degradation of pesticides whose accumulation otherwise causes biomagnification is possible with the help of this microbe. Actinobacterial genera have also proven versatile to bring about the degradation of xenobiotic pollutants in the nutrient starvation conditions in the soil and their capability of using these toxic compounds as their nutrient source, mainly a source of carbon is something that speeds up the process of degradation. Cocktail of microbes of this genus is effective in causing faster degradation. Hence, this microbe is quite adaptive for maintaining the environment eco-friendly.

Conflict of interest

None to declare.

List of abbreviations

ACC	1-aminocyclopropane-1-carboxylate
AFA	anti-*Fusarium* activity
BCAs	biocontrol agents
HCN	hydrogen cyanide
HeLa	human cervical cancer cell line
IAA	indole-acetic acid
ISR	induced systemic resistance
MBCAs	microbial biocontrol agents
PAHs	poly-aromatic hydrocarbons
PGPR	plant growth-promoting rhizobacteria
PGRs	plant growth regulators
ROS	reactive oxygen species
AuNP	gold nanoparticles

AgNP	silver nanoparticles
SACs	surface-active compounds
SAR	systemic acquired resistance
VOCs	volatile organic compounds

Author details

Saloni Jain[1†], Ishita Gupta[1†], Priyanshu Walia[2†] and Shalini Swami[1*]

1 Department of Microbiology, Ram Lal Anand College, University of Delhi, New Delhi, India

2 Department of Botany, Institute of Science, Banaras Hindu University, Varanasi, Uttar Pradesh, India

*Address all correspondence to: dr.shaliniswami@gmail.com

† Contributed equally.

IntechOpen

References

[1] Barka EA, Vatsa P, Sanchez L, Gaveau-Vaillant N, Jacquard C, Meier-Kolthoff JP, et al. Taxonomy, physiology, and natural products of Actinobacteria. Microbiology and Molecular Biology Reviews. 2016;**80**(1):1-43

[2] Anandan R, Dharumadurai D, Manogaran GP. An introduction to Actinobacteria. In: Dhanasekaran D, Jiang Y, editors. Actinobacteria. Rijeka: IntechOpen; 2016

[3] Salwan R, Sharma V. Molecular and biotechnological aspects of secondary metabolites in actinobacteria. Microbiological Research. 2020;**231**: 126374

[4] Shivlata L, Satyanarayana T. Thermophilic and alkaliphilic Actinobacteria: Biology and potential applications. Frontiers in Microbiology. 2015;**6**:1014

[5] Sharma M, Dangi P, Choudhary M. Actinomycetes: Source, identification, and their applications. International Journal of Current Microbiology and Applied Sciences. 2014;**3**(2):801-832

[6] Singh R, Dubey AK. Diversity and applications of endophytic actinobacteria of plants in special and other ecological niches. Frontiers in Microbiology. 2018;**9**:1767

[7] Bhatti AA, Haq S, Bhat RA. Actinomycetes benefaction role in soil and plant health. Microbial Pathogenesis. 2017;**111**:458-467

[8] Poomthongdee N, Duangmal K, Pathom-aree W. Acidophilic actinomycetes from rhizosphere soil: Diversity and properties beneficial to plants. Journal of Antibiotics. 2015;**68**(2):106-114

[9] Abdelshafy Mohamad OA, Li L, Ma J-B, Hatab S, Rasulov BA, Musa Z, et al. Halophilic Actinobacteria biological activity and potential applications. In: Egamberdieva D, Birkeland N-K, Panosyan H, Li W-J, editors. Extremophiles in Eurasian Ecosystems: Ecology, Diversity, and Applications. Singapore: Springer Singapore; 2018. pp. 333-364

[10] Lewin GR, Carlos C, Chevrette MG, Horn HA, McDonald BR, Stankey RJ, et al. Evolution and ecology of Actinobacteria and their bioenergy applications. Annual Review of Microbiology. 2016;**70**:235-254

[11] Goudjal Y, Zamoum M, Meklat A, Sabaou N, Mathieu F, Zitouni A. Plant-growth-promoting potential of endosymbiotic actinobacteria isolated from sand truffles (Terfezia leonis Tul.) of the Algerian Sahara. Annales de Microbiologie. 2015;**66**(1):91-100

[12] Gandhimathi R, Arunkumar M, Selvin J, Thangavelu T, Sivaramakrishnan S, Kiran GS, et al. Antimicrobial potential of sponge associated marine actinomycetes. Journal of Medical Mycology. 2008;**18**(1):16-22

[13] Polti MA, Aparicio JD, Benimeli CS, Amoroso MJ. 11—Role of Actinobacteria in bioremediation. In: Das S, editor. Microbial Biodegradation and Bioremediation. Oxford: Elsevier; 2014. pp. 269-286

[14] Sathya A, Vijayabharathi R, Gopalakrishnan S. Plant growth-promoting actinobacteria: A new strategy for enhancing sustainable production and protection of grain legumes. 3 Biotech. 2017;**7**(2):102

[15] Shanthi V. Actinomycetes: Implications and prospects in sustainable agriculture. Biofertilizers: Study and Impact. Jul 20, 2021:335-370

[16] Haukka K, Lindström K, Young JP. Three phylogenetic groups of nodA and

nifH genes in Sinorhizobium and Mesorhizobium isolates from leguminous trees growing in Africa and Latin America. Applied and Environmental Microbiology. 1998;**64**(2):419-426

[17] Gifford I, Vance S, Nguyen G, Berry AM. A stable genetic transformation system and implications of the type IV restriction system in the nitrogen-fixing plant endosymbiont Frankia alni ACN14a. Frontiers in Microbiology. 2019;**10**:2230

[18] Yadav N, Yadav AN. Actinobacteria for sustainable agriculture [Internet]. Journal of Applied Biotechnology and Bioengineering. 2019;**6**:38-41. DOI: 10.15406/jabb.2019.06.00172

[19] Sharma SB, Sayyed RZ, Trivedi MH, Gobi TA. Phosphate solubilizing microbes: Sustainable approach for managing phosphorus deficiency in agricultural soils. Springerplus. 2013;**2**:587

[20] Sharma V, Salwan R. Biocontrol potential and applications of Actinobacteria in agriculture. In: New and Future Developments in Microbial Biotechnology and Bioengineering. Elsevier; Jan 1, 2018. pp. 93-108

[21] Swarnalakshmi K, Senthilkumar M, Ramakrishnan B. Endophytic Actinobacteria: Nitrogen fixation, phytohormone production, and antibiosis. In: Subramaniam G, Arumugam S, Rajendran V, editors. Plant Growth Promoting Actinobacteria: A New Avenue for Enhancing the Productivity and Soil Fertility of Grain Legumes. Singapore: Springer Singapore; 2016. pp. 123-145

[22] Köhl J, Kolnaar R, Ravensberg WJ. Mode of action of microbial biological control agents against plant diseases: Relevance beyond efficacy. Frontiers in Plant Science. 2019;**10**:845

[23] Law JW-F, Ser H-L, Khan TM, Chuah L-H, Pusparajah P, Chan K-G,

et al. The potential of Streptomyces as biocontrol agents against the Rice blast fungus, Magnaporthe oryzae (Pyricularia oryzae). Frontiers in Microbiology. 2017;**8**:3

[24] Kasuga K, Sasaki A, Matsuo T, Yamamoto C, Minato Y, Kuwahara N, et al. Heterologous production of kasugamycin, an aminoglycoside antibiotic from Streptomyces kasugaensis, in Streptomyces lividans and Rhodococcus erythropolis L-88 by constitutive expression of the biosynthetic gene cluster. Applied Microbiology and Biotechnology. 2017;**101**(10):4259-4268

[25] Doolotkeldieva T, Bobusheva S. Fire blight disease caused by Erwinia amylovora on Rosaceae plants in Kyrgyzstan and biological agents to control this disease. Advances in Microbiology. 2016;**6**(11):831

[26] Wang W, Qiu Z, Tan H, Cao L. Siderophore production by actinobacteria. Biometals. 2014;**27**(4):623-631

[27] Hazarika SN, Thakur D. Actinobacteria. In: Beneficial Microbes in Agro-Ecology. Academic Press; Jan 1, 2020. pp. 443-476

[28] Manivasagan P, Venkatesan J, Sivakumar K, Kim S-K. Actinobacteria mediated synthesis of nanoparticles and their biological properties: A review. Critical Reviews in Microbiology. 2016;**42**(2):209-221

[29] Singh MJ. Green nano actinobacteriology—An interdisciplinary study. In: Actinobacteria—Basics and Biotechnological Applications. Intech; Feb 11, 2016. pp. 377-387

[30] Manimaran M, Kannabiran K. Actinomycetes-mediated biogenic synthesis of metal and metal oxide nanoparticles: Progress and challenges.

Letters in Applied Microbiology. 2017;**64**(6):401-408

[31] Sharma P, Dutta J, Thakur D. Future prospects of actinobacteria in health and industry. In: New and Future Developments in Microbial Biotechnology and Bioengineering. Elsevier; Jan 1, 2018. pp. 305-324

[32] Chau C-F, Wu S-H, Yen G-C. The development of regulations for food nanotechnology [Internet]. Trends in Food Science & Technology. 2007;**18**:269-280. DOI: 10.1016/j.tifs.2007.01.007

[33] Pattekari P, Zheng Z, Zhang X, Levchenko T, Torchilin V, Lvov Y. Top-down and bottom-up approaches in production of aqueous nanocolloids of low solubility drug paclitaxel [Internet]. Physical Chemistry Chemical Physics. 2011;**13**:9014. DOI: 10.1039/c0cp02549f

[34] Edison LK, Pradeep NS. Actinobacterial nanoparticles: Green synthesis, evaluation and applications. In: Nanotechnology in the Life Sciences. Cham: Springer International Publishing; 2020. pp. 371-384

[35] Koul B, Poonia AK, Yadav D, Jin JO. Microbe-mediated biosynthesis of nanoparticles: Applications and future prospects. Biomolecules. Jun 15, 2021;**11**(6):886

[36] Song JY, Kim BS. Biological synthesis of bimetallic Au/Ag nanoparticles using Persimmon (Diopyros kaki) leaf extract [Internet]. Korean Journal of Chemical Engineering. 2008;**25**:808-811. DOI: 10.1007/s11814-008-0133-z

[37] Otari SV, Patil RM, Ghosh SJ, Thorat ND, Pawar SH. Intracellular synthesis of silver nanoparticle by actinobacteria and its antimicrobial activity. Spectrochimica Acta. Part A, Molecular and Biomolecular Spectroscopy. 2015;**136**(Pt B): 1175-1180

[38] Beveridge TJ, Hughes MN, Lee H, Leung KT, Poole RK, Savvaidis I, et al. Metal-microbe interactions: Contemporary approaches. Advances in Microbial Physiology. 1997;**38**:177-243

[39] Bruins MR, Kapil S, Oehme FW. Microbial resistance to metals in the environment [Internet]. Ecotoxicology and Environmental Safety. 2000;**45**:198-207. DOI: 10.1006/eesa.1999.1860

[40] Sharma NC, Sahi SV, Nath S, Parsons JG, Gardea-Torresdey JL, Pal T. Synthesis of plant-mediated gold nanoparticles and catalytic role of biomatrix-embedded nanomaterials. Environmental Science & Technology. 2007;**41**(14):5137-5142

[41] Ahmad A, Senapati S, Islam Khan M, Kumar R, Ramani R, Srinivas V, et al. Intracellular synthesis of gold nanoparticles by a novel alkalotolerant actinomycete, Rhodococcusspecies [Internet]. Nanotechnology. 2003;**14**:824-828. DOI: 10.1088/0957-4484/14/7/323

[42] Alani F, Moo-Young M, Anderson W. Biosynthesis of silver nanoparticles by a new strain of Streptomyces sp. compared with *Aspergillus fumigatus* [Internet]. World Journal of Microbiology and Biotechnology. 2012;**28**:1081-1086. DOI: 10.1007/s11274-011-0906-0

[43] Balagurunathan R, Radhakrishnan M, Rajendran RB, Velmurugan D. Biosynthesis of gold nanoparticles by actinomycete Streptomyces viridogens strain HM10. Indian Journal of Biochemistry & Biophysics. 2011;**48**(5):331-335

[44] Mohanpuria P, Rana NK, Yadav SK. Biosynthesis of nanoparticles: Technological concepts and future applications [Internet]. Journal of Nanoparticle Research. 2008;**10**:507-517. DOI: 10.1007/s11051-007-9275-x

[45] Ahmad A, Senapati S, Islam Khan M, Kumar R, Sastry M. Extracellular biosynthesis of monodisperse gold nanoparticles by a novel extremophilic actinomycete,thermomonosporasp [Internet]. Langmuir. 2003;**19**:3550-3553. Available from: http://dx.doi.org/10.1021/la026772l

[46] Chauhan R, Kumar A, Abraham J. A biological approach to the synthesis of silver nanoparticles with Streptomyces sp JAR1 and its antimicrobial activity. Scientia Pharmaceutica. 2013;**81**(2):607-621

[47] Gopal JV, Thenmozhi M, Kannabiran K, Rajakumar G, Velayutham K, Rahuman AA. Actinobacteria mediated synthesis of gold nanoparticles using Streptomyces sp. VITDDK3 and its antifungal activity [Internet]. Materials Letters. 2013;**93**:360-362. DOI: 10.1016/j.matlet.2012.11.125

[48] Otari SV, Patil RM, Nadaf NH, Ghosh SJ, Pawar SH. Green biosynthesis of silver nanoparticles from an actinobacteria Rhodococcus sp [Internet]. Materials Letters. 2012;**72**:92-94. DOI: 10.1016/j.matlet.2011.12.109

[49] Tsibakhashvili NY, Kirkesali EI, Pataraya DT, Gurielidze MA, Kalabegishvili TL, Gvarjaladze DN, et al. Microbial synthesis of silver nanoparticles by *Streptomyces glaucus* and *Spirulina platensis* [Internet]. Advanced Science Letters. 2011;**4**:3408-3417. DOI: 10.1166/asl.2011.1915

[50] Chen X-J, Sanchez-Gaytan BL, Qian Z, Park S-J. Noble metal nanoparticles in DNA detection and delivery. Wiley Interdisciplinary Reviews. Nanomedicine and Nanobiotechnology. 2012;**4**(3):273-290

[51] Doria G, Conde J, Veigas B, Giestas L, Almeida C, Assunção M, et al. Noble metal nanoparticles for biosensing applications. Sensors. 2012;**12**(2):1657-1687

[52] Ramesh S, Mathivanan N. Screening of marine actinomycetes isolated from the Bay of Bengal, India for antimicrobial activity and industrial enzymes [Internet]. World Journal of Microbiology and Biotechnology. 2009;**25**:2103-2111. DOI: 10.1007/s11274-009-0113-4

[53] Jensen PR, Williams PG, Oh D-C, Zeigler L, Fenical W. Species-specific secondary metabolite production in marine actinomycetes of the genus Salinispora. Applied and Environmental Microbiology. 2007;**73**(4):1146-1152

[54] Nel AE, Mädler L, Velegol D, Xia T, EMV H, Somasundaran P, et al. Understanding biophysicochemical interactions at the nano–bio interface [Internet]. Nature Materials. 2009;**8**:543-557. DOI: 10.1038/nmat2442

[55] Krutyakov YA, Kudrinskiy AA, Yu Olenin A, Lisichkin GV. Synthesis and properties of silver nanoparticles: Advances and prospects [Internet]. Russian Chemical Reviews. 2008;**77**:233-257. DOI: 10.1070/rc2008v077n03abeh003751

[56] Lin Y-SE, Vidic RD, Stout JE, McCartney CA, Yu VL. Inactivation of *Mycobacterium avium* by copper and silver ions [Internet]. Water Research. 1998;**32**:1997-2000. DOI: 10.1016/s0043-1354(97)00460

[57] Morones JR, Elechiguerra JL, Camacho A, Holt K, Kouri JB, Ramírez JT, et al. The bactericidal effect of silver nanoparticles. Nanotechnology. 2005;**16**(10):2346-2353

[58] Lok C-N, Ho C-M, Chen R, He Q-Y, Yu W-Y, Sun H, et al. Silver nanoparticles: Partial oxidation and antibacterial activities. Journal of Biological Inorganic Chemistry. 2007;**12**(4):527-534

[59] Kumar VS, Siva Kumar V, Nagaraja BM, Shashikala V, Padmasri AH,

Shakuntala Madhavendra S, et al. Highly efficient Ag/C catalyst prepared by electro-chemical deposition method in controlling microorganisms in water [Internet]. Journal of Molecular Catalysis A: Chemical. 2004;**223**:313-319. DOI: 10.1016/j.molcata.2003.09.047

[60] Prakasham RS, Buddana SK, Yannam SK, Guntuku GS. Characterization of silver nanoparticles synthesized by using marine isolate Streptomyces albidoflavus. Journal of Microbiology and Biotechnology. 2012;**22**(5):614-621

[61] Sivalingam P, Antony JJ, Siva D, Achiraman S, Anbarasu K. Mangrove Streptomyces sp. BDUKAS10 as nanofactory for fabrication of bactericidal silver nanoparticles. Colloids and Surfaces. B, Biointerfaces. 2012;**98**:12-17

[62] Samundeeswari A, Dhas SP, Nirmala J, John SP, Mukherjee A, Chandrasekaran N. Biosynthesis of silver nanoparticles using actinobacterium Streptomyces albogriseolus and its antibacterial activity. Biotechnology and Applied Biochemistry. 2012;**59**(6):503-507

[63] Lee PC, Meisel D. Adsorption and surface-enhanced Raman of dyes on silver and gold sols [Internet]. The Journal of Physical Chemistry. 1982;**86**:3391-3395. DOI: 10.1021/j100214a025

[64] Sanjenbam P, Gopal JV, Kannabiran K. Anticandidal activity of silver nanoparticles synthesized using Streptomyces sp.VITPK1. Journal of Medical Mycology. 2014;**24**(3):211-219

[65] Bray F, Ferlay J, Soerjomataram I, Siegel RL, Torre LA, Jemal A. Global cancer statistics 2018: GLOBOCAN estimates of incidence and mortality worldwide for 36 cancers in 185 countries. CA: A Cancer Journal for Clinicians. 2018;**68**(6):394-424

[66] Vickers A. Alternative cancer cures: "unproven" or "disproven"? CA: A Cancer Journal for Clinicians. 2004;**54**(2):110-118

[67] Manivasagan P, Venkatesan J, Sivakumar K, Kim S-K. Pharmaceutically active secondary metabolites of marine actinobacteria. Microbiological Research. 2014;**169**(4):262-278

[68] Shanmugasundaram T, Radhakrishnan M, Gopikrishnan V, Pazhanimurugan R, Balagurunathan R. A study of the bactericidal, anti-biofouling, cytotoxic and antioxidant properties of actinobacterially synthesised silver nanoparticles. Colloids and Surfaces. B, Biointerfaces. 2013;**111**:680-687

[69] Chapman J, Weir E, Regan F. Period four metal nanoparticles on the inhibition of biofouling. Colloids and Surfaces. B, Biointerfaces. 2010;**78**(2):208-216

[70] Kannabiran K. Actinobacteria are better bioremediating agents for removal of toxic heavy metals: An overview [Internet]. International Journal of Environmental Technology and Management. 2017;**20**:129. DOI: 10.1504/ijetm.2017.10010678

[71] Alvarez A, Saez JM, Davila Costa JS, Colin VL, Fuentes MS, Cuozzo SA, et al. Actinobacteria: Current research and perspectives for bioremediation of pesticides and heavy metals. Chemosphere. 2017;**166**:41-62

[72] Alkorta I, Epelde L, Garbisu C. Environmental parameters altered by climate change affect the activity of soil microorganisms involved in bioremediation [Internet]. FEMS Microbiology Letters. 2017;**364**. DOI:10.1093/femsle/fnx200

[73] Polti MA, Amoroso MJ, Abate CM. Chromate reductase activity in

Streptomyces sp. MC1. The Journal of General and Applied Microbiology. 2010;**56**(1):11-18

[74] Malkin R, Malmström BG. The state and function of copper in biological systems. Advances in Enzymology and Related Areas of Molecular Biology. 1970;**33**:177-244

[75] Benimeli CS, Polti MA, Albarracín VH, Abate CM, Amoroso MJ. Bioremediation potential of heavy metal–resistant actinobacteria and maize plants in polluted soil. In: Biomanagement of Metal-Contaminated Soils. Dordrecht: Springer; 2011. pp. 459-477

[76] Mercer JF. The molecular basis of copper-transport diseases. Trends in Molecular Medicine. Feb 1, 2001;**7**(2):64-69

[77] Albarracín VH, Avila AL, Amoroso MJ, Abate CM. Copper removal ability by Streptomyces strains with dissimilar growth patterns and endowed with cupric reductase activity. FEMS Microbiology Letters. 2008;**288**(2):141-148

[78] Albarracín VH, Winik B, Kothe E, Amoroso MJ, Abate CM. Copper bioaccumulation by the actinobacterium Amycolatopsis sp. AB0. Journal of Basic Microbiology. 2008;**48**(5):323-330

[79] Bagwell CE, Hixson KK, Milliken CE, Lopez-Ferrer D, Weitz KK. Proteomic and physiological responses of Kineococcus radiotolerans to copper. PLoS One. 2010;**5**(8):e12427

[80] Costa JSD, Albarracín VH, Abate CM. Cupric reductase activity in copper-resistant Amycolatopsis tucumanensis [Internet]. Water, Air, & Soil Pollution. 2011;**216**:527-535. DOI: 10.1007/s11270-010-0550-6

[81] Poopal AC, Laxman RS. Studies on biological reduction of chromate by

Streptomyces griseus. Journal of Hazardous Materials. 2009;**169**(1-3):539-545

[82] Das S, Chandra AL. Chromate reduction in Streptomyces. Experientia. 1990;**46**(7):731-733

[83] Laxman RS, More S. Reduction of hexavalent chromium by *Streptomyces griseus* [Internet]. Minerals Engineering. 2002;**15**:831-837. DOI: 10.1016/s0892-6875(02)00128-0

[84] Pattanapipitpaisal P, Brown NL, Macaskie LE. Chromate reduction and 16S rRNA identification of bacteria isolated from a Cr(VI)-contaminated site. Applied Microbiology and Biotechnology. 2001;**57**(1-2):257-261

[85] Colin VL, Pereira CE, Villegas LB, Amoroso MJ, Abate CM. Production and partial characterization of bioemulsifier from a chromium-resistant actinobacteria. Chemosphere. 2013;**90**(4):1372-1378

[86] Amoroso MJ, Benimeli CS, Cuozzo SA, editors. Actinobacteria: Application in Bioremediation and Production of Industrial Enzymes. CRC Press; Mar 12, 2013

[87] Camargo FAO, Bento FM, Okeke BC, Frankenberger WT. Hexavalent chromium reduction by an actinomycete, arthrobacter crystallopoietes ES 32. Biological Trace Element Research. 2004;**97**(2):183-194

[88] Focardi S, Pepi M, Landi G, Gasperini S, Ruta M, Di Biasio P, et al. Hexavalent chromium reduction by whole cells and cell free extract of the moderate halophilic bacterial strain Halomonas sp. TA-04 [Internet]. International Biodeterioration & Biodegradation. 2012;**66**:63-70. DOI: 10.1016/j.ibiod.2011.11.003

[89] Sugiyama T, Sugito H, Mamiya K, Suzuki Y, Ando K, Ohnuki T. Hexavalent

chromium reduction by an actinobacterium Flexivirga alba ST13T in the family Dermacoccaceae. Journal of Bioscience and Bioengineering. 2012;**113**(3):367-371

[90] Schumann P, Prauser H, Rainey FA, Stackebrandt E, Hirsch P. Friedmanniella antarctica gen. nov., sp. nov., an LL-diaminopimelic acid-containing actinomycete from Antarctic sandstone. International Journal of Systematic Bacteriology. 1997;**47**(2):278-283

[91] Liu H, Wang H, Wang G. Intrasporangium chromatireducens sp. nov., a chromate-reducing actinobacterium isolated from manganese mining soil, and emended description of the genus Intrasporangium. International Journal of Systematic and Evolutionary Microbiology. 2012;**62**(Pt 2):403-408

[92] Nabi S. Toxic Effects of Mercury. New Delhi, India: Springer India; Jul 25, 2014

[93] Ravel J, Amoroso MJ, Colwell RR, Hill RT. Mercury-resistant actinomycetes from the Chesapeake Bay. FEMS Microbiology Letters. 1998;**162**(1):177-184

[94] Sanjenbam P, Saurav K, Kannabiran K. Biosorption of mercury and lead by aqueous Streptomyces VITSVK9 sp. isolated from marine sediments from the bay of Bengal, India [Internet]. Frontiers of Chemical Science and Engineering. 2012;**6**:198-202. DOI: 10.1007/s11705-012-1285-2

[95] Flora G, Gupta D, Tiwari A. Toxicity of lead: A review with recent updates. Interdisciplinary Toxicology. 2012;**5**(2):47-58

[96] El Baz S, Baz M, Barakate M, Hassani L, El Gharmali A, Imziln B. Resistance to and accumulation of heavy metals by actinobacteria isolated from abandoned mining areas. ScientificWorldJournal. 2015;**2015**:761834

[97] Chopra AK, Sharma MK, Chamoli S. Bioaccumulation of organochlorine pesticides in aquatic system—An overview [Internet]. Environmental Monitoring and Assessment. 2011;**173**:905-916. DOI: 10.1007/s10661-010-1433-4

[98] Yadav IC, Devi NL, Syed JH, Cheng Z, Li J, Zhang G, et al. Current status of persistent organic pesticides residues in air, water, and soil, and their possible effect on neighboring countries: A comprehensive review of India. Science of the Total Environment. 2015;**511**:123-137

[99] Becerra-Castro C, Prieto-Fernández Á, Kidd PS, Weyens N, Rodríguez-Garrido B, Touceda-González M, et al. Improving performance of *Cytisus striatus* on substrates contaminated with hexachlorocyclohexane (HCH) isomers using bacterial inoculants: Developing a phytoremediation strategy [Internet]. Plant and Soil. 2013;**362**:247-260. DOI: 10.1007/s11104-012-1276-6

[100] Sullivan Jr JB, Krieger GB, Thomas RJ. Hazardous materials toxicology: Clinical principles of environmental health. Journal of Occupational and Environmental Medicine. Apr 1, 1992;**34**(4):365-371

[101] Wang Z, Gerstein M, Snyder M. RNA-Seq: A revolutionary tool for transcriptomics. Nature Reviews. Genetics. 2009;**10**(1):57-63

[102] Kumar M, Vidya Lakshmi C, Khanna S. Biodegradation and bioremediation of endosulfan contaminated soil [Internet]. Bioresource Technology. 2008;**99**:3116-3122. DOI: 10.1016/j.biortech.2007.05.057

[103] Devers-Lamrani M, Pesce S, Rouard N, Martin-Laurent F. Evidence

for cooperative mineralization of diuron by Arthrobacter sp. BS2 and Achromobacter sp. SP1 isolated from a mixed culture enriched from diuron exposed environments. Chemosphere. 2014;**117**:208-215

[104] Chen S, Lai K, Li Y, Hu M, Zhang Y, Zeng Y. Biodegradation of deltamethrin and its hydrolysis product 3-phenoxybenzaldehyde by a newly isolated Streptomyces aureus strain HP-S-01. Applied Microbiology and Biotechnology. 2011;**90**(4):1471-1483

[105] Sagarkar S, Bhardwaj P, Storck V, Devers-Lamrani M, Martin-Laurent F, Kapley A. s-triazine degrading bacterial isolate Arthrobacter sp. AK-YN10, a candidate for bioaugmentation of atrazine contaminated soil. Applied Microbiology and Biotechnology. 2016;**100**(2):903-913

[106] Kolekar PD, Phugare SS, Jadhav JP. Biodegradation of atrazine by Rhodococcus sp. BCH2 to N-isopropylammelide with subsequent assessment of toxicity of biodegraded metabolites. Environmental Science and Pollution Research International. 2014;**21**(3):2334-2345

[107] Benimeli CS, Castro GR, Chaile AP, Amoroso MJ. Lindane removal induction byStreptomyces sp. M7 [Internet]. Journal of Basic Microbiology. 2006;**46**:348-357. DOI: 10.1002/jobm.200510131

[108] Poopal AC, Seeta LR. Chromate reduction by PVA-alginate immobilized *Streptomyces griseus* in a bioreactor [Internet]. Biotechnology Letters. 2009;**31**:71-76. DOI: 10.1007/s10529-008-9829-8

[109] Saez JM, Benimeli CS, Amoroso MJ. Lindane removal by pure and mixed cultures of immobilized actinobacteria. Chemosphere. 2012;**89**(8):982-987

[110] Saez JM, Aparicio JD, Amoroso MJ, Benimeli CS. Effect of the acclimation

of a Streptomyces consortium on lindane biodegradation by free and immobilized cells [Internet]. Process Biochemistry. 2015;**50**:1923-1933. DOI: 10.1016/j.procbio.2015.08.014

[111] Yemashova NA, Murygina VP, Zhukov DV, Zakharyantz AA, Gladchenko MA, Appanna V, et al. Biodeterioration of crude oil and oil derived products: A review [Internet]. Reviews in Environmental Science and Bio/Technology. 2007;**6**:315-337. DOI: 10.1007/s11157-006-9118-8

[112] Lal B, Khanna S. Degradation of crude oil by *Acinetobacter calcoaceticus* and *Alcaligenes odorans*. The Journal of Applied Bacteriology. 1996;**81**(4):355-362

[113] Hua Z, Chen J, Lun S, Wang X. Influence of biosurfactants produced by *Candida antarctica* on surface properties of microorganism and biodegradation of n-alkanes [Internet]. Water Research. 2003;**37**:4143-4150. DOI: 10.1016/s0043-1354(03)00380-4

[114] Herman DC, Lenhard RJ, Miller RM. Formation and removal of hydrocarbon residual in porous media: Effects of attached bacteria and biosurfactants [Internet]. Environmental Science & Technology. 1997;**31**:1290-1294. DOI: 10.1021/es960441b

[115] Churchill PF, Dudley RJ, Churchill SA. Surfactant-enhanced bioremediation [Internet]. Waste Management. 1995;**15**:371-377. DOI: 10.1016/0956-053x(95)00038-2

[116] Willumsen PA, Arvin E. Kinetics of degradation of surfactant-solubilized fluoranthene by a *Sphingomonas paucimobilis* [Internet]. Environmental Science & Technology. 1999;**33**:2571-2578. DOI: 10.1021/es981022c

[117] Doong R-A, Lei W-G. Solubilization and mineralization of

polycyclic aromatic hydrocarbons by pseudomonas putida in the presence of surfactant. Journal of Hazardous Materials. 2003;**96**(1):15-27

[118] Volkering F, Breure AM, van Andel JG. Effect of micro-organisms on the bioavailability and biodegradation of crystalline naphthalene [Internet]. Applied Microbiology and Biotechnology. 1993;**40**:535, 10.1007/bf00175745-540

[119] Straube WL, Jones-Meehan J, Pritchard PH, Jones WR. Bench-scale optimization of bioaugmentation strategies for treatment of soils contaminated with high molecular weight polyaromatic hydrocarbons [Internet]. Resources, Conservation and Recycling. 1999;**27**:27-37. DOI: 10.1016/s0921-3449(98)00083-4

[120] Kwok C-K, Loh K-C. Effects of Singapore soil type on bioavailability of nutrients in soil bioremediation [Internet]. Advances in Environmental Research. 2003;**7**:889-900. DOI: 10.1016/s1093-0191(02)00084-9

[121] Holden PA, LaMontagne MG, Bruce AK, Miller WG, Lindow SE. Assessing the role of *Pseudomonas aeruginosa* surface-active gene expression in hexadecane biodegradation in sand. Applied and Environmental Microbiology. 2002;**68**(5):2509-2518

[122] Pernetti M, Di Palma L. Experimental evaluation of inhibition effects of saline wastewater on activated sludge. Environmental Technology. 2005;**26**(6):695-703

[123] Onwurah INE, Nwuke C. Enhanced bioremediation of crude oil-contaminated soil by a Pseudomonas species and mutually associated adapted *Azotobacter vinelandii* [Internet]. Journal of Chemical Technology & Biotechnology. 2004;**79**:491-498. DOI: 10.1002/jctb.1009

[124] Ghazali FM, Rahman RNZ, Salleh AB, Basri M. Biodegradation of hydrocarbons in soil by microbial consortium [Internet]. International Biodeterioration & Biodegradation. 2004;**54**:61-67. DOI: 10.1016/j.ibiod.2004.02.002

[125] Wang J, Yan G, An M, Liu J, Zhang H, Chen Y. Study of a plugging microbial consortium using crude oil as sole carbon source [Internet]. Petroleum Science. 2008;**5**:367-374. DOI: 10.1007/s12182-008-0061-x

[126] Franzetti A, Tamburini E, Banat IM. Applications of biological surface active compounds in remediation technologies. Advances in Experimental Medicine and Biology. 2010;**672**:121-134

[127] Chaturvedi S, Khurana SM. Importance of actinobacteria for bioremediation. In: Plant Biotechnology: Progress in Genomic Era. Singapore: Springer; 2019. pp. 277-307

[128] Bordoloi NK, Konwar BK. Bacterial biosurfactant in enhancing solubility and metabolism of petroleum hydrocarbons. Journal of Hazardous Materials. 2009;**170**(1):495-505

[129] Khopade A, Biao R, Liu X, Mahadik K, Zhang L, Kokare C. Production and stability studies of the biosurfactant isolated from marine Nocardiopsis sp. B4 [Internet]. Desalination. 2012;**285**:198-204. DOI: 10.1016/j.desal.2011.10.002

Chapter 9

Biosurfactant Production by Mycolic Acid-Containing Actinobacteria

Fiona M. Stainsby, Janki Hodar and Halina Vaughan

Abstract

The *Actinobacteria* produce an array of valuable metabolites including biosurfactants which are gaining increased attention in the biotechnology industries as they are multifunctional, biorenewable and generally superior to chemically syn- thesized compounds. Biosurfactants are surface-active, amphipathic molecules pre- sent at the microbial cell-surface or released extracellularly and in a variety of chemical forms. The mycolic acid-containing actinobacteria (MACA), classified in the order *Corynebacteriales*, represent a potentially rich source of biosurfactants for novel applications and undiscovered biosurfactant compounds. Members of the mycolate genus *Rhodococcus* produce various well-characterised glycolipids. However, other mycolate genera including *Corynebacterium*, *Dietzia*, *Gordonia* and *Tsukamurella* although less extensively investigated also possess biosurfactant-producing strains. This chapter captures current knowledge on biosurfactant production amongst the MACA, including their chemical structures and producer organisms. It also provides an overview of approaches to the recovery of biosurfactant producing MACA from the environment and assays available to screen for biosurfactant production. Meth- odologies applied in the extraction, purification, and structural elucidation of the different types of biosurfactants are also summarised. Potential future applications of MACA-derived biosurfactants are highlighted with particular focus on biomedical and environmental possibilities. Further investigation of biosurfactant production by MACA will enable the discovery of both novel producing strains and compounds with the prospect of biotechnological exploitation.

Keywords: actinobacteria, antimicrobial, bioemulsifiers, bioremediation, biosurfactants, biotechnology, *Corynebacteriales*, mycolic acids, *Rhodococcus*

1. Introduction

Members of the class *Actinobacteria* produce an impressive range of bioactive metabolites that are of commercial importance and many more that have the potential for future exploitation. This includes biosurfactants which are synthesised by many actinobacterial species. Microbial biosurfactants are gaining increased attention in the biotechnology industries as they are multifunctional, enabling diverse applications. Biosurfactants can also claim strong green credentials as not only are they biorenewable with the possibility of production on various substrates including wastes, but they may also be applied to environmental remediation [1]. Further, biosurfactants are generally considered superior to their chemically synthesized counterparts. Amongst the most common biosurfactant producers are members of the mycolic acid-containing (mycolate) genus *Rhodococcus* which have received considerable attention. However, other related mycolate genera including *Corynebacterium*, *Dietzia*, *Gordonia* and *Tsukamurella* also possess biosurfactant-producing strains but have not been explored to the same extent. Additionally, there are several other mycolate genera that have received little or no investigation in this respect that may produce novel biosurfactant compounds.

Membership of the mycolic acid-containing actinobacterial (MACA) group has expanded considerably over the past 20 years with revisions to the classification of existing species and the publication of copious new mycolate species and genera [2]. This substantial and metabolically diverse group therefore warrants further attention in the search for valuable biosurfactants. This chapter provides an overview of the current knowledge on biosurfactants produced by members of this group and describes approaches to the recovery, screening and biosurfactant-producing strains from the environment and their growth requirements. Methodologies applied to screen for biosurfactant production and for extraction, purification, and structural elucidation of biosurfactant compounds are also described. Current and potential future applications of biosurfactants derived from MACA are examined with particular focus on potential biomedical and environmental possibilities.

1.1 Biosurfactant properties

Microbial biosurfactants are amphipathic compounds, with both hydrophilic (polar) and hydrophobic (non-polar) moieties. The hydrophobic portion has saturated, unsaturated, or hydroxylated long-chain fatty acids and the hydrophilic portion can contain amino acids, carbohydrate, carboxyl acid, peptides, phosphate, or alcohol [3]. Biosurfactants may be categorised according to molecular weight (low or high), ionic charge (anionic, cationic, neutral, or non-ionic) or according to chemical composition and structure. The main classes of biosurfactants include fatty acids, glycolipids, lipopeptides, lipoproteins, neutral lipids, phospholipids, and polymeric biosurfactants. Their amphipathic nature enables biosurfactants to partition at water-air, oil-air, or oil-water interfaces thereby reducing surface and/or interfacial tension. They exhibit many other useful properties including de-/emulsification, dispersion, foaming, lubrication, softening, stabilisation, viscosity reduction and wetting [4].

Biosurfactants may be located intracellularly, on the cell surface (cell-bound) or excreted extracellularly (free) [5] and are produced during growth on both hydrophilic and hydrophobic substrates, to reduce surface or interfacial properties of the microbial cell or the surrounding environment. Biosynthesis of these compounds is required for gliding, motility, swarming, and biofilm formation. Biosurfactants also mediate between cells and hydrophobic compounds, enabling enhanced solubilisation and uptake across the cell membrane for utilisation as a substrate for growth and energy (**Figure 1**).

Many microbially derived biosurfactants are already used in diverse industries including agriculture, bioremediation, cosmetics, food, healthcare and medicine, and the petrochemical industry (**Figure 2**). In addition to being multifunctional, biosurfactants have several advantages over chemically synthesised surfactants. They are less/non-toxic and biodegradable, have higher surface activity and lower critical micelle concentrations (CMC), greater biocompatibility and selectivity, they function over wide pH, salinity, and temperature ranges, and can be produced using renewable and waste substrates [6]. These unique eco-friendly features make biosurfactants particularly attractive options as industries focus on longer-term sustainability and working towards a circular economy.

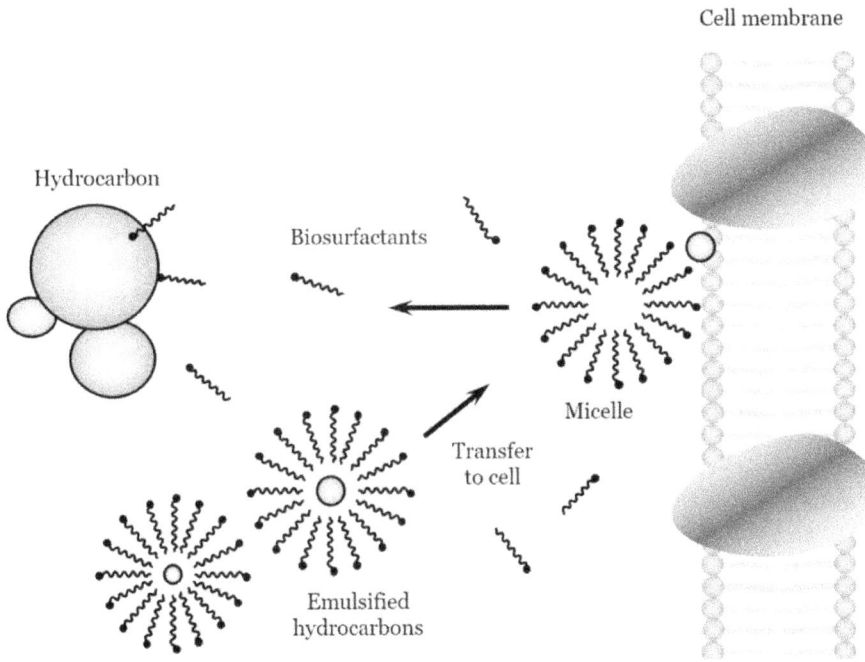

Figure 1.
Emulsification of hydrocarbons by microbial biosurfactants to enhance bioavailability.

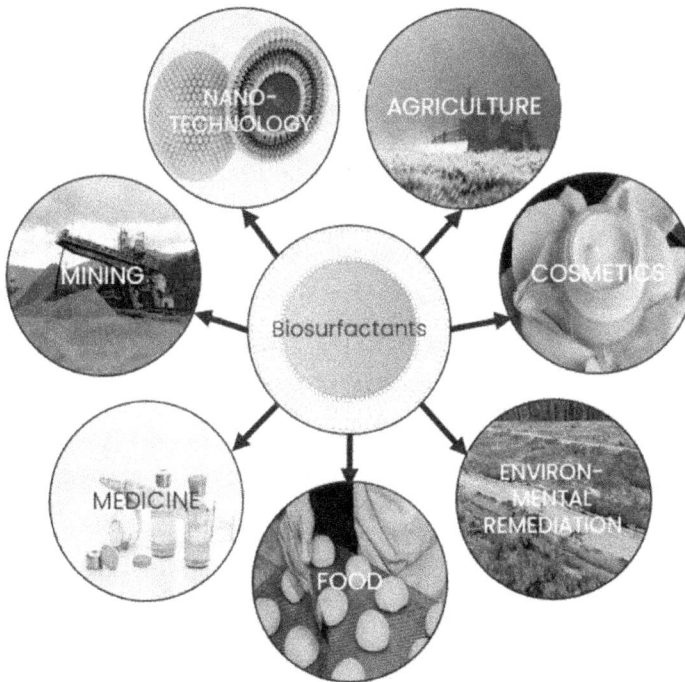

Figure 2.
Various sectors of application for microbial biosurfactants.

1.2 Mycolic acid-containing actinobacteria

The MACA form a phylogenetically coherent group that resides in the order *Corynebacteriales* based on 16S rRNA gene sequence analysis. The members are Gram-positive with high guanine-plus-cytosine (G + C) content in their genomic DNA. They currently comprise more than 400 species classified in 15 genera, namely *Corynebacterium, Dietzia, Gordonia, Hoyosella, Lawsonella, Millisia, Mycobacterium, Nocardia, Rhodococcus, Segniliparus, Skermania, Smarigdococcus, Tomitella, Tsukamurella* and *Williamsia* [2]. The almost universal production of mycolic acids by members of this group is a synapomorphic trait that is unique to this phylogenetic lineage [7]. However, several members of this order appear to have lost the ability to produce mycolic acids over the course of evolution, including several species of the genus *Corynebacterium* and *Hoyosella*. It was recently proposed that the single species belonging to the genus *Turicella*, also characterised by the absence of mycolic acids, be reclassified in the genus *Corynebacterium* [8].

Mycolic acids, which are high molecular weight 3-hydroxy fatty acids with a long alkyl branch in the 2-position, represent the major lipid constituents of the cell envelope of these organisms. They show structural variations from relatively simple mixtures of saturated and unsaturated compounds in corynebacteria to highly complex mixtures in mycobacteria. Mycolic acids also vary in the number of carbons on the 2-alkyl-branch from C22–C38 in corynebacteria to C60–C90 in mycobacteria [9]. Mycolic acids play an essential role in the architecture and functions of the cell envelope, where attached to the cell wall arabinogalactan they help to form a barrier that contributes to impermeability and resilience and conveys hydrophobicity to the cell surface. Trehalose mycolates, also termed cord factors, play an important role in pathogenicity in mycobacterial species that cause infection [9]. The presence and carbon chain length of mycolic acids can be used as taxonomic markers for the identification and classification of actinobacteria to the order *Corynebacteriales* [2].

Members of order *Corynebacteriales* can usually be distinguished from one another and from corresponding taxa in the phylum Actinobacteria based on 16S rRNA phylogeny supported by phenotypic (cell wall chemistry and morphology) features. Cell morphology amongst the MACA varies from simple rods and cocci to branched filaments that fragment to pleomorphic forms (**Table 1**). Members of the species *Skermania piniformis* are micromorphologically unique in this group as they form pine tree-like acute-angle branched filaments [10]. Colonies growing on agar plates are normally visible within several days of inoculation (**Figure 3**) although slow-growing mycobacteria take considerably longer. Species vary widely in colony appearance and are often colourful however it is usually not possible to unambiguously assign strains to a genus based on this feature alone.

Chemotaxonomy is the study of the distribution of various cell wall components to classify and identify strains and is particularly useful to differentiate between the various mycolic acid-containing genera. Cell wall markers typically used to differentiate between MACA genera are summarised in **Table 2**. Some of the methods used to analyse these chemotaxonomic markers provide quantitative or semi-quantitative data, as in the case of fatty acids, whereas other techniques provide only qualitative data as in the case of muramic acid type and phospholipid pattern.

Reliable identification of MACA strains to species level depends upon phylogenetic analysis of the gene encoding 16S rRNA and DNA:DNA homology determination provides definitive delineation of species with 70% homology and above signifying membership of same species [11]. Increasingly, whole-genome sequencing (WGS) is becoming a standard technique and comparative genomic analysis is providing useful insights to the relatedness and divergence of MACA species [11].

Genus	Micro-morphology	Acid-fastness	Aerial hyphae	Visible colonies (days)	Strictly aerobic
Corynebacterium	Pleomorphic rods, often club-shaped in palisade or angular arrangements	Some weakly acid-fast	Absent	1–2	No
Dietzia	Short rods and cocci	No	Absent	1–3	Yes
Gordonia	Rods, cocci and/or moderately branching hyphae	Partially acid-alcohol fast	Absent	1–3	Yes
Hoyosella	Cocci occur singly, in pairs, tetrads or in groups	Slightly acid–alcohol-fast	Absent	2	Yes
Lawsonella	Pleomorphic bacilli and cocci	Partially acid-fast	Absent	5–7	No
Millisia	Short rods	Acid-alcohol fast	Absent	1–3	Yes
Mycobacterium	Rods, occasionally branched filaments that fragment to rods and cocci	Strongly acid-fast	Rare	2–40	Yes
Nocardia	Mycelia that fragment into rods and cocci	Partially acid-fast	Present	1–5	Yes
Rhodococcus	Rods to extensive substrate mycelia that fragment to irregular rods and cocci	Partially acid-fast	Absent	1–3	Yes
Segniliparus	Rods	Acid-alcohol fast	Absent	3–4	Yes
Skermania	Acute angled branched mycelia	No	Only visible under the microscope	10–21	No
Smaragdicoccus	Coccoid	ND	Absent	7–14	Yes
Tomitella	Irregular rods	ND	Absent	ND	Yes
Tsukamurella	Single rods or in pairs or masses, sometimes rudimentary filaments and coccobacillary forms	Partially alcohol-acid fast	Absent	1–3	Yes
Williamsia	Thin rods or cocci in pairs or clusters	ND	Present	1–4	Yes

ND, not determined.
Adapted from [2].

Table 1.
General phenotypic features of mycolate genera classified in the order Corynebacteriales.

Protein sequences from *Corynebacteriales* genomes have revealed many conserved signature indels (CSIs) conserved signature proteins (CSPs) that are specific for members of this order [12].

2. Biosurfactants produced by MACA

In addition to *Rhodococcus*, diverse members of the order *Corynebacteriales* have been reported to synthesise extra-cellular and cell-bound biosurfactants, including

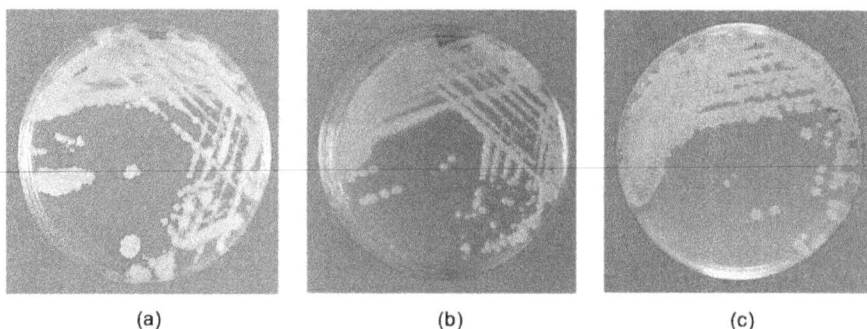

(a) (b) (c)

Figure 3.
The appearance of (a) Gordonia amarae, *(b)* Rhodococcus erythropolis *and (c)* Tsukamurella spumae *on glucose yeast-extract agar after 7 days incubation at 30°C.*

Genus	Mycolic acids (chain length)	Fatty acids*	Phospholipid type	Major menaquinone(s)	Muramic acid type	gDNA G + C (mol%)
Corynebacterium	22–38	S,U	I	MK-8(H$_2$)	Acetylated	51–67
Dietzia	34–38	S,U,T	II	MK-8(H$_2$)	Acetylated	65.5–73
Gordonia	46–66	S,U,T	II	MK-9(H$_2$)	Glycolated	63–69
Hoyosella	30–38		II	MK-8	Acetylated	49.3–61.8
Lawsonella	α+-mycolate	S,U	I	MK-9	Acetylated	58.6
Millisia	44–52	S,U, T	II	MK-8(H$_2$)	Glycolated	64.7
Mycobacterium	60–90	S,U,T	II	MK-9(H$_2$)	Glycolated	57–73
Nocardia	48–60	S,U,T	II	MK-8(H$_4$, ⍵-cycl)	Glycolated	63–72
Rhodococcus	30–54	S,U,T	II	MK-8(H$_2$)	Glycolated	63–73
Segniliparus	α+-mycolate	T	ND	ND	ND	68–72
Skermania	58–64	S,U,T	II	MK-8(H$_4$, ⍵-cycl)	Glycolated	67.5
Smaragdicoccus	43–49	S,U	II	SQA-8(H$_4$, ⍵-cycl) SQB(H$_4$, dicycl)	Glycolated	63.7
Tomitella	42–52	S,U	II	MK-9(H$_2$)	Glycolated	67.5–71.6
Tsukamurella	64–78	S,U,T	II	MK-9	Glycolated	67–78
Williamsia	50–56	S,U,T	II	MK-9(H$_2$)	Glycolated	64–65

*S, straight-chain saturated fatty acids; U, straight-chain unsaturated fatty acids; T, tuberculostearic acid.
Adapted from [2].

Table 2.
Chemotaxonomic features of mycolate genera classified in the order Corynebacteriales.

members of the genera *Corynebacterium*, *Dietzia*, *Gordonia*, *Mycobacterium*, *Nocardia*, and *Tsukamurella*. Species belonging to the genus *Rhodococcus* have been most extensively investigated and are known to produce different chemical types, including a variety of glycolipids. However, an interesting array of biosurfactant structures are synthesized by MACA including lipopeptides, oligosaccharide lipids, polymeric glycolipids, terpenoid glycosides, trehalose corynemycolates, trehalose mycolates and dimycolates, and trehalose lipid (THL) esters [13]. Example structures of the different types of biosurfactants produced by MACA are shown in **Figure 4**. The chemical structure of trehalose-containing glycolipids have perhaps been studied in most detail. Several structural types have been reported including mono-, di- and tri-corynemycolates which have been characterised for species such as *Rhodococcus erythropolis*, *Rhodococcus ruber* and *Rhodococcus wratislaviensis* [14] and trehalose di-nocardiomycolates which have been characterised for *Rhodococcus opacus* [13]. The mycobacterial trehalose mycolates or di-mycolates (cord factors) are also thoroughly investigated given their role as modulators of mycobacterial pathogenesis and host immune response.

3. Habitats, recovery, and growth requirements of MACA

MACA are widely distributed in the environment including natural habitats such as mangroves, soil, freshwater, and deep ocean sediments as well as man-made sites such as activated sludge foams, biofilters, industrial wastewater and indoor building materials. Although predominantly saprophytic, many species are opportunistic pathogens forming parasitic associations with plants and animals, including humans, notably immunocompromised individuals. Several members of the genus *Mycobacterium* cause a plethora of diseases most notably tuberculosis caused by *Mycobacterium bovis* and *Mycobacterium tuberculosis*.

Mycolic and corynemycolic containing trehalose lipids

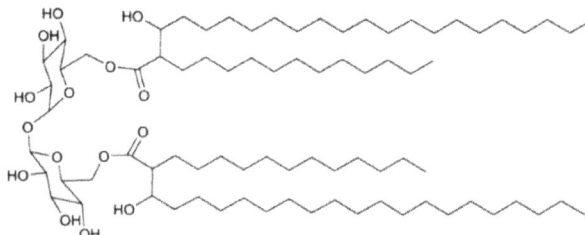

Trehalose dicorynemycolate produced by *Rhodococcus erythropolis*

Trehalose lipid esters

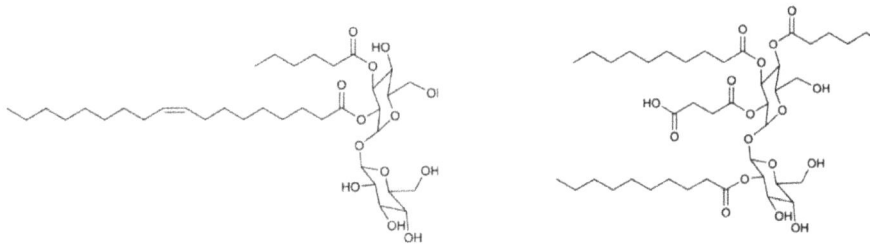

Trehalose diester produced by *Tsukamurella spumae*   Succinic trehalose tetra-ester produced by *Nocardia farcinia*

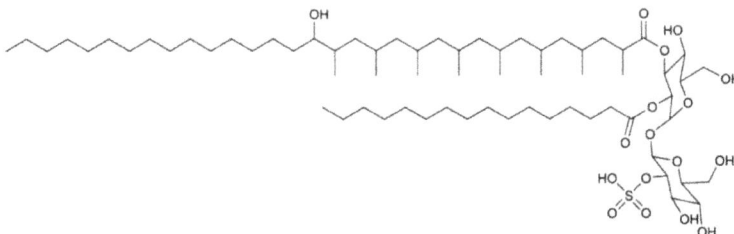

Diacetylated trehalose sulfolipid produced by *Mycobacterium tuberculosis*

Oligosaccharide lipids.

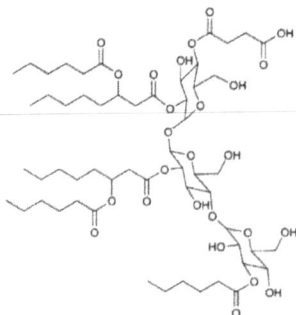

Succinic trisaccharide lipid from *Rhodococcus fascians*

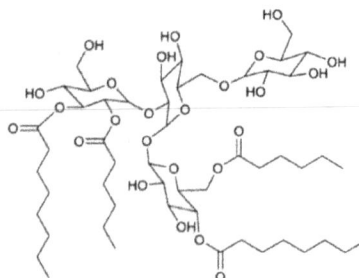

Tetrasaccharide lipid from *Tsukamurella tyrosinosolvens*

Macrocyclic glycosides   **Macrocyclic dilactones**

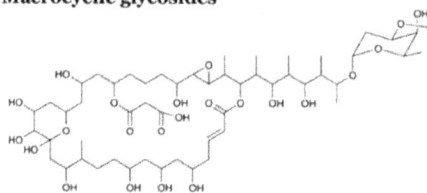

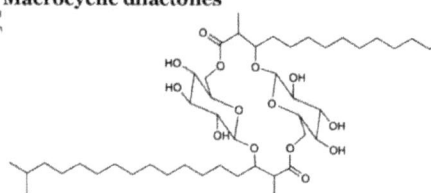

Brasilinoide A produced by *Nocardia brasiliensis*

Glucolypsin A produced by *Nocardia vaccinii*

Terpene-containing biosurfactants

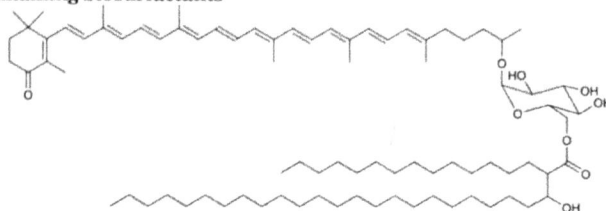

Carotenoid glycoside esterified with a rhodococcus type mycolic acid produced by *Rhodococcus rhodochrous*

Figure 4.
Types and key structural features of various biosurfactants produced by MACA. (Adapted from [13]).

MACA capable of producing various biosurfactants have been isolated from environments (**Table 3**) including oil-contaminated soils [24, 25], water from oil wells [26], wastewater from the rubber industry [21], activated sludge, and effluent and sediment from pesticide manufacturing facilities [23]. The ability of MACA to produce biosurfactants in these habitats appears to be driven by the environmental conditions to which they are exposed whereby the biosurfactants act as mediators for the biodegradation of hydrophobic carbon substrates. Genes involved in biosynthesis of rhamnolipids by *Dietzia maris* for example have been shown to be upregulated in the presence of hydrophobic substrates including *n*-hexadecane, *n*-tetradecane and pristane [15]. However, the true distribution of biosurfactant-producing MACA in the environment may not solely depend on the presence of hydrophobic substrates.

MACA species	Source of isolation	Biosurfactant type	References
D. maris As-13-3	Deep-sea hydrothermal field	Di-rhamnolipid (DRL)	[15]
Dietzia cinnamea KA-1	Water and sediments collected from oil-polluted seasonal ponds	Methylated ester	[16]
Nocardia otitidiscaviarum MTCC 6471	Oil contaminated seawater	Rhamnolipid	[17]
T. spumae DSM44113, *T. spumae* DSM44114 and *T. pseudospumae* DSM44117	Activated sludge foam	THL	[18]
Gordonia amicalis HS-11	Oil contaminated soil	Glycolipid	[19]
Gordonia westfalica GY40	Agricultural soil	Glycolipid	[20]
R. erythropolis 16 LM. USTHB	Water polluted by rejections of 2-mercaptobenzothiazole and its derivatives used in the rubber industry	Fatty acid methyl esters	[21]
Rhodococcus fascians strain A-3	Fell field soil	Rhamnose-containing glycolipid	[22]
Rhodococcus pyridinivorans NT2	Effluent-sediment collected from a pesticide manufacturing facility	THLs	[23]

Table 3.
Various environmental sources of biosurfactant-producing MACA.

Isolation of biosurfactant producers largely relies on selective isolation strategies, utilising hydrophobic compounds as sole carbon sources for energy and growth. Typically, strains are isolated and cultivated using mineral salt medium containing essential trace elements supplemented with a hydrocarbon substrate such as crude oil, diesel, n-alkanes, n-hexadecane, paraffin, polyaromatic hydrocarbons (PAHs), or vegetable oils such as olive oil and rapeseed oil, as the sole carbon source. These may be incorporated into the liquid or solid medium, spread across the agar surface or soaked onto a filter in the lid of petri dishes. Besides the selectivity of the culture medium, pre-enrichment techniques utilising hydrophobic compounds as the sole carbon source, can be used [27]. The principle of enrichment is to provide growth conditions that are favourable for the organisms of interest but not for competing organisms. This selective advantage allows target populations to expand through a series of passages, maximising the chances of successful recovery at the isolation stage. Incorporating antibiotics into the isolation media may provide a useful additional selective pressure to eliminate or reduce unwanted fungi and bacteria.

The ability of an organism to grow on hydrophobic compounds is a good indicator of biosurfactant production but is not a guarantee. It is therefore important that isolates of interest are tested in pure culture for biosurfactant production using further screening assays. It is also possible that biosurfactant-producing organisms may be present in an environment but not enriched by in the conditions provided or indeed producers may be recovered from the environment but not synthesize biosurfactants under the culture conditions imposed. Mining genomes for cryptic biosurfactant biosynthesis pathways, and metagenomic screening of DNA from environmental samples promise an alternative approach to biosurfactant discovery that may circumvent some of the issues associated with culture-dependent strategies [28].

4. Detection and characterisation of biosurfactants

4.1 Biosurfactant screening methods

A variety of methods, both qualitative and quantitative, have been applied to screen microbial cultures and cell-free media for total (intracellular, surface-bound, and freely released) and freely released biosurfactants, respectively. As biosurfactants are structurally diverse, complex molecules, most of these methods are indirect, reliant on physico-chemical properties such as emulsification, surface activity or hydrophobicity. Commonly reported screening methods used to detect biosurfactant production amongst MACA strains are listed in **Table 4**. Besides the bacterial adhesion to hydrocarbons (BATH) assay [37] other tests based on cell surface hydrophobicity include salt aggregation [38] and hydrocarbon overlay [39] assays. The atomized oil assay [40] may be used to directly screen colonies growing on primary isolation plates and is therefore useful as an initial screen for novel-producing strains recovered from the environment. The microplate assay [41] which relies on the wetting properties of biosurfactants and the penetration assay [42], which relies on the reduction of interfacial tension are also considered useful for screening large numbers of strains. Recently, a rapid, high throughput assay that utilises Victoria pure blue BO dye, and is based on surface-active properties, has been developed for quantitative screening, but has not yet been applied to MACA [43].

Detection property	Screening method	MACA species	Reference
Surface activity	Oil spreading	*R. erythropolis*	[29]
		Gordonia spp.	[30]
		Dietzia spp.	[31]
	Drop collapse/ modified drop collapse	*G. westfalica*	[20]
		R. erythropolis	[29]
	Surface and interfacial tension measurement	*Rhodococcus* spp.	[14, 26, 29]
		Gordonia spp.	[20, 30, 32]
		Tsukamurella spp.	[18]
		Dietzia sp.	[31, 33]
		Nocardia sp.	[17]
Emulsification	Emulsification assay	*G. amicalis*	[34]
	Emulsification index	*Rhodococcus* sp.	[14, 22, 35]
		Gordonia spp.	[30, 32]
		Tsukamurella spp.	[18]
		Dietzia spp.	[31, 33]
		Nocardia spp.	[17]
Cell-surface hydrophobicity	Microbial adhesion to hydrocarbons (MATH)/BATH assay	*R. fascians*	[22]
		Gordonia alkanivorans	[36]
		D. maris	[33]
		N. otitidiscaviarum	[17]

Table 4.
Examples of screening methods used to detect biosurfactant production by MACA.

These assays are simpler and more rapid than chemical analytical procedures, and most enable larger-scale screening for biosurfactant production. However, perhaps owing to the general and indirect nature of these assays and various limitations associated with some, test results between assays are not always congruent and no one assay is considered definitive for biosurfactant production. It is thus advisable to use several methods in combination, adopting simple methods to undertake preliminary screening of large strains collections prior to further investigation of those found to be most promising. The development of high-throughput screening, metabolic profiling technologies, and whole-genome analysis promise a more thorough investigation of potential biosurfactant producing strain in the future [28].

4.2 Extraction and structural analysis of biosurfactants

Crude biosurfactant extracts may be obtained from cell cultures (cell-associated and free surfactants) or cell-free broth (free surfactant only) by acidification and solidification followed by solvent extraction of the precipitate. In the case of MACA commonly used solvents include MTBE, dichloromethane, or varying ratios of chloroform–methanol or MTBE–chloroform [44]. Various analytical techniques are used in combination to detect, quantify, and characterise biosurfactants. Thin layer chromatography (TLC) is a straightforward method to separate biosurfactant fractions present in crude extracts. Samples are spotted at the base of a silica plate before development in a solvent system, then air-dried and sprayed with a particular reagent to detect certain chemical groups based on spot colour and/or R_f values. Orcinol, for example, allows detection and differentiation of glycolipids and can distinguish mono-rhamnolipid (MRL) and DRL congeners [45]. However, TLC provides little further detail on congener structure, and it is not generally considered suitable for quantitative analysis although densiometry has been used for this purpose [46]. Biosurfactants may be further separated by silica gel column chromatography.

High-performance liquid chromatography-mass spectrometry (HPLC-MS) allows more precise and accurate characterisation and quantitation of biosurfactant compounds. Isocratic HPLC-UV has been reported for structural and yield determination of THLs produced by *R. erythropolis* strain MTCC 2794 from semi-purified extractions of whole-cell broth [47]. Nuclear magnetic resonance spectroscopy (NMR) is considered the gold standard method to characterise the chemical structure of novel biosurfactants. This has been used in combination with matrix-assisted laser desorption/ionization-time-of-flight mass spectrometry (MALDI-ToF/MS) to elucidate the structure of two novel extracellular THLs TL A and TL B from *Tsukamurella* spp. [18].

A combination of Fourier transform infrared spectroscopy (FTIR), NMR, and liquid chromatography-mass spectrometry (LC-MS) enabled structural characterisation of a novel cyclic lipopeptide, Coryxin, produced by *Corynebacterium xerosis* NS5 [48]. Multiple-Stage Linear Ion-Trap Mass Spectrometry with Electrospray Ionization has been used to determine the structure of trehalose monomycolate (TMM) and trehalose dimycolate (TDM) in the cell wall of *Rhodococcus equi* and *R. opacus* [49]. Ultra-performance liquid chromatography-tandem mass spectrometry (UPLC-MS/MS) has been utilised successfully for the purification and characterisation of sophorolipids and rhamnolipids in *Pseudomonas aeruginosa* [50] and could be applied to similar compounds produced by mycolate species. Gas chromatography-mass spectrometry (GC-MS) is used to characterisation of the fatty acid and mycolic acid components and for the carbohydrate portion of THLs.

5. Potential applications of biosurfactants from MACA

Biosurfactants produced by rhodococci and related MACA have been investigated primarily for their potential application in oil remediation but are otherwise under-studied and under-exploited. However, research studies reveal various potential applications for these molecules, including in environmental and medical fields as summarised in **Figure 5**.

5.1 Biomedical applications

Biosurfactants produced by microorganisms are reported to have various potential biomedical and pharmaceutical applications which have been reviewed widely [1, 51, 52]. This stems from an array of biological properties including anti-adhesion and antibiofilm, anti-inflammatory, antimicrobial (anti-bacterial, anti-fungal and anti-viral), antioxidant, anti-tumour, and wound healing activities. Other potential applications include adjuvants for antigens in vaccines, pulmonary surfactants, drug delivery systems, enhanced vehicles for gene therapy and in dermatological care. Biosurfactants also have several applications in therapeutic dentistry [53]. Daptomycin, a cyclic lipopeptide produced by the actinobacterium *Streptomyces filamentosus*, is used as an antibiotic to treat serious blood and skin infections caused by Gram-positive pathogens [54] and there are other examples of actinobacteria that produce surfactants with potential biomedical applications, such as *Nocardiopsis* strains [55]. Only limited investigation has focused on the biomedical potential of biosurfactants from MACA, except for TDM or cord factors synthesised by intracellular pathogens of the genera *Mycobacterium*. Nevertheless, as shown in **Table 5**, studies over the past two decades reveal that various biosurfactants produced by members of the genera *Corynebacterium*, *Nocardia*, *Rhodococcus*, and *Tsukamurella* demonstrate a range of promising properties.

The amphipathic nature of biosurfactants makes them suitable for anti-adhesion and anti-biofilm applications such as the development of anti-adhesive coatings for intra-urinary devices that are prone to the formation of intractable biofilms, to prevent or delay the onset of biofilm growth by pathogens such as *Escherichia coli* and *Proteus mirabilis*. *C. xerosis* strain NS5, *Nocardia vaccinii* K-8 and various *Rhodococcus* strains demonstrate anti-adhesion, biofilm inhibition and/or biofilm disruption effects against various clinically significant pathogens (**Table 5**). Some

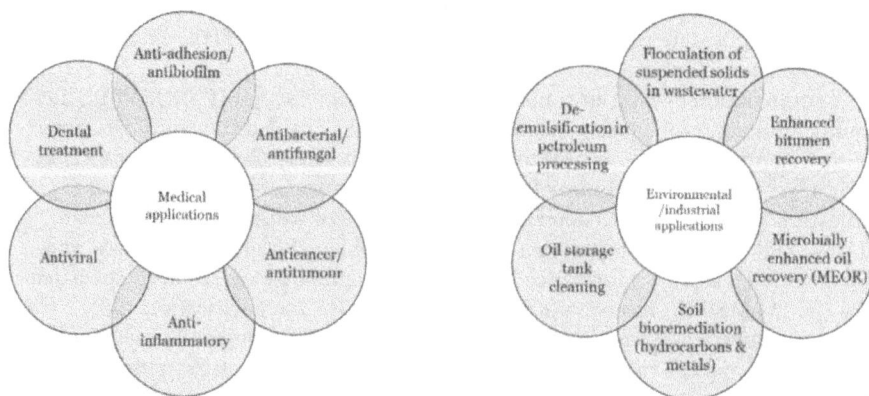

Figure 5.
Promising medical and environmental applications for biosurfactants produced by MACA.

also exhibit antimicrobial properties although in the case of *R. ruber* strain IEGM 231 the trehalolipids had no effect on cell viability despite preventing adhesion of various bacteria to polystyrene [63]. Oligosaccharides produced by *Tsukamurella tyrosinosolvens* (DSM 44370) showed some activity against Gram-positive bacteria, although the pathogenic strain *Staphylococcus aureus* was not affected. *Rhodococcus* strain I2R shows anti-viral activity against herpes simplex virus 1 (HSV-1) and human coronavirus HcoV-OC43 [62].

Nocardia farcinica BN26 produces a THL with anti-cancer effects, showing cytotoxicity against human tumour and promyelocytic leukaemia (HL60) cell lines [57]. *Rhodococcus erythropolis* SD-74 and *Rhodococcus* sp. TB-43 also cause the induction of HL60 cells [59, 60]. *R. ruber* has been studied in some detailed and reported to show immunomodulatory effects, including both *in vitro* induction of Th1-polarizing factors IL-12 and IL-18 by human mononuclear cells and monocytes and *in vivo* induction of IL-1β by mouse peritoneal macrophages [64, 65, 68, 69]. Two succinoyl trehalose lipids, STL-1 and STL-3, produced by *R. erythropolis* SD-74 inhibit growth and induce cell differentiation into monocytes instead of cell proliferation when tested on the HL60 cell line.

Glycolipid bearing mycolic acids, such as trehalose dimycolate (TDM) have attracted extensive investigation as they play a central role in pathogenesis during infection by intracellular pathogens such as *M. tuberculosis* and *R. equi*. TDM's have been researched as a possible tuberculosis vaccine and as an adjuvant. In addition, modification of mycobacterial TDM has been shown to reduce virulence and suppress the host immune response [9]. Interestingly, TDM also possesses biological activities that point towards medical and pharmaceutical applications, such as antitumor activity and immunomodulating functions. Despite this, the potential for TDM is perhaps limited by relatively high toxicity and the pathogenic nature of the species that produce them.

Although biologics including surfactants are generally regarded as less toxic than synthesized pharmaceuticals not much work has focussed on this with respect to MACA surfactants. However, a THL from *R. erythropolis* strain 51T7 has been reported to be suitable for use in cosmetic preparations as it was less irritating than SDS when tested on mouse fibroblast and human keratinocyte lines [70]. Further investigation into the potential biomedical and pharmaceutical applications of biosurfactants produced by members of the MACA, including toxicity testing, is certainly warranted. The high costs and technical challenges associated with production and downstream extraction of biosurfactants may not be a barrier to their commercial application in biomedical fields given that smaller-scale productions would likely be required.

5.2 Environmental applications

Biosurfactants have a range of promising, and increasingly important, applications in the environmental, industrial, and agricultural sectors (**Table 6**). These include bioremediation of both organic pollutants (especially hydrocarbons) and metals, microbial enhanced oil recovery (MEOR), cleaning and maintenance of tanks and pipelines in the petroleum industry, wastewater treatment, and agricultural applications such as promotion of plant growth/health and inhibition of phytopathogenic fungi [1, 78]. MACA-derived surfactants have been investigated in some of these contexts, although the focus is on well-known species such as *R. ruber* and *R. erythropolis*. Members of *Gordonia, Corynebacterium, Nocardia* and *Dietzia* have also been investigated but there is likely to be much unexplored potential within the group [79]. This is supported by the promising results obtained with rhamnolipids produced by other bacteria, most notably *P. aeruginosa*, and their

Strain (origin)	Biosurfactant	Biomedical properties	Reference
C. xerosis NS5 (human axilla)	Purified Coryxin (lipopeptide)	Antibacterial activity, biofilm inhibition and disruption of pre-formed biofilms of Gram-positive *S. aureus* and *Streptococcus mutans* and Gram-negative *E. coli* and *P. aeruginosa* strains	[48]
Nocardia brasiliensis IFM-0406 (patient with lung nocardiosis)	Aliphatic macrolide (Brasilinolide)	Moderately antifungal against *Aspergillis*, *Candida*, *Cryptococcus* and *Paecilomyces* species	[56]
N. farcinica BN26 (hydrocarbon polluted soil)	THL	Anti-tumour activity: cytotoxic effects on human tumour cell lines BV-173 and SKW-3, and to a lesser extent, HL-60. Mediated cell death by the induction of partial apoptotic DNA laddering	[57]
N. vaccinii IMB B7405 (K-8) (oil-contaminated soil)	Complex of amino lipids; neutral lipids (mycolic and *n*-alkanic acids); trehalose di-acelates and di-mycolates (surfactant solution and supernatant)	Anti-adhesive activity against Gram-negative bacteria *E. coli*, *Proteus vulgaris*, *P. aeruginosa* and *Enterobacter cloaceae* and the yeast *Candida albicans* on silicon urogenital catheters. Anti-adhesive activity against fungus *C. albicans* and bacterium *E. coli* on treated acrylic dental material and against Gram-positive *Bacillus subtilis* and micromycete *Aspergillis niger* when coated on various abiotic substrates	[58]
R. erythropolis SD-74 (alkaline soil)	Purified STL-1	*In vitro* induction of human monocytoid leukemic cell line U937 differentiation and cytotoxicity against human lung carcinoma cell line A549	[59, 60]
		In vitro induction of human promyelocytic leukaemia (HL60) cell line differentiation into monocytes and inhibition of protein kinase C	
R. erythropolis IMB Ac-5017 (EK-1) (oil-contaminated soil)	Complex of trehalose mono- and di-mycolates; neutral lipids (cetyl alcohol, palmitic acid, methyl ether of *n*-pentadecanoic acid, mycolic acids); phospholipids (phosphatidylglycerol, phosphatidylethanol-amine) (surfactant solution and supernatant)	Antibacterial activity against Gram-positive bacteria *B. subtilis* and *S. aureus* and Gram-negative *E. coli* and *Pseudomonas* sp., and anti-fungal activity against *C. albicans*, *C. utilis* and *C. tropicalis*	[58]
		Anti-adhesive activity against Gram-negative bacteria and fungus *C. albicans* on silicon urogenital catheters. Anti-adhesive activity against *B. subtilis* on various abiotic substrates, against *C. albicans* and *E. coli* on acrylic dental material and *S. aureus* and *P. aeruginosa* on plastic and steel	
R. fascians BD8 (Arctic soil	THL	Antibacterial activity against *Vibrio harveyi* and *P. vulgaris*, and partial inhibition of other Gram-positive	[61]

Strain (origin)	Biosurfactant	Biomedical properties	Reference
polluted with hydrocarbons		and negative bacteria and fungus *C. albicans*. Anti-adhesion properties on polystyrene against various Gram-positive and negative strains and fungal strains of *C. albicans*. Biofilm inhibition on glass, polystyrene, and silicone urethral catheters against Gram-positive *Enterococcus hirae* and *E. faecalis*, Gram-negative *E. coli*, and fungus *C. albicans*	
Rhodococcus sp. I2R (marine sediment)	Extracellular complex of glycolipids (crude extracts and purified fractions)	Antiviral activity against HSV-1 and human coronavirus HCoV-OC43. Antiproliferation activity against human prostatic carcinoma cell line PC3	[62]
R. ruber IEGM 231 (spring water, oil-extracting enterprise)	Crude trehalolipids	Anti-adhesive activity against exponentially growing Gram-positive bacteria *Arthrobacter simplex*, *B. subtilis*, *Brevibacterium linens*, *Corynebacterium glutamicum*, and *Micrococcus luteus* and against Gram-negative bacteria *E. coli* and *P. fluorescens* on polystyrene.	[63]
	Mixture of TDM, diacyltrehalose and monoacyltrehalose isolated by column chromatography	*In vitro* induction of IL-1β, IL-6, and TNF-α cytokine secretion by human monocytes	[64]
		In vitro induction of Th1-polarizing factors IL-12 and IL-18 by human mononuclear cells and monocytes and reactive oxygen species (ROS) by peripheral blood leukocytes	[65]
	Glycolipid	*In vivo* induction of IL-1β by mouse peritoneal macrophages	[66]
	Monoacyltrehalose fraction (MAT)	*In vivo* suppression of bactericidal activity and proinflammatory cytokine IL-1β of mouse peritoneal macrophages, antibody production by splenocytes and stimulates the production of IL-10	[67]
Rhodococcus sp. TB-42 (soil)	Analogues of STL-3	Inhibited growth and induced the differentiation of human HL-60 promyelocytic leukaemia cell line	[66]
T. tyrosinosolvens DSM 44370 (oil containing soil)	Purified oligosaccharide lipids	Antimicrobial activity against Gram-positive bacterial strain of *Bacillus megaterium*, Gram-negative *E. coli* and fungal strain *Ustilago violacea*	[67]

Table 5.
Biomedical research on biosurfactants produced by MACA.

commercialisation [80]. It is not unreasonable to expect that rhamnolipids produced by MACA may also exhibit such properties. Indeed, the search for non-pathogenic producers is important for further development of biosurfactant production at industrial scale [81].

Pollution of soils with organic and inorganic chemical compounds is a major environmental issue. Biosurfactants are used to improve the solubility of hydrocarbon organic compounds, either to make them available for subsequent biodegradation or to facilitate removal by soil washing. A remediation agent called JE1058BS containing biosurfactant from *Gordonia* sp. strain JE-1058 was evaluated as an oil spill dispersant using the baffled flask test recommended by the US Environmental Protection Agency and performed better than commercially available dispersants. It also enhanced the bioremediation of crude oil by indigenous marine bacteria and significantly improved removal of crude oil from contaminated sea sand by washing compared with the use of seawater alone [73]. Various *Dietzia*, *Gordonia* and *Rhodococcus* strains have been shown to degrade hydrocarbon compounds and many studies show that the production of surface-active compounds makes an important contribution. In a recent study, *G. amicalis* HS-11 was able to remove 92.85% of the diesel oil provided as the sole carbon source after 16 days of incubation, with a corresponding reduction in surface tension due to the production of extracellular surfactants. Microscopy suggested that these surfactants play a role in the emulsification and uptake of the hydrocarbons. Plant-based bioassays also showed that toxicity of the diesel oil decreased. This illustrates the potential of this strain and perhaps other gordoniae for use in the bioremediation of contaminated environments, or industrial wastewaters [82].

The properties and actions of biosurfactants make them particularly relevant to the petroleum industry. MEOR is perhaps the most well-known application in this area. Biosurfactants, or biosurfactant-producing microorganisms, are used to extract some of the oil remaining in reservoirs after primary and secondary processing has been carried out. Mechanisms include reduction of capillary forces holding the oil in porous rock, stabilisation of desorbed oil in water and increased viscosity of oil for easier removal [83]. *Dietzia* sp. ZQ-4, a hydrocarbon-degrading, surfactant-producing MACA isolated from an oil reservoir, demonstrated potential for use in *ex situ* oil recovery. Fermentation broth significantly increased oil displacement efficiency by 18.82% in rock cores and performed well within the range reported for other strains. However, injection of the strain itself was not so successful, and field trials testing nutrient injection did not always result in an increase in the population of *Dietzia* sp. ZQ-4, indicating that an *in-situ* approach may not be viable although it may be possible to optimise this strategy further [72]. Biosurfactants produced by various rhodococci strains recovered from oil-polluted soils have been shown to be effective at recovering trapped oil from oil-saturated sand packs. Glycolipids produced by strain ST-5 recovered up to 86% [84] and a mix of glycolipids and extracellular lipids produced by strain TA6 up to 86% [24] using the sand pack column method. Studies on biosurfactant produced by *R. ruber* IEGM 231 showed that 2.5 times greater washing activity could be achieved than with synthetic surfactant Tween-60 in soil columns spiked with polyaromatic carbons (PAHs) and alkanes. The biosurfactant maintained activity at a high (5% w/w) contamination level and consistently removed 0.3–0.5 g PAHs per kg dry soil in a single run of washing [71].

Biosurfactants may also be used to de-emulsify water–oil emulsions that form during oil production in the oilfields, as well as during transportation, and processing and offer a more ecologically friendly solution than chemically synthesized de-emulsifiers. A lipopeptide bio-demulsifier produced by *Dietzia* sp. strain S-JS-1 grown on waste frying oil achieved 88.3% of oil separation ratio in water/oil emulsion and 76.4% of water separation ratio in oil/water emulsion [75].

Biosurfactants have been shown to reduce phytotoxicity of heavy metals, and pre-treatment of seeds could allow plants to be grown successfully in contaminated soil, facilitating phytoremediation of the environment. Crude biosurfactant from *R. ruber* IEGM 231 mitigated the toxic effects of high concentrations of molybdenum on oat, white mustard, and vetch seeds. Germination increased up to 4.5 times and

Application	Examples of MACAs	Reference/s
Bioremediation: enhanced hydrocarbon solubility and degradation	*D. maris* As-13-3	[15]
	D. maris WR3	[33]
	G. amicalis HS-11	[34]
	Gordonia cholesterolivorans AMP 10	[32]
	N. otitidiscaviarum	[17]
	R. erythropolis 3C-9	[29]
	R. pyridinivorans NT2	[23]
Bioremediation: soil washing	*Gordonia* sp. strain BS29	[30]
	R. ruber (IEGM 231)	[71]
MEOR	*Dietzia* sp. ZQ-4	[72]
	Gordonia sp. JE-1058	[73]
	R. ruber Z25	[74]
	Rhodococcus sp. strain TA6	[24]
Bio-demulsification: treatment of water-oil emulsions generated during processing of petroleum	*Dietzia* sp. S-JS-1	[75]
Paraffin control in oil transport pipelines	*G. amicalis* LH3	[76]
Bioflocculation (e.g., for oil recovery from wastewater)	*R. erythropolis* S-1	[77]

Table 6.
Various potential environmental applications of biosurfactants produced by MACA.

shoot and/or root length up to 2.5 times when seeds were pre-treated with a biosurfactant emulsion and grown under conditions of molybdenum contamination [85]. Similar results have been recorded for other heavy metals such as copper [86].

The use of biosurfactants in environmental and industrial applications is limited by the current high costs of production, and the large amounts of biosurfactant required. However, using waste and/or renewable substrates would be cheaper, and a highly purified product is not essential so costs of downstream processing can also be reduced. In addition, different approaches such as selective stimulation of biosurfactant producers *in situ*, and inoculation of biosurfactant-producing cultures, are being explored [87]. This could potentially overcome some of the challenges associated with accessing the cell-bound biosurfactants produced by MACA such as *Rhodococcus* spp.

5.3 Challenges to commercialisation

Currently, commercial production of biosurfactants is not economically competitive with chemical surfactant production as there are various challenges to overcome. Bioprocesses presently achieve low biosurfactant productivity and yield and substrates are expensive [6]. Foam formation can cause serious operational issues and downstream biosurfactant recovery can be technically involved and costly. Development work to optimise bioprocesses should focus on enhancing biosurfactant yield and potency. Approaches include the search and discovery of novel biosurfactant-producing organisms and strain improvement by various genetic engineering methods and/or stress-fermentation including co-cultivation [84]. Yield can also be enhanced through the optimisation of culture conditions and costs reduced through the introduction of renewable or waste products [6, 28, 77] as cheaper feed stocks. The effects of biosurfactants on human health and the environment also require further assessment to ensure safe production and use.

6. Conclusions

Biosurfactants offer an attractive proposition for biotechnological application across various sectors and are considered superior to synthetic surfactants. Diverse MACA produce biosurfactants with interesting properties that have been explored in the context of biomedicine and environmental remediation. However, many MACA have not yet been investigated for biosurfactant production and various potential applications are yet to receive significant research. Rapid, reliable methods for high throughput screening for biosurfactant production are essential as are robust standard methods for biosurfactant purification and characterisation. Efforts to evaluate and expand the knowledge of structural characteristics and gene regulation of biosurfactants are warranted to improve their effectiveness and productivity. Commercial-scale production will need to employ various existing and new strategies to become economic and sustainable. Cutting-edge technologies such high-throughput omics-based tools should accelerate the development of commercial production of biosurfactants. Furthering our understanding of biosurfactants produced by MACA will facilitate their commercial exploitation thereby contributing to a sustainable bio-based economy.

Conflict of interest

The authors declare that there is no conflict of interest.

Author details

Fiona M. Stainsby*, Janki Hodar and Halina Vaughan
Department of Life Sciences, School of Applied Sciences, Edinburgh Napier University, Edinburgh, UK

*Address all correspondence to: f.stainsby@napier.ac.uk

IntechOpen

References

[1] Mnif I, Ghribi D. Lipopeptides biosurfactants: Mean classes and new insights for industrial, biomedical, and environmental applications. Biopolymers. 2015;**104**:129-147. DOI: 0.1002/bip.22630

[2] Goodfellow M, Jones AL. *Corynebacteriales* ord. nov. In: Whitman WB, editor. Bergey's Manual of Systematics of Archaea and Bacteria. New Jersey: Wiley; 2015. p. 14. DOI: 10.1002/9781118960608.obm00009

[3] Bognolo G. Biosurfactants as emulsifying agents for hydrocarbons. Colloids and Surfaces A: Physicochemical and Engineering Aspects. 1999;**152**(1-2):41-52. DOI: 10.1016/S0927-7757(98)00684-0

[4] Banat I, Franzetti A, Gandolfi I, Bestetti G, Martinotti M, Fracchia L, et al. Microbial biosurfactants production, applications, and future potential. Applied Microbiology and Biotechnology. 2010;**87**(2):427-444. DOI: 10.1007/s00253-010-2589-0

[5] Mao X, Jiang R, Xiao W, Yu J. Use of surfactants for the remediation of contaminated soils: A review. Journal of Hazardous Materials. 2015;**285**:419-435. DOI: 10.1016/j.jhazmat.2014.12

[6] Makkar R, Cameotra S, Banat I. Advances in utilization of renewable substrates for biosurfactant production. AMB Express. 2011;**1**(1):5. DOI: 10.1186/2191-0855-1-5

[7] Embley TM, Stackebrandt E. Phylogeny and systematics of the actinomycetes. Annual Review of Microbiology. 1994;**48**:257-289. DOI: 10.1146/annurev.mi.48.100194.001353

[8] Baek I, Kim M, Lee I, Na S-I, Goodfellow M, Chun J. Phylogeny trumps chemotaxonomy: A case study involving *Turicella otitidis*. Frontiers in

Microbiology. 2018;**9**:834-844. DOI: 10.3389/fmicb.2018.00834

[9] Marrakchi H, Laneelle MA, Daffe M. Mycolic acids: Structures, biosynthesis, and beyond. Chemistry and Biology. 2014;**21**:67-85. DOI: 10.1016/j. chembiol.2013.11.011

[10] Eales KL, Nielsen JL, Seviour EM, Nielsen PH, Seviour RJ. The *in situ* physiology of *Skermania piniformis* in foams in Australian activated sludge plants. Environmental Microbiology. 2006;**8**:1712-1720. DOI: 10.1111/ j.1462-2920.2006.01107.x

[11] Sangal V, Goodfellow M, Jones AL, Schwalbe EC, Blom J, Hoskisson PA, et al. Next-generation systematics: An innovative approach to resolve the structure of complex prokaryotic taxa. Scientific Reports. 2016;**2016**(6): 383-392. DOI: 10.1038/srep38392

[12] Gao B, Gupta RS. Phylogenetic framework and molecular signatures for the main clades of the phylum Actinobacteria. Microbiology and Molecular Biology Reviews. 2012;**76**(1): 66-112. DOI: 10.1128/mmbr.05011-11

[13] Kügler JH, Le Roes-Hill M, Syldatk C, Hausmann R. Surfactants tailored by the class actinobacteria. Frontiers in Microbiology. 2015;**6**:212. DOI: 10.3389/fmicb.2015.00212

[14] Tuleva B, Christova N, Cohen R, Stoev G, Stoineva I. Production and structural elucidation of trehalose tetraesters (biosurfactants) from a novel alkanothrophic *Rhodococcus wratislaviensis* strain. Journal of Applied Microbiology. 2008;**104**:1703-1710. DOI: 10.1111/j.1365-2672.2007.03680.x

[15] Wang W, Cai B, Shao Z. Oil degradation and biosurfactant production by the deep sea bacterium *Dietzia maris* As-13-3. Frontiers in

Microbiology. 2014;**5**:711. DOI: 10.3389/
fmicb.2014.00711

[16] Kavyanifard A, Ebrahimipour G,
Ghasempour A. Structure
characterization of a methylated ester
biosurfactant produced by a newly
isolated *Dietzia cinnamea* KA1.
Microbiology. 2016;**85**(4):430-435. DOI:
10.1134/S0026261716040111

[17] Vyas TK, Dave BP. Production of
biosurfactant by *Nocardia
otitidiscaviarum* and its role in
biodegradation of crude oil.
International Journal of Environmental
Science and Technology. 2011;**8**:
425-432. DOI: 10.1007/BF03326229

[18] Kügler J, Muhle-Goll C, Kühl B,
Kraft A, Heinzler R, Kirschhöfer F, et al.
Trehalose lipid biosurfactants produced
by the actinomycetes *Tsukamurella
spumae* and *T. pseudospumae*. Applied
Microbiology and Biotechnology. 2014;
98(21):8905-8915. DOI: 10.1007/
s00253-014-5972-4

[19] Sowani H, Mohite P, Munot H,
Shouche Y, Bapat T, Kumar A, et al.
Green synthesis of gold and silver
nanoparticles by an actinomycete
Gordonia amicalis HS-11: Mechanistic
aspects and biological application.
Process Biochemistry. 2016;**51**(3):
374-383. DOI: 10.1016/j.
procbio.2015.12.013

[20] Laorrattanasak S,
Rongsayamanont W, Khondee N,
Paorach N, Soonglerdsongpha S,
Pinyakong O, et al. Production and
application of *Gordonia westfalica* GY40
biosurfactant for remediation of fuel oil
spill. Water Air and Soil Pollution. 2016;
227(9):325. DOI: 10.1007/s11270-016-
3031-8

[21] Sadouk Z, Hacene H, Tazerouti A.
Biosurfactants production from low cost
substrate and degradation of diesel oil
by a *Rhodococcus* strain. Oil & Gas
Science and Technology–Revue de l'IFP.

2008;**63**(6):747-753. DOI: ff10.2516/
ogst:2008037ff.ffhal-02002052f

[22] Gesheva V, Stackebrandt E,
Vasileva-Tonkova E. Biosurfactant
production by halotolerant *Rhodococcus
fascians* from Casey Station, Wilkes
Land, Antarctica. Current Microbiology.
2010;**61**(2):112-117. DOI: 10.1007/
s00284-010-9584-7

[23] Kundu D, Hazra C, Dandi N,
Chaudhari A. Biodegradation of 4-
nitrotoluene with biosurfactant
production by *Rhodococcus
pyridinivorans* NT2: Metabolic pathway,
cell surface properties and toxicological
characterization. Biodegradation. 2013;
24(6):775-793. DOI: 10.1007/
s10532-013-9627-4

[24] Shavandi M, Mohebali G, Haddadi A,
Shakarami H, Nuhi A. Emulsification
potential of a newly isolated biosurfactant-
producing bacterium, *Rhodococcus* sp.
strain TA6. Colloids and Surfaces B:
Biointerfaces. 2011;**82**(2):477-482. DOI:
10.1016/j.colsurfb.2010.10.005

[25] Kumari B, Singh SN, Singh DP.
Characterization of two biosurfactant
producing strains in crude oil
degradation. Process Biochemistry.
2012;**47**(12):2463-2471. DOI: 10.1016/j.
procbio.2012.10.010

[26] Zheng C, Li S, Yu L, Huang L,
Wu Q. Study of the biosurfactant-
producing profile in a newly isolated
Rhodococcus ruber strain. Annals of
Microbiology. 2009;**59**:771-776. DOI:
10.1007/BF03179222

[27] Domingues PM, Louvado A,
Oliveira V, Coelho FJCR, Almeida A,
Gomes NCM, et al. Selective cultures for
the isolation of biosurfactant producing
bacteria: Comparison of different
combinations of environmental inocula
and hydrophobic carbon sources.
Biochemistry & Biotechnology. 2013;
43(3):237-255. DOI: 10.1080/
10826068.2012.719848

[28] Kubicki S, Bollinger A, Katzke N, Jaeger KE, Loeschcke A, Thies S. Marine biosurfactants: Biosynthesis, structural diversity and biotechnological applications. Marine Drugs. 2019;**17**(7): 408. DOI: 10.3390/md17070408

[29] Peng F, Liu Z, Wang L, Shao Z. An oil-degrading bacterium: *Rhodococcus erythropolis* strain 3C-9 and its biosurfactants. Journal of Applied Microbiology. 2007;**102**:1603-1611. DOI: 10.1111/j.1365-2672.2006.03267.x

[30] Franzetti A, Bestetti G, Caredda P, La Colla P, Tamburini E. Surface-active compounds and their role in the access to hydrocarbons in *Gordonia* strains. FEMS Microbiology Ecology. 2008; **63**(2):238-248. DOI: 10.1111/ j.1574-6941.2007.00406.x

[31] Hvidsten I, Mjøs SA, Holmelid B, Bødtker G, Barth T. Lipids of *Dietzia* sp. A14101. Part I: A study of the production dynamics of surface-active compounds. Chemistry and Physics of Lipids. 2017;**208**:19-30. DOI: 10.1016/j. chemphyslip.2017.08.00

[32] Kurniati TH, Rusmana I, Suryani A, Mubarik NR. Degradation of polycyclic aromatic hydrocarbon pyrene by biosurfactant-producing bacteria *Gordonia cholesterolivorans* AMP 10. Biosaintifika: Journal of Biology and Biology Education. 2016;**8**(3): 336-343

[33] Nakano M, Kihara M, Iehata S, Tanaka R, Maeda H, Yoshikawa T. Wax ester-like compounds as biosurfactants produced by *Dietzia maris* from *n*-alkane as a sole carbon source. Journal of Basic Microbiology. 2011;**51**:490-498. DOI: 10.1002/jobm.201000420

[34] Jackisch-Matsuura AB, Santos LS, Eberlin MN, de Faria AF, Matsuura T, Grossman MJ, et al. Production and characterization of surface-active compounds from *Gordonia amicalis*. Environmental Sciences–Brazilian Archives of Biology and Technology. 2014;**57**(1):138-144. DOI: 10.1590/ S1516-89132014000100019

[35] Suryanti V, Hastuti S, Andriani D. Optimization of biosurfactant production in soybean oil by *Rhodococcus rhodochrous* and its utilization in remediation of cadmium-contaminated solution. IOP Conference Series: Materials Science and Engineering. 2016;**107**(1):012018. DOI: 10.1088/1757-899X/107/1/012018

[36] Mohebali G, Ball A, Kaytash A, Rasekh B. Stabilization of water/gas oil emulsions by desulfurizing cells of *Gordonia alkanivorans* RIPI90A. Microbiology. 2007;**153**(5):1573-1581. DOI: 10.1099/mic.0.2006/002543-0

[37] Rosenberg M, Gutnick D, Rosenberg E. Adherence to bacteria to hydrocarbons: A simple method for measuring cell-surface hydrophobicity. Federation of European Microbiological Societies Microbiology Letter. 1980;**9**: 29-33. DOI: 10.1111/j.1574-6968.1980. tb05599.x

[38] Lindahl M, Faris A, Wadström T, Hjertén S. A new test based on 'salting out' to measure relative hydrophobicity of bacterial cells. Biochimica et Biophysica Acta (BBA)–General Subjects. 1981;**677**(3-4):471-476. DOI: 10.1016/0304-4165(81)90261-0

[39] Morikawa M, Ito M, Imanaka T. Isolation of a new surfactin producer *Bacillus pumilus* A-1, and cloning and nucleotide sequence of the regulator gene, psf-1. Journal of Fermentation and Bioengineering. 1992;**74**(5):255-261. DOI: 10.1016/0922-338X(92)90055-Y

[40] Burch A, Shimada B, Browne P, Lindow S. Novel high-throughput detection method to assess bacterial surfactant production. Applied and Environmental Microbiology. 2010; **76**(16):5363-5372. DOI: 10.1128/ AEM.00592-10

[41] Cottingham M, Bain C, Vaux D. Rapid method for measurement of surface tension in multiwell plates. Lab Investigation. 2004;**84**:523-529. DOI: 0.1038/labinvest.3700054

[42] Maczek J, Junne S, Götz P. Examining biosurfactant producing bacteria–an example for an automated search for natural compounds. Application Note CyBio AG. 2007

[43] Kubicki S, Bator I, Jankowski S, Schipper K, Tiso T, Feldbrügge M, et al. A straightforward assay for screening and quantification of biosurfactants in microbial culture supernatants. Frontiers in Bioengineering and Biotechnology. 2020;**8**:958. DOI: 10.3389/fbioe.2020.00958 8

[44] Kuyukina MS, Ivshina IB, Philp JC, Christofi N, Dunbar SA, Ritchkova MI. Recovery of *Rhodococcus* biosurfactants using methyl tertiary-butyl ether extraction. Journal of Microbiological Methods. 2001;**46**(2):149-156. DOI: 10.1016/S0167-7012(01)00259-7

[45] Christova N, Tuleva B, Lalchev Z, Jordanova A, Jordanov B. Rhamnolipid biosurfactants produced by *Renibacterium salmoninarum* 27BN during growth on *n*-hexadecane. Zeitschrift für Naturforschung C. 2004; **59**(1-2):70-74. DOI: 10.1515/znc-2004-1-215

[46] Das P, Mukherjee S, Sen R. Improved bioavailability and biodegradation of a model polyaromatic hydrocarbon by a biosurfactant producing bacterium of marine origin. Chemosphere. 2008;**72**(9):1229-1234. DOI: 10.1016/j.chemosphere.2008. 05.015

[47] Patil HI, Pratap AP. Production and quantitative analysis of trehalose lipid biosurfactants using high-performance liquid chromatography. Journal of Surfactants and Detergents. 2018;**21**: 553-564. DOI: 10.1002/jsde.12158

[48] Dalili D, Amini M, Faramarzi MA, Fazeli MR, Khoshayand MR, Samadi N. Isolation and structural characterization of Coryxin, a novel cyclic lipopeptide from *Corynebacterium xerosis* NS5 having emulsifying and anti-biofilm activity. Colloids and Surfaces B: Biointerfaces. 2015;**135**:425-432. DOI: 10.1016/j.colsurfb.2015.07.005

[49] Frankfater C, Henson WR, Juenger-Leif A, Foston M, Moon TS, Turk J, et al. Structural determination of a new peptidolipid family from *Rhodococcus opacus* and the pathogen *Rhodococcus equi* by multiple stage mass spectrometry. Journal of the American Society for Mass Spectrometry. 2020; **31**(3):611-623. DOI: 10.1021/ jasms.9b00059

[50] Dardouri M, Mendes RM, Frenzel J, Costa J, Ribeiro IAC. Seeking faster, alternative methods for glycolipid biosurfactant characterization and purification. Analytical and Bioanalytical Chemistry. 2021;**413**: 4311-4320. DOI: 10.1007/s00216-021-03387-4

[51] Markande AR, Patel D, Varjani S. A review on biosurfactants: Properties, applications and current developments. Bioresource Technology. 2021;**330**: 124963. DOI: 10.1016/j. biortech.2021.124963

[52] Naughton P, Marchant R, Naughton V, Banat I. Microbial biosurfactants: Current trends and applications in agricultural and biomedical industries. Journal of Applied Microbiology. 2019;**127**:12-28. DOI: 10.1111/jam.14243

[53] Patil R, Ishrat S, Chaurasia A. Biosurfactants - A new paradigm in therapeutic dentistry. Saudi Journal of Medicine. 2021;**6**:20-28. DOI: 10.36348/ sjm.2021.v06i01.005

[54] Tedesco KL, Rybak MJ. Daptomycin. Pharmacotherapy: The

Journal of Human Pharmacology and Drug Therapy. 2004;**24**:41-57. DOI: 10.1592/phco.24.1.41.34802

[55] Rajivgandhi G, Vijayan R, Maruthupandy M, Vaseeharan B, Manoharan N. Antibiofilm effect of *Nocardiopsis* sp. GRG 1 (KT235640) compound against biofilm forming Gram negative bacteria on UTIs. Microbial Pathogenesis. 2018;**118**: 190-198. DOI: 10.1016/j. micpath.2018.03.011

[56] Mikami Y, Komaki H, Imai T, Yazawa K, Nemoto A, Tanaka Y, et al. New antifungal macrolide component, brasilinolide B, produced by *Nocardia brasiliensis*. The Journal of Antibiotics. 2000;**53**:70-74. DOI: 10.7164/ antibiotics.53.70

[57] Christova N, Lang S, Wray V, Kaloyanov K, Konstantinov S, Stoineva I. Production, structural elucidation, and in vitro antitumor activity of trehalose lipid biosurfactant from *Nocardia farcinica* strain. Journal of Microbiology and Biotechnology. 2015;**25**(4):439-447. DOI: 10.4014/ jmb.1406.06025

[58] Pirog TP, Konon AD, Beregovaya KA, Shulyakova MA. Antiadhesive properties of the surfactants of *Acinetobacter calcoaceticus* IMB B-7241, *Rhodococcus erythropolis* IMB Ac-5017, and *Nocardia vaccinii* IMB B-7405. Microbiology. 2014;**83**:732-739. DOI: 10.1134/S0026261714060150

[59] Isoda H, Shinmoto H, Matsumura M, Nakahara T. Succinoyl trehalose lipid induced differentiation of human monocytoid leukemic cell line U937 into monocyte-macrophages. Cytotechnology. 1995;**19**:79-88. DOI: 10.1007/BF00749758

[60] Isoda H, Shinmoto H, Kitamoto D, Matsumura M, Nakahara T. Differentiation of human promyelocytic leukemia cell line HL60 by microbial extracellular glycolipids. Lipids. 1997;**32**: 263-271. DOI: 10.1007/s11745-997- 0033-0

[61] Janek T, Krasowska A, Czyżnikowska Ż, Łukaszewicz M. Trehalose lipid biosurfactant reduces adhesion of microbial pathogens to polystyrene and silicone surfaces: An experimental and computational approach. Frontiers in Microbiology. 2018;**9**:2441. DOI: 10.3389/ fmicb.2018.02441

[62] Palma Esposito F, Giugliano R, Della Sala G, Vitale GA, Buonocore C, Ausuri J, et al. Combining OSMAC approach and untargeted metabolomics for the identification of new glycolipids with potent antiviral activity produced by a marine *Rhodococcus*. International Journal of Molecular Sciences. 2021;**22**: 9055. DOI: 10.3390/ijms22169055

[63] Kuyukina MS, Ivshina IB, Korshunova IO, Stukova GI, Krivoruchko AV. Diverse effects of a biosurfactant from Rhodococcus ruber IEGM 231 on the adhesion of resting and growing bacteria to polystyrene. AMB Express. 2016;**6**:14. DOI: 10.1186/ s13568-016-0186-z

[64] Kuyukina MS, Ivshina IB, Gein SV, Baeva TA, Chereshnev VA. In vitro immunomodulating activity of biosurfactant glycolipid complex from *Rhodococcus ruber*. Bulletin of Experimental Biology and Medicine. 2007;**144**(3):326-330. DOI: 10.1007/ s10517-007-0324-3 PMID: 18457028

[65] Baeva TA, Gein SV, Kuyukina MS, Ivshina IB, Kochina OA, Chereshnev VA. Effect of glycolipid *Rhodococcus* biosurfactant on secretory activity of neutrophils in vitro. Bulletin of Experimental Biology and Medicine. 2014;**157**(2):238-242. DOI: 10.1007/ s10517-014-2534-9

[66] Sudo T, Zhao X, Wakamatsu Y, Shibahara M, Nomura N, Nakahara T, et

al. Induction of the differentiation of human HL-60 promyelocytic leukemia cell line by succinoyl trehalose lipids. Cytotechnology. 2000;**33**(1-3): 259-264. DOI: 10.1023/A: 1008137817944

[67] Vollbrecht E, Rau U, Lang S. Microbial conversion of vegetable oils into surface-active di-, tri-, and tetrasaccharide lipids (biosurfactants) by the bacterial strain *Tsukamurella* spec. European Journal of Lipid Science and Technology. 1999;**101**:389-394. DOI: 10.1002/(SICI)1521-4133(199910) 101:10<389::AID-LIPI389>3.0.CO;2-9

[68] Gein SV, Kochina OA, Kuyukina MS, Ivshina IB. Effects of glycolipid Rhodococcus biosurfactant on innate and adaptive immunity parameters in vivo. Bulletin of Experimental Biology and Medicine. 2018;**165**:368-372. DOI: 10.1007/s10517-018-4172-0

[69] Gein SV, Kochina OA, Kuyukina MS, Klimenko DP, Ivshina IB. Effects of monoacyltrehalose fraction of *Rhodococcus* biosurfactant on the innate and adaptive immunity parameters *in vivo*. Bulletin of Experimental Biology and Medicine. 2020;**169**(4):474-477. DOI: 10.1007/s10517-020-04912-8

[70] Marqués AM, Pinazo A, Farfan M, Aranda FJ, Teruel JA, Ortiz A, et al. The physicochemical properties and chemical composition of trehalose lipids produced by *Rhodococcus erythropolis* 51T7. Chemistry and Physics of Lipids. 2009;**158**(2):110-117. DOI: 10.1016/j. chemphyslip.2009.01.001

[71] Ivshina I, Kostina L, Krivoruchko A, Kuyukina M, Peshkur T, Anderson P, et al. Removal of polycyclic aromatic hydrocarbons in soil spiked with model mixtures of petroleum hydrocarbons and heterocycles using biosurfactants from *Rhodococcus ruber* IEGM 231. Journal of Hazardous Materials. 2016; **312**:8-17. DOI: 10.1016/j.jhazmat. 2016.03.007

[72] Gao P, Wang H, Li G, Ma T. Low-abundance *Dietzia* inhabiting a water-flooding oil reservoir and the application potential for oil recovery. BioMed Research International. 2019. DOI: 10.1155/2019/2193453

[73] Saeki H, Sasaki M, Komatsu K, Miura A, Matsuda H. Oil spill remediation by using the remediation agent JE1058BS that contains a biosurfactant produced by *Gordonia* sp. strain JE-1058. Bioresource Technology. 2009;**100**(572-577). DOI: 10.1016/j. biortech.2008.06.046

[74] Zheng LY, Huang L, Xiu J, Huang Z. Investigation of a hydrocarbon-degrading strain, *Rhodococcus ruber* Z25, for the potential of microbial enhanced oil recovery. Journal of Petroleum Science and Engineering. 2012;**81**:49-56. DOI: 10.1016/j.petrol.2011.12.019

[75] Liu J, Huang X-F, Lu L-J, Xu J-C, Wen Y, Yang D-H, et al. Comparison between waste frying oil and paraffin as carbon source in the production of biodemulsifier by *Dietzia* sp. S-JS-1. Bioresource Technology. 2009;**100**(24): 6481-6487. DOI: 10.1016/j. biortech.2009.07.006.7

[76] Hao DH, Lin JQ, Song X, Lin J-Q, Su Y-J, Qu Y-B. Isolation, identification, and performance studies of a novel paraffin-degrading bacterium of *Gordonia amicalis* LH3. Biotechnology and Bioprocess Engineering. 2008;**13**:61-68. DOI: 10.1007/s12257-007-0168-8 2008

[77] Peng L, Yang C, Zeng G, Wang L, Dai C, Long Z, et al. Characterization and application of bioflocculant prepared by *Rhodococcus erythropolis* using sludge and livestock wastewater as cheap culture media. Applied Microbiology and Biotechnology. 2014;**98**:684-658. DOI: 10.1007/s00253-014-5725-4

[78] Franzetti A, Gandolfi I, Bestetti G, Smyth TJP, Banat IM. Production and

applications of trehalose lipid biosurfactants. European Journal of Lipid Science and Technology. 2010;**112** (6):617-627. DOI: 10.1002/ ejlt.200900162

[79] Pacwa-Plociniczak M, Plaza GA, Piotrowska-Seget Z, Cameotra SS. Environmental applications of biosurfactants: Recent advances. International Journal of Molecular Sciences. 2011;**12**:633-654. DOI: 10.3390/ijms12010633

[80] Geys R, Soetaert W, Van Bogaert I. Biotechnological opportunities in biosurfactant production. Current Opinion in Biotechnology. 2014;**30**:66-72. DOI: 10.1016/j.copbio.2014.06.002

[81] Henkel M, Muller MM, Kugler JH, Lovaglio RB, Contiero J, Syldatk C, et al. Rhamnolipids as biosurfactants from renewable resources: Concepts for next-generation rhamnolipid production. Process Biochemistry. 2012;**47**:1207-1219. DOI: 10.1016/j. procbio.2012.04.018

[82] Sowani H, Deshpande A, Gupta V, Kulkarni M, Zinjarde S. Biodegradation of squalene and *n*-hexadecane by *Gordonia amicalis* HS-11 with concomitant formation of biosurfactant and carotenoids. International Biodeterioration and Biodegradation. 2019;**142**:172-181. DOI: 10.1016/j. ibiod.2019.05.005

[83] Khire JM. Bacterial biosurfactants, and their role in microbial enhanced oil recovery (MEOR). In: Sen R, editor. Biosurfactants. Advances in Experimental Medicine and Biology. Vol. 672. New York: Springer; 2010. pp. 146-157. DOI: 10.1007/978-1-4419-5979-9_11

[84] Abu-Ruwaida AS, Banat IM, Haditirto S, Salem A, Kadri M. Isolation of biosurfactant-producing bacteria, product characterization, and evaluation. Engineering in Life Sciences.

1991;**11**(4):315-324. DOI: 10.1002/ abio.370110405

[85] Tishchenko AV, Litvinenko LV, Shumikhin SA. Effects of *Rhodococcus*-biosurfactants on the molybdenum ion phytotoxicity. IOP Conference Series: Materials Science and Engineering. 2019;**487**(1):012021. DOI: 10.1088/1757-899X/487/1/012021

[86] Litvinenko LV, Tishchenko AV, Ivshina IB. Reduction of copper ion phytotoxicity using *Rhodococcus*-biosurfactants. Biology Bulletin of the Russian Academy of Sciences. 2019;**46** (10):1333-1338. DOI: 10.1134/ S1062359019100200

[87] Pirog T, Kluchka L, Skrotska O, Stabnikov V. The effect of co-cultivation of *Rhodococcus erythropolis* with other bacterial strains on biological activity of synthesized surface-active substances. Enzyme and Microbial Technology. 2020;**142**:109677. DOI: 10.1016/j.enzmictec.2020.109677

A Laboratory-Scale Study: Biodegradation of Bisphenol A (BPA) by Different Actinobacterial Consortium

Adetayo Adesanya and Victor Adesanya

Abstract

The unique diversity of microbes makes them ideal for biotechnological purposes. In this present study, 16 actinobacterial isolates were screened on media supplemented with Bisphenol A (BPA). Three out of 16 isolates exhibited high biocapacity to degrade BPA as *a* carbon source. Four different mixed actinobacterial consortia were developed using the above strains and the effect of each consortium on biomass growth; laccase production and BPA degradation were examined. At 100-mg/L BPA concentration, the three-member consortium grew well with maximum laccase activity as well as maximal degradation rate of Bisphenol A than the other two-member consortium. The consortium of *Actinomyces naeslundii*, *Actinomyces bovis*, and *Actinomyces israelii* degraded 93.1% with maximum laccase activity of 15.9 U/mL, followed by *A. naeslundii* and *A. israelii* with 87.3% and 9.5 U/mL. This was followed by *A. naeslundi* and *A. bovis* with 80.4% and 8.7 U/mL, while *A. bovis* and *A. israelii* degraded 76.0% with laccase activity of 7.0. The gas chromatography–mass spectrometry (GC–MS) analysis of biodegraded BPA showed the presence of oxalic acid and new products like 1,2,4-trimethylbenzene and 2,9-dimethyldecane.

Keywords: Bisphenol A, laccase, biodegradation, biocapacity, gas chromatography–mass spectrometry (GC–MS), *Actinomyces naeslundii*, *Actinomyces bovis*, *Actinomyces israelii*

1. Introduction

One of the most pressing health and environmental issues in today's world is the generation of cumbersome waste with toxic organic substances like Bisphenol A (BPA) from home and industrial sector. BPA poses threats to both aquatic and terrestrial animals [1, 2]. BPA (4,4-isopropylidenediphenol) is an industrial chemical that is produced through the condensation of acetone and phenol using acid or alkaline as catalyst [3, 4].

It is one of the highest production volume chemicals [5], which is widely used as an intermediate in the synthesis of polycarbonate plastics, epoxy resins, and flame retardants [6, 7]. BPA is a monomer of polycarbonate plastics and a constituent of epoxy and polystyrene resins, which are used in the food packing industry [6, 8]. Despite these relevant usages, it has a strong estrogenic property, thus, classified as part of endocrine disruptive compounds (EDCs) [9].

The intensification of anthropogenic activities in manufacturing industries has contributed to the direct or indirect release of a wide range of toxic compounds into the environment and BPA has the potential of causing significant threats to flora, fauna, and human [1, 10, 11]. Microbial degradation is described as a major approach and a natural mechanism, by which one-can clean-up pollutants from the environment in an eco-friendly manner [12–14].

Most of the exploited biodegradation research processes rely mostly on enzymes from different strain of plankton, fungi, or bacteria [15–18]. It has been reported that BPA bioremediation by fungi and bacteria is mediated mainly through lignin-degrading enzymes, such as laccase and manganese peroxidase (MnP), which are produced extracellularly [19–21]. At present, actinobacteria are relatively less explored for biodegradation processes that utilize laccase for remediating BPA.

Laccase is a multicopper oxidase and catalyzes one-electron oxidation of phenolic compounds by reducing oxygen to water [22, 23]. Laccase typically contains 15–30% carbohydrate. It usually has an acidic isoelectric point and a molecular mass of 60–90 kDa [24–26]. Laccase is encoded by a family of genes and produced in the form of multiple isozymes [27, 28]. It has been proven that genes encoding laccase isozymes were differentially regulated [29, 30].

Laccase is an important industrial enzyme. It can be applied extensively in many fields, which include textile dye transformation, waste detoxification and demineralization, and production of biofuels [31–33]. In the case of laccase, BPA metabolism is faster in the presence of mediators such as 1-hydroxybenxotriaxzole (HBT) and 2,2-azino-bis (3-ethylbenzthiazoline-6-sulfonate) than in laccase alone. Thus, the objective of this research work was to investigate the microbial growth and laccase activity from the different actinobacterial consortium during the biodegradation process. The metabolites from the BPA biodegradation were also analyzed using gas chromatography–mass spectrometry (GC–MS).

2. Material and methods

2.1 Chemicals and reagents

BPA, hydrochloric acid, sodium acetate, sodium hydroxide, calcium chloride, iron sulphate, potassium phosphate, hydrogen peroxide, ammonium nitrate, and 2,2′azino-di-(3 ~ ethyl benzothiazoline-6-sulphonic acid) (ABTS) were products of Sigma-Aldrich, Germany. Ethanol and yeast extract were ordered from BDH Chemicals Limited, England, Scharlaun Chemie S.A. Barcelona, Spain, and Biomark Laboratories Pune, India. All the other chemicals and reagents used were of analytical grade.

2.2 Microorganism

Sixteen actinobacterial isolates that were obtained from the culture collection of the Enzyme and Microbial Technology laboratory, Department of Biochemistry, Federal University of Technology, Akure, Nigeria, were screening for this study. Three (*A. naeslundii*, *A. bovis*, and *A. israelii*) of the isolates with best-growing ability on media supplemented with 100 mg/L of BPA were selected from the culture bank for further studies.

2.2.1 Seed culture preparation

Seed culture for each actinobacterial strain was prepared by growing a loopful portion from the slant in sterile media containing (5 g/L), yeast extract (1.5 g/L), beef extract (1.5 g/L), and NaCl (5 g/L) with pH adjusted to 7.4. The media were incubated in an orbital incubator at 30°C for 24 h at 160 r·min^{-1} in a shaking incubator (Stuart, UK). After 24 h, the seed culture was used as inoculum for the production media. Each seed inoculum (constituting 5% v/v) was transferred into 500-mL Erlenmeyer flask containing 100 mL of production media. Three different actinobacterial consortia were developed as follows: *A. naeslundii*, *A. israelii*, and *A. bovis* used as inoculant for the degradation processes.

2.2.2 Preparation of mineral salt medium

The mineral salt medium used in this study was composed of KH_2PO_4 (0.2 g/L), K_2HPO_4 (0.2 g/L), $CaCl_2.H_2O$ (0.1 g/L), NaCl (0.8 g/L), $MgSO_4.7H_2O$ (0.2 g/L), $MnSO_4.7H_2O$ (0.01 g/L), $FeSO_4.7H_2O$ (0.02 g/L), and yeast (2.0 g/L) with pH adjusted to 7.0. The basal mineral media was supplemented with 100 mg/L of BPA. These were sterilized in an autoclave at 15 psi and (121°C) for 20 minutes.

2.2.3 Actinobacterial growth during the biodegradation process

Biomass concentration was estimated at every 24 h from the absorbance of appropriately diluted culture medium at 620 nm according to the predetermined correlation between optical density and dry weight of biomass [34, 35]. The media were incubated at room temperature for 312 h at 160 r·min^{-1}.

2.3 Effect of actinobacterial consortium on laccase production and BPA degradation

The effect of actinobacterial consortium on laccase production and BPA degradation was investigated for a period of 312 h in sterilized mineral salt media supplemented with 100 mg/L of BPA at pH 7.0. The consortia were developed using equal volume of individual seed culture. BPA degradation efficiency and laccase production were monitored, as described below, at 24-h interval throughout the incubation period.

2.3.1 Enzyme assays

The activity of laccase was measured using the common substrate, ABTS ($E_{420} = 36,000 \ M^{-1} \ cm^{-1}$) a modified method of Bourbonnais and Paice [36]. This was done by monitoring spectrophotometrically change in absorbance at 420 nm (A_{420}) correlated with the rate of oxidation of 1-mM ABTS in 1-mM Tris–HCl buffer pH 7.0. The experiment was performed in 1-mL cuvettes at 30°C. Reaction mixture contained 750-µL ABTS and 250 µL of enzyme solution.

At an interval of 1 minute for 5 minutes, the absorbance of the mixture was measured. One unit of laccase activity was defined as the amount of enzyme that oxidized 1 mM of ABTS per minute under standard assay conditions. Laccase activity was expressed as U/mL. The enzyme activity was calculated using the Eq. (1).

$$Enzyme \ activity = \frac{\frac{Absorbance}{minute} \times Total \ volume \ of \ mixture}{Total \ time \times Extinction \ coefficient \times volume \ of \ enzyme} \qquad (1)$$

2.3.2 Determination of BPA degradation

The percentage removal of BPA was determined using Folin–Ciocalteu reagent according to the method of Yordanova *et al.* [37]. The residual of BPA supplemented in the mineral salt media was determined at 24-h intervals throughout the degradation period. Aliquots of the culture media were withdrawn at intervals and centrifuged for 10 min at 3500 r·min⁻¹. One milliliter of the supernatant was added to 10 mL of distilled H_2O and 1 mL of Folin–Ciocalteu reagents. The mixture was left for 5 min, and 2 mL of 20% Na_2CO_3 (w/v) was added to the mixture. The solution was kept in the dark for 60 min, and therefore, absorbance at 750 nm was measured [37]. The degradation rate was expressed as the difference between the initial and final absorbance. This was estimated in percentage as follows (Eq. (2)):

$$BPA\ Degradation\,(\%) = \frac{Initial\,(BPA)\,concentration - Final\,(BPA)}{Initial\,(BPA)\,concentration} \times 100 \quad (2)$$

2.4 Gas chromatography: mass spectrometry (GC: MS) analysis of BPA degradation metabolites

From the quantitative confirmation analysis employing a Shimadzu gas chromatograph GC-2010 series connected to a Shimadzu spectrometer GCMS-QP2010 PLUS (Japan), the separation of the compounds was achieved by employing a DB5MS capillary column (60 m) (Supelco), the carrier gas was helium and maintained at constant flow of (0.9 mL·min⁻¹). A sample volume of 1 μL was injected in the splitless mode at an inlet temperature of 280°C. The MS transfer line temperature was maintained at 280°C, whereas the ion source temperature was 180°C.

3. Results and discussion

3.1 Actinobacterial growth

Of all the six actinobaterial isolates examined, only three display the potential to grow on BPA. All the three actinobacterial isolates (*A. naeslundii*, *A. bovis*, and *A. israelii*) grew in the presence of BPA indicating their ability to metabolize BPA as a carbon and energy source. Likewise, the actinobacterial growth suggests that the necessary enzymes are required for the degradation of BPA; thus, the production of laccase in basal salt mineral medium confirmed the BPA degrading activity.

3.2 Influence of consortium on enzyme production and BPA degradation

3.2.1 Effect of consortium on laccase production

The influence of actinobacterial consortium on the production of laccase was studied over the degradation period of 312 h, as shown in **Figure 1**. The consortium of *A. naeslundii*, *A. bovis*, and *A. israelii* supported a maximum laccase yield of 15.9 U/mL followed by the consortium of *A. naeslundii* and *A. israelii* (9.5 U/mL), *A. naeslundii* and *A. bovis* (8.7 U/mL), while *A. bovis* and *A. israelii* (7.0 U/mL). Initially, at the incubation time of 24 h, the laccase activity observed for the three-member consortium had the lowest enzyme activity; this may be because the three individual strains are adjusting for one another to coexist in the culture media. Interestingly, at this period, the laccase activity increases steadily up to 168 h where maximal laccase

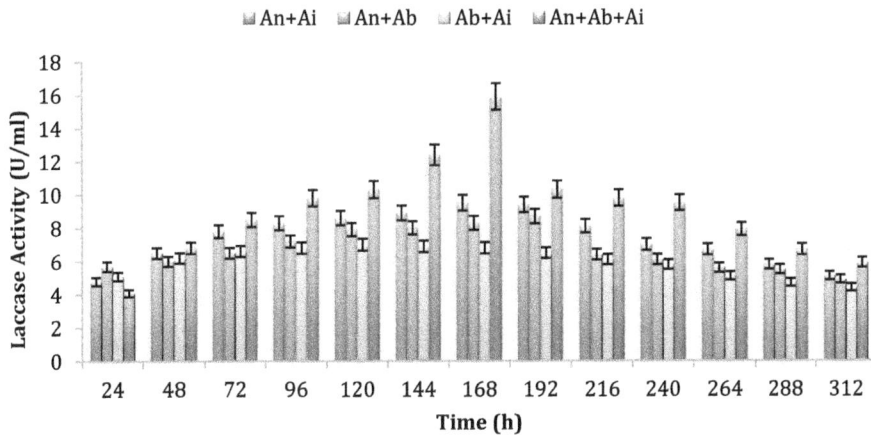

Figure 1.
Laccase activity of the different consortia. Key: An+Ai = Actinomyces naeslundii and Actinomyces israelii; An+Ab = A. naeslundii and A. bovis; Ab+Ai = A. bovis and A. israelii; and An+Ab+Ai = A. naeslundii, A. bovis, and A. israelii.

activity was recorded. This result suggested that laccase production increases proportionately with the growth of the actinobacterial consortium. This result is similar to the findings of Bogan and Lamar [38], who observed that extracellular enzymes of organisms are produced in response to their growth phases. Tsioulpas et al. reported that maximum laccase activity was measured in the growth medium, while 69–76% of phenolic compounds were removed by Pleurotus spp. [39]. This present work recognized that enzyme secretion also depends on the physical factor, nutritional, physiological, and biochemical nature of the microorganism.

3.2.2 Effect of consortium on BPA degradation

The influence of the actinobacterial consortium on BPA degradation was studied over the degradation period of 312 h, as shown in **Figure 2**. The consortium of *Actinomyces naeslundii, A. bovis*, and *Actinomyces israelii* supported a maximum BPA degradation of 93.1% followed by that of *A. naeslundii* and *A. israelii* (87.3%), *A. naeslundi* and *A. bovis* (80.4%), and *A. bovis* and *A. israelii* (76.0%). Microbial consortia have the capability of degrading a wide range of hydrocarbons. This research was able to link actinobacterial growth and laccase activity to the rate of BPA degradation because it can be deduced that at 24 h, all the two-member consortium growth gradually increases until 120 h when a steady decrease set in. However, a different growth pattern was observed for the three-member consortium where the growth steadily decreased at 24 h and then sharply increases at 48 h, and the growth increases until 168 h when there was a gradual decline to the incubation period of 312 h. The decrease in growth and laccase activity might be due to the depletion of nutrients or the production of waste or toxic substances into the culture media during this period. BPA degradation rate hardly increases after the 168-h incubation period. However, the enhanced BPA degradation performance displayed by the three-member actinobacterial consortium is due to synergic in the secretion of laccase compared to the two-member consortium. Although all the actinobacterial strains under study exhibited promising potentials to degrade BPA, their interactive or compatibility test was not investigated.

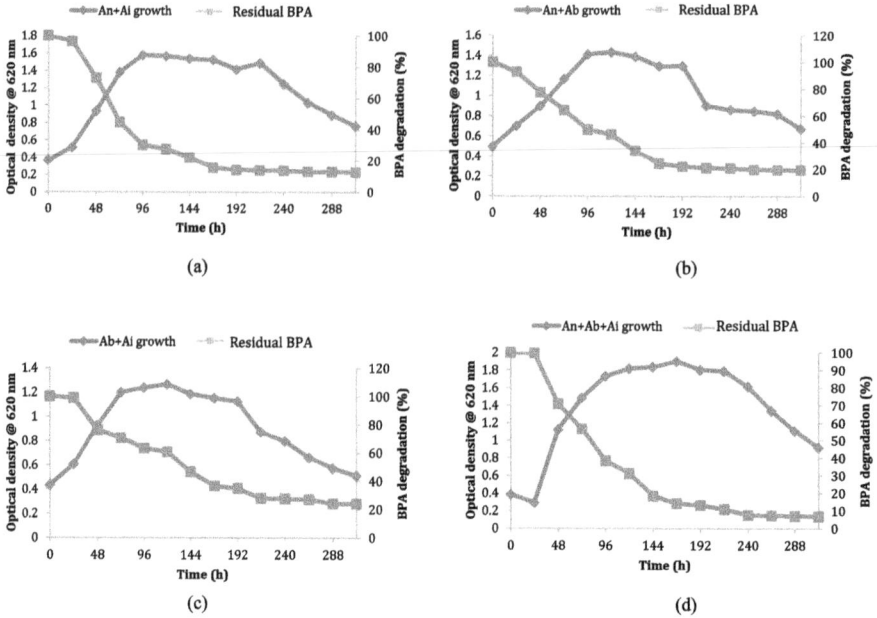

Figure 2.
Degradation of BPA by different consortia (a) An+Ai; (b) An+Ab; (c) Ab+Ai; and (d) An+Ab+Ai.

3.3 Degradation by-product of BPA

The mineral-salt-medium-supplemented BPA as a sole carbon source at pH 7.0 inoculated in each actinobacterial consortium was analyzed after an incubation period of 312 h. Cultures were extracted for GC–MS analysis to determine the biodegradation products of BPA. The by-products of BPA degradation were investigated, and each metabolite produced during the biodegradation process is shown in **Figure 3**. The GC–MS analysis showed that the metabolites could be identified as

Isolate (SAMPLE-ABC)

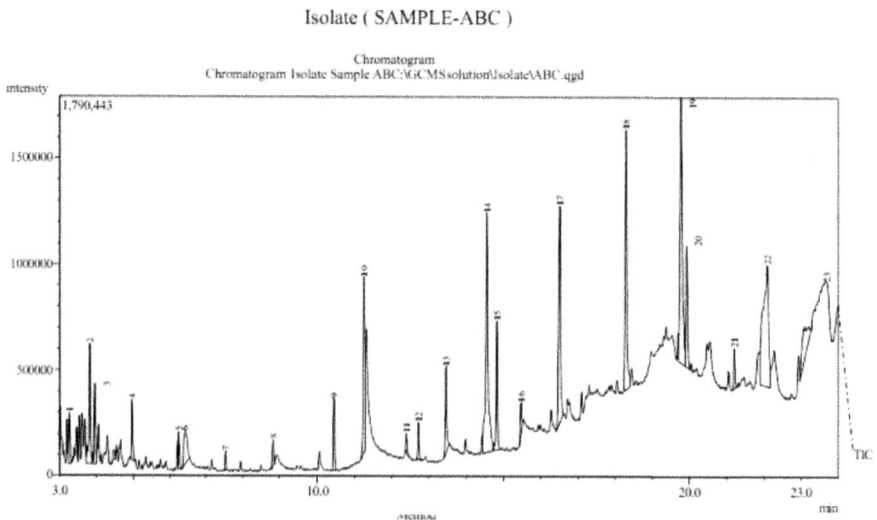

Figure 3.
Chromatogram generated by GC–MS.

concerned compounds by comparisons with known authentic compounds using the NIST Chemistry library. The mass peak is found at 38 for 1,2,4-trimethylbenzene, and its relative molecular mass is 120 at the retention time of 4.0 min, while the base peak value was observed at 105. In addition, 2,9-dimethyldecane was identified at mass peak of 21 at the retention time of 4.9 min. The relative molecular mass of the compound 2,9-dimethyldecane was observed as 170 and base peak was noticed at 43. Oxalic acid was also identified as one of the intermediate products, the relative molecular mass of the compound was 216 with a retention time of 7.0 min, and the mass and base peak values were recorded at 12 and 57, respectively. According to Kusvuran and Yildrim [40], oxalic acid was identified as organic intermediate from the degradation of BPA, which is similar to our observation in this study. However, intermediates such as p-hydroxyacetophone, hydroquinone, p-hydroxybenzaldehyde, and p-hydroxbenzoic identified by previous studies [41] were not detected.

4. Conclusion

This study focused on the actinobacterial isolates identified as *Actinomyces naeslundii*, *A. bovis*, and *Actinomyces israelii*, which showed adaptive and biocapacity mechanisms to survive on culture media supplemented with BPA. A direct relationship was found between the microbial growth, laccase activity of the actinobacterial consortium, and BPA degradation. From the evidence presented in this research work, it can be concluded that the investigated actinobacterial strains could be considered as good prospects for their application in the bioremediation of BPA-contaminated environments as they revealed promising potential.

Author details

Adetayo Adesanya[1]* and Victor Adesanya[2]

1 Enzyme Biotechnology and Environmental Health Unit, Department of Biochemistry, Federal University of Technology, Akure, Ondo State, Nigeria

2 Department of Chemistry, Federal University of Petroleum, Effurun, Delta State, Nigeria

*Address all correspondence to: adesanya88@gmail.com

IntechOpen

References

[1] Sidorkiewicz I, Czerniecki J, Jarząbek K, Zbucka-Krętowska M, Wołczyński S. Cellular, transcriptomic and methylome effects of individual and combined exposure to BPA, BPF, BPS on mouse spermatocyte GC-2 cell line. Toxicology and Applied Pharmacology. 2018;**359**:1-11

[2] Yang Q, Yang X, Liu J, Chen Y, Shen S. Effects of exposure to BPF on development and sexual differentiation during early life stages of zebrafish (Danio rerio). Comparative Biochemistry and Physiology Part C: Toxicology & Pharmacology. 2018;**210**:44-56

[3] Karak N. Modification of epoxies. In: Sustainable Epoxy Thermosets and Nanocomposites. Washington, DC, USA: American Chemical Society; 2021. pp. 37-68

[4] Wei N, Zhang Y, Zhang H, Chen T, Wang G. The effect of modifying group on the acid properties and catalytic performance of sulfonated mesoporous polydivinylbenzene solid acid in the Bisphenol-A synthesis reaction. Reactive and Functional Polymers. 2022:105156

[5] Burridge E. Bisphenol A product profile. European Chemical News. 2003:14-20

[6] Idowu GA, David TL, Idowu AM. Polycarbonate plastic monomer (Bisphenol-A) as emerging contaminant in Nigeria: Levels in selected rivers, sediments, well waters and dumpsites. Marine Pollution Bulletin. 2022;**176**:113444

[7] Staples CA, Dorn PB, Klecka GM, O'Block ST, Harris LR. A review of the environmental fate, effects, and exposures of Bisphenol-A. Chemosphere. 1998;**36**:2149-2173

[8] Cooper JE, Kendig EL, Belcher SM. Assessment of Bisphenol A released from reusable plastic, aluminium and stainless steel water bottles. Chemosphere. 2011;**85**(6):943-947

[9] Palacios-Arreola MI, Moreno-Mendoza NA, Nava-Castro KE, Segovia-Mendoza M, Perez-Torres A, Garay-Canales CA, et al. The endocrine disruptor compound Bisphenol-A (BPA) regulates the intra-tumoral immune microenvironment and increases lung metastasis in an experimental model of breast cancer. International Journal of Molecular Sciences. 2022;**23**(5):2523

[10] Oehlmann J, Schulte-Oehlmann U, Kloas W, Jagnytsch O, Lutz I, Kusk KO, et al. A critical analysis of the biological impacts of plasticizers on wildlife. Philosophical Transactions of the Royal Society B. 2009;**364**:2047-2062

[11] Zhang W, Xia W, Liu W, Li X, Hu J, Zhang B, et al. Exposure to Bisphenol A substitutes and gestational diabetes mellitus: A prospective cohort study in China. Frontiers in Endocrinology. 2019;**262**

[12] Atlas RM. Petroleum microbiology, in Encyclopedia of Microbiology. Baltimore: Academic Press; 1992. pp. 363-369

[13] Chatterjee S, Kumari S, Rath S, Das S. Prospects and scope of microbial bioremediation for the restoration of the contaminated sites. In: Microbial Biodegradation and Bioremediation. 2022. pp. 3-31

[14] Lal B, Khanna S. Degradation of crude oil by Acinetobacter calcoaceticus and Alcaligenes odorans. Journal of Applied Bacteriology. 1996;**81**(4):355-362

[15] Etaware PM. Biomineralization of toxicants, recalcitrant and radioactive wastes in the environment using genetically modified organisms. Journal

of Agricultural Research Advances. 2021;**3**(2):28-36

[16] Hirooka T, Akiyama Y, Tsuji N, Nakamura T, Nagase H, Hirata K, et al. Removal of hazardous phenols by microalgae under photoautotrophic conditions. Jpirnal of Bioscience and Bioenginerring. 2003;**95**:200-203

[17] Lee SM, Koo BW, Choi JW, Chai DH, An BS, Jeung EB, et al. Degradation of BPA by white rot fungi, Stereum hirsutum and Heterobasidium insulare, and reduction of its estrogenic activity. Biological & Pharmaceutical Bulletin. 2005;**28**:201-207

[18] Sasaki M, Maki JI, Oshiman KI, Matsumura Y, Tsuchido T. Biodegradation of BPA by cells and cell lysate from Sphingomonas sp. strain AO1. Biodegradation. 2005;**16**:449-459

[19] González-Rodríguez S, Lu-Chau TA, Trueba-Santiso A, Eibes G, Moreira MT. Bundling the removal of emerging contaminants with the production of ligninolytic enzymes from residual streams. Applied Microbiology and Biotechnology. 2022:1-13

[20] Kabiersch G, Rajasarkka J, Ullrich R, Tuomela M, Hofrichter M, Virta M, et al. Fate of Bisphenol A during treatment with the litter-decomposing fungi Stropharia rugosoannulata and Stropharia coronilla. Chemosphere. 2011;**83**:226-232

[21] Rampinelli JR, Melo MP, Arbigaus A, Silveira ML, Wagner TM, Gern RM, et al. Production of Pleurotus sajor-caju crude enzyme broth and its applicability for the removal of Bisphenol A. Anais da Academia Brasileira de Ciências. 2021;**93**:1-16

[22] Dixit M, Gupta GK, Usmani Z, Sharma M, Shukla P. Enhanced bioremediation of pulp effluents through improved enzymatic treatment strategies: A greener approach.

Renewable and Sustainable Energy Reviews. 2021;**152**:111664

[23] Reinhammar B. Laccase. In: Lontie R, editor. Copper Proteins and Copper Enzymes. Vol. 3. Boca Raton, FL: CRC Press; 1984. pp. 1-35

[24] Akram F, Ashraf S, Shah FI, Aqeel A. Eminent industrial and biotechnological applications of laccases from bacterial source: A current overview. Applied Biochemistry and Biotechnology. 2022:1-21

[25] Tülek A, Karataş E, Çakar MM, Aydın D, Yılmazcan Ö, Binay B. Optimisation of the production and bleaching process for a new laccase from Madurella mycetomatis, expressed in Pichia pastoris: From secretion to yielding prominent. Molecular Biotechnology. 2021;**63**(1):24-39

[26] Xiao YZ, Chen Q, Hang J, Shi YY, Xiao YZ, Wu J, et al. Selective induction, purification and characterization of a laccase isozyme from the basidiomycete Trametes sp. AH28-2. Mycologia. 2004;**96**(1):26-35

[27] Chatterjee A, Chakraborty P, Abraham J. Microbial Enzymes for the Mineralization of Xenobiotic Compounds. In: Bioprospecting of Microorganism-Based Industrial Molecules. 2021. pp. 319-336

[28] Zhang Y, Wu Y, Yang X, Yang E, Xu H, Chen Y, et al. Alternative splicing of heat shock transcription factor 2 regulates expression of the laccase gene family in response to copper in *Trametes trogii*. Applied and Environmental Microbiology. 2021;**87**(8):e00055-e00021

[29] Durán-Sequeda D, Suspes D, Maestre E, Alfaro M, Perez G, Ramírez L, et al. Effect of nutritional factors and copper on the regulation of laccase enzyme production in *pleurotus ostreatus*. Journal of Fungi. 2021;**8**(1):7

[30] Soden DM, Dobson AD. Differential regulation of laccase gene expression in *Pleurotus sajorcaju*. Microbiology. 2001;**147**:1755-1763

[31] Abadulla E, Tzanov T, Costa S, Robra KH, Cavaco PA, Guebitz GM. Decolorization and detoxification of textile dyes with a laccase from *Trametes hirsuta*. Applied and Environmental Microbiology. 2000;**66**:3357-3362

[32] Brazkova M, Koleva R, Angelova G, Yemendzhiev H. Ligninolytic enzymes in Basidiomycetes and their application in xenobiotics degradation. In: BIO Web of Conferences. 2022

[33] Fukuda T, Uchida H, Takashima Y, Uwajima T, Kawabata T, Suzuki M. Degradation of Bisphenol A by purified laccase from *Trametes villsoa*. Biochemical and Biophysical Research Communications. 2001;**284**:704-706

[34] Gangola S, Kumar R, Sharma A, Singh H. Bioremediation of petrol engine oil polluted soil using microbial consortium and wheat crop. Journal of Pure and Applied Microbiology. 2017;**11**(3):1583-1588

[35] Muthuswamy S, Arthur RB, Sang-Ho B, Sei-Eok Y. Biodegradation of crude oil by individual bacterial strains and a mixed bacterial consortium isolated from hydrocarbon contaminated areas. Clean. 2008;**36**:92-96

[36] Bourbonnais R, Paice MG. Oxidation of non-phenolic substrates: An expanded role for laccase in lignin biodegradation. FEBS Letters. 1990;**267**(1):99-102

[37] Yordanova G, Godjevargova T, Nenkova R, Ivanova D. Biodegradation of phenol and phenolic derivatives by a mixture of immobilized cells of Aspergillus awamori and Trichosporon cutaneum. Biotechnology and Biotechnological Equipment. 2013;**27**(2):3681-3688

[38] Bogan BW, Lamar RT. Polycyclic aromatic hydrocarbon-degrading capabilities of *Phanerochaete laevis* HHB-1625 and its extracellular ligninolytic enzymes. Applied and Environmental Microbiology. 1996;**62**(5):1597-1603

[39] Tsioulpas A, Dimou D, Iconomou D, Aggelis G. Phenolic removal in olive oil mill wastewater by strains of *Pleurotus* spp. in respect to their phenol oxidase (laccase) activity. Bioresource Technology. 2002;**84**:251-257. DOI: 10.1016/S0960-8524(02)00043-3

[40] Kusvuran E, Yildirim D. Degradation of bisphenol A by ozonation and determination of degradation intermediates by gas chromatography–mass spectrometry and liquid chromatography–mass spectrometry. Chemical Engineering Journal. 2013;**220**:6-14

[41] Lu N, Lu Y, Liu F, Zhao K, Yuan X, Zhao Y, et al. H3PW12O40/TiO$_2$ catalyst-induced photodegradation of Bisphenol A (BPA): Kinetics, toxicity and degradation pathways. Chemosphere. 2013;**91**(9):1266-1272

Actinobacteria and Diseases

Chapter 11

Actinomycosis: Diagnosis, Clinical Features and Treatment

Onix J. Cantres-Fonseca, Vanessa Vando-Rivera,

Vanessa Fonseca-Ferrer, Christian Castillo Latorre

and Francisco J. Del Olmo-Arroyo

Abstract

Actinomycosis is a filamentous bacterium that forms part of the normal human flora of the gastrointestinal, oropharynx and female genitalia. This indolent infection is characterized by abscess formation, widespread granulomatous disease, fibrosis, cavitary lung lesions and mass-like consolidations, simulating an active malignancy or systemic inflammatory diseases. It is subacute, chronic and variable presentation may delay diagnosis due to its capability to simulate other conditions. An accurate diagnostic timeline is relevant. Early diagnosis of pulmonary actinomycosis decreases the risk of indolent complications. Proper treatment reduces the need for invasive surgical methods. Actinomycosis can virtually involve any organ system, the infection spread without respecting anatomical variables as metastatic disease does, making malignancy an important part of the differential diagnosis. As it is normal gastrointestinal florae, it is difficult to cultivate, and share similar morphology to other organisms such as Nocardia and fungus. It is often difficult to be identified as the culprit of disease. Its true imitator capability makes this infectious agent a remarkable organism within the spectra of localized and disseminated disease. In this chapter, we will discuss different peculiarities of actinomycosis as an infectious agent, most common presentation in different organ systems, and challenging scenarios.

Keywords: actinomycosis, pulmonary, systemic, disseminated, imitator

1. Introduction

Few infectious agents can cause a broad variety of clinical presentations, including scenarios that mimic other infections, systemic inflammatory diseases, and even localized and metastatic cancer. The pathogenic characteristics of *Actinomycosis*, its capability to disseminate and grow in different environments and tissues, and its indolent and resistant presentation, has allowed this organism to confuse the clinician before obtaining a diagnosis and start treatment. Also, the invasiveness of the actinobacteria, many times require complicated procedures to obtain tissue to rule out cancer and to culture and visualize the organism, many times delaying identification. Finally, this organism can cause infection in a variety of hosts, including immunocompetent patients, in which suspicion of infection is not an initial consideration.

Actinomyces is a prokaryotic bacterium that belongs to the family *Actinomyceatacea*. Initially identified as a fungi in the 19th century, this bacteria has unique characteristics that have allowed to be distinguished as a group from other organisms. Unicellular, but elongated that can mimic the physical structure of a fungal hyphae, this bacteria has the capability to live in different organic and inorganic substrates, making this organism unique and fascinating. Actinobacteria can survive using different environmental substrates found in soil and water, and even in the human body. The word "actinomycosis" is derived from the Greek aktino, which refers to the radiating appearance of the sulfur granules produced by the bacteria. The human form was first described in the year 1857, and the first thoracic case was identified approximately 25 years later [1]. In 1891, the *A. israelli,* the most common human infecting specie, was identified [2]. Incidence of the disease is not updated, but the most recent data reports that it affects 1 in 300,000 individuals yearly, and that incidence hast decreased in the last 3 decades [3]. Men are more commonly affected, and several risk factors have been identified as chronic alcohol abuse, diabetes, cancer, temporary immunosuppressive therapy, organ transplant and end stage renal disease [4]. However, immunocompetent patients with poor oral hygiene, and an event of aspiration, have also been documented. The most common form of Actinomycosis is the cervicofacial invasion, but the organism can affect virtually any organ or body site, most commonly the thorax and the abdomen. Delay in diagnosis occurs in many cases as the signs and symptoms are nonspecific, and are often similar to those seen in malignancy and other systemic diseases and infections. There are cases in which the actinomyces infection can coexist with other diseases, as for example lung cancer, as it tends to colonize devitalized tissue, making diagnosis even more challenging [5]. Even in developed countries, the incidence of the disease is underestimate and diagnosis is difficult. It many cases patients are receiving treatment for other conditions and inadvertently been cured for actinomycosis infection. The use of antibiotic therapy has significantly improved the prognosis of patient with actinomycosis. Morbidity and mortality is significatively decreased with appropriate identification and treatment.

2. Epidemiology and pathogenesis

As this organism constitutes part of the human flora, disease manifestation depends on the site of invasion. Several anatomical sites have been identified and include cervicofacial, pulmonary, abdominal, genitourinary, cutaneous, extra facial and joint, central nervous system and disseminated infection. More than 30 species of Actinomycosis have been identified and the most common human pathogen is *Actinomyces Israelli* [6].

Actinomycosis has a worldwide distribution and usually affects middle age individuals. It is two to four times more common in males than in females [1]. The condition is usually considered as infrequent, however the epidemiologic data is very limited and probable underreported [7]. Infection involving the cervicofacial area is the most commonly known, representing 60% of the cases [8]. This occurs after dental or oral procedure, or in a patient with poor dental care. The second most common presentation of actinomycosis is abdominal involvement, constituting 20% of the reported cases [6]. The organism can invade any abdominal tissue, most commonly the bowel, causing appendicitis in 65% of the cases, but also, gastrointestinal perforation and upper and lower bowel obstruction have been reported. The third most common presentation is thoracic actinomycosis (15–20%

of the reported cases) [6]. Infection develop after aspiration of the oropharyngeal secretions or perforation of the esophagus. Some cases develop from progression from cervicofacial or abdominal actinomycosis, or after hematogenous spread in disseminated disease [1].

The rarest cases of actinomycosis include musculoskeletal, central nervous system or joint actinomycosis. Central nervous system actinomycosis infection develops from hematogenous spread or extension of a cervicofacial actinomycosis [9]. In one study, the distribution of presentations included brain abscess (67%), meningitis or meningoencephalitis (13%), actinomycoma (7%), subdural empyema (6%), and epidural abscess (6%) [10]. Musculoskeletal actinomycotic infections of hip and knee prostheses have been described, with early presentation suggesting introduction of the organism perioperatively, and late presentation usually indicating hematogenous spread from an extra-articular site [6]. Disseminated Actinomycosis, is a rare presentation, but can be seen after extensive use of antibiotics.

The pathological damage cause by these organisms are secondary to direct tissue invasion and damage by the bacterial population. Also, actinomyces is often found cohabitating with other bacteria as *Eikenella, Enterobacteriaceace*, and species of *Fusobacterium, Bacteroides, Capnocytophagia, Staphylococci*, and *Streptococci*. It has been always a question of how those other organisms contribute to the pathogenesis of actinomycosis. It is believed that the cohabitant bacterias create an ecosystem that facilitates bacterial spread and growth of the actinomyces [11].

3. Risks factors

Actinomycosis is considered an endogenous infection. This organism can affect immunocompromise and immunocompetent hosts. Risk factors associated with the acquisition of actinomycosis include: male sex between 20 and 60 years old, diabetics, poor oral hygiene, implanted foreign bodies as occurs during aspiration, or intrauterine implantable contraceptive device (IUD) [4]. Also, the use of immunosuppressive therapy, as systemic steroids or chemotherapy, has also being identified as possible risk factor for localized and disseminated infection. Other identified factors to acquire the infection include history of HIV, active hematogenous and solid malignancy, organ transplant, alcohol abuse, and accidental or intentional tissue trauma, as tissue radiation or surgery [4].

4. Diagnosis

Diagnosis of Actinomycosis is quite challenging and requires high clinical suspicion. Prolonged time for work up and invasive testing is often part of the diagnostic process in an actinomyces infection. A definitive diagnosis is made from identification of the organism from a collected specimen.

Blood tests are often unspecific, with normochromic anemia and mild leukocytosis with predominance of polymorphonuclear cells as most common finding [4]. Others nonspecific laboratory findings include elevation of alkaline phosphate (ALP) and erythrocyte sedimentation rate (ERS), representing the nonspecific inflammatory nature of the illness. The most common laboratory pattern of the infection is related to the organ and tissue involved.

Imaging studies such as CT scan and MRI are nonspecific and nondiagnostic, specially at the earlier stages of the disease as other inflammatory diseases and neoplastic processes such as lung and intestinal cancer may show similar findings.

In cases of thoracic actinomyces infections, the findings are more predominant at the peripheral and lower lobes, reflecting the role of aspiration in the pathogenesis of the disease [3]. Common findings include nodules, masses, cavitations, infected bronchiectasis, segment collapses, tissue calcifications and foreign bodies [3]. All those findings mimicking other infectious or malignant diseases. At latter stages, infiltration to surrounding tissues at different planes and sinus tract formation and fistulas may be identified in radiological images. Mucosal involvement as occurs in the bowel track can mimic cancer and other inflammatory conditions such as Ulcerative Colitis and Chron's disease.

Histopathological findings strongly support the diagnosis of Actinomycosis. The use of Hematoxylin stain (H&E stain) will show the sulfur granules which represent colonies of eosinophilic oval or round basophilic masses, which charac-terized actinomycosis infection [4]. Giemsa, Gram Stain and Gomori methanamine silver staining are need for identification of Gram-positive filamentous branching organisms. This filamentous organisms can be confused with other bacterias as Nocardia u other branching organisms as fungus. Actinomyces is distinguished from Nocardia as nocardia is acid fast positive and aerobic, and actino is acid fast negative and grow in anaerobic medium.

Identification of the organism from a sterile specimen confirms diagnosis, but failure rates are high due to inhibition of Actinomycosis growth by a coexistent/contaminant organism, the previous use of antibiotic therapy or inadequate incu-bation period. Culture growth is slow, with the earliest identification done as a minimum of 5 days, but may take up to 20 days to be identified. Allowing cultures to take at least 10 days is necessary to confirm a negative diagnosis. Actinomycosis culture should be performed on chocolate agar media at 37 degrees Celsius, as well as brain heart infusion broth and Brucella Blood Agar with vitamin K and hemin [4]. Serological assays have a low clinical yield and are still under investigation.

5. Systemic clinical features

Several anatomical areas have been identified where invasion with actinomyco-sis can led to acute/chronic pathological findings. Most of the presenting symptoms are not specific, making diagnosis challenging. The following presentations will describe the clinical manifestations, diagnosis and general treatment of actinomy-cosis according to anatomical site.

5.1 Cervicofacial actinomycosis

Cervicofacial actinomycosis is the most frequent clinical form of actinomy-cosis. Actinomyces is part of the normal flora of the mouth, which tends to cause infections when the oral mucosa barrier is breach, and poor oral hygiene exists. This clinical presentation usually involves tissues surrounding the upper and lower mandible, cheek, chin, and submaxillary area [6]. Patients may present with a fever and slowly progressive painless indurated mass, in the peri-mandibular region, that can evolve to abscess formation and osteomyelitis. A typical thick yellow exudate can be seen in tissues [4]. This mass lesions usually are misdiagnosed as a malignant lesion, and is some cases as a granulomatous disease [12]. Head and neck lymphadenopathy is usually seen related to the infection. In a retrospective study of 317 patient with cervicofacial actinomycosis, disease progress to bone and muscle leading to bone infection [4]. Diagnosis of the infection is mainly thorough fine needle aspiration of the mass like structure with identification of actinomyces.

5.2 Bone and joint actinomycosis

Extra facial and joint actinomycosis is not as common as cervicofacial actinomycosis. Few data is available on the incidence of the disease. Reported cases are
mostly secondary to hematogenous spread. Other causes include polymicrobial
bone infection after bone exposition, continuous spread to the spine after pulmonary actinomycosis [6]. Patients who develop extra facial and joint actinomycosis
usually present symptoms months after suspected bacteremia, and symptoms are
usually insidious and similar to other chronic bone infections or malignancies [6].

5.3 Pulmonary actinomycosis

Pulmonary actinomycosis is a rare and challenging disease to diagnose. It mainly
results from aspiration of oropharyngeal or gastrointestinal secretions. Individuals
with history of poor oral hygiene, preexisting dental disease, and alcoholism
have an increased risk for developing pulmonary actinomycosis. Patients with
pulmonary disease like COPD and bronchiectasis are at increased risk of infection
[6]. Other risk factors include disease that have high risk of aspirations, such as
neurological and psychiatric diseases, drug abuse and severe reflux with hernias. In
the 1950s pulmonary actinomycosis presented similar to tuberculosis, as a empyema necessitans, with a sharp chest pain and cutaneous fistulation [3]. In the post
antibiotic era, the most common presentation is a focal pulmonary consolidation,
mimicking persistent bacterial pneumonia or lung malignancy. It is recommended
that in patients with persistent symptoms of cough or hemoptysis refractory to
antibiotic therapy, for actinomycosis to become a possible differential diagnosis.

Most cases of pulmonary actinomycoses are misdiagnosed. Differential diagnosis includes other lung pathologies such lung malignancy, pneumonia, mycobacterium infection, aspergillosis and lung abscess. Pulmonary actinomycosis is
commonly subdivided depending on the radiological findings, into the airway type
including bronchiectasis, the endobronchial, mediastinum and that involving the
chest wall [3]. Imaging allows for proper localization and extension of the disease.
Patchy air space consolidation, ground glass opacity, cavitation, pleural thickening, atelectasis, pleural effusion, nodular and the appearance of mediastinal and
hilar lymphadenopathies may be seen in CT imaging of pulmonary actinomycosis
[13]. Parenchymal involvement have an extensive range of presentations from a
pulmonary nodule to extensive space consolidation, to the most invasive spreading
involving pleural and chest wall invasion with pleural effusions and empyema [13].

Colonization and infection of preexisting bronchial dilations or causing the
bronchial changes to develop bronchiectasis, is another presentation. This form may
coexist with parenchymal actinomycosis infection. Aspergillus infection can cause
similar findings and should also be ruled out. Finally, endobronchial actinomycosis,
develops after colonization of broncholiths or foreign bodies in the airway, by
actinomyces. These broncholiths are form from eroded and calcified lymph nodes
into the airway, which are commonly seen in patient suffering from granulomatous
infections such as *Histoplasmosis Capsulatum* or *Mycobacterium Tuberculosis* [14]. On
CT imaging, endobronchial nodules with associated obstructive pneumonia, may
be observed. Most endobronchial actinomycosis may present as tumor, with irregu-
lar thickening or an intrabronchial mass associated to post-obstructive pneumonia,
resembling lung cancer and its complications [3]. Some cases of chest wall invasion have been noted, and produce similar findings as those seen in blastomycosis,
cryptococcosis, nocardiosis and invasive aspergillosis. Other possible differential
diagnosis include lymphoma, malignant mesothelioma and chest wall tumors. Vocal
cord can be also involved, and may mimic papilloma, or cancer.

Fiberoptic bronchoscopy help in diagnosis when an endobronchial lesion is biopsied and the organism colonies are identified in the tissue obtained, but it may just represent colonization, so clinical correlation must be done and other diagnosis as malignancy should be rule out.

5.4 Central nervous system

The central nervous system can also be affected by actinomyces. It is most seen as brain and epidural abscess, meningitis, meningoencephalitis, and as subdural empyema after hematogenous spread most likely from lung or cervicofacial infection or from a penetrating head injury. Its signs and symptoms are non-specific but tend to be similar to other central nervous system infections.

5.5 Cutaneous actinomycosis

Cutaneous Actinomycosis has been poorly describe in literature. Most cases present as a soft tissue inflammation that develops into an abscess, nodules mass like lesions.

5.6 Abdominal actinomycosis

Abdominal Actinomycosis usually affects the appendix and ileus [15]. Factors that predispose to infection, include trauma, neoplasia, recent surgery, perforated viscus, as well as the use of intrauterine contraceptive devices [16]. Patient with presents with an indolent course of symptoms such as fever weight loss, fatigue and associated abdominal pain. Physical findings may show a palpable mass, fistula or sinus tract [6]. Laboratory work up often is not specific. Clinical presentation can be similar to clinical Chron's disease, malignancy and extrapulmonary tuberculosis, making diagnosis difficult. The lack of extra intestinal manifestation, as well as the lack of improvement with anti-inflammatory or immunosuppressive drugs may help differentiate form Chron's disease and consider other possible diagnosis, as actinomycosis infection [17].

Radiological findings are nonspecific for the diagnosis of abdominal actinomycosis. Ct Scan may help localized the infection and extension of the disease for tissue diagnosis. Endoscopic evaluation are also nonspecific and remarkable for engrossment and inflammation of mucosa, and bowel wall ulcer formation and nodular and mass like lesions. Definitive diagnosis is obtained with biopsies showing and growing the organism.

5.7 Genitourinary actinomycosis

After cervicofacial actinomycosis, the genitourinary tract is the system most affected by actinomycosis [6]. It is most seen in women using an intrauterine implantable devices [18]. This is most likely due to disruption of the endothelium of the uterus, facilitating microbial invasion. It is recommended to change the intrauterine device every 5 years as it is mostly seen with prolonged use [19].

Bladder actinomycosis can be confused with bladder carcinoma, and it is important to diagnosed accordingly to avoid unnecessary treatment and procedure thought to be from carcinoma. Findings are usually the identification of a genital mass associated with abdominal pain, and genital infections. Macroscopic hematuria can be seen if bladder wall is invaded by actinomyces. Imaging usually show a mass like lesion with associated lymphadenopathy. An abscess of the tubulo-ovarian structures include actinomycosis in the differential diagnosis [6].

5.8 Disseminated actinomycosis

Hematogenous spread to distant multiple organs is uncommon with Actinomycosis infection, but has been reported. Disseminated actinomycosis have been documented in 15.9% of the reported cases [20]. In many cases, the presenting symptoms do not correlate with the extension of the disease. According to Weese, the diagnosis is properly identified at admission in only 7% of the cases [21]. Disseminated disease may involve any organ, but the most common include lung, skin, brain, liver, bone and muscle.

6. Treatment for actinomycosis

The management of actinomycosis infection consists of prolong antimicrobial therapy, but surgical debridement and resection may be indicated in some cases. The use of antimicrobials has greatly improved the prognosis. Drug resistance is not considered a problem in actinomycosis and tends to be susceptible to beta-lactams antibiotics [22]. For patients with monomicrobial infections, treatment can be divided base on mild versuss severe disease. If the infection involves an organ causing a life threatening disease or multiple organs, it is considered severe. For mild actinomycosis, initial oral therapy with penicillin V (divided in four daily doses) is recommended [23]. For severe infection, initial course of 10 to 20 million units daily of intravenous penicillin G (divided into four to six hours) is recommended [22]. If the patient has penicillin allergy, a cephalosporin or doxycycline can be use [24].

As mentioned before, Actinomyces can grow with other organisms in tissue and sample cultures in almost 75 to 95% of cases [10]. The other organisms are usually anaerobic from the oral flora, and they can produce beta-lactamases that can protect actinomyces from penicillin.In those cases a combination of a beta-lactam plus beta-lactamase inhibitor is recommended as treatment.

Antimicrobial treatment should be continued until resolution of infection, usually between 6 and 12 months [25]. Actinomycosis infection can recur, especially in thoracic infections without surgical debridement [25]. Therapy duration of less than 3 months should be avoided in those cases. When infection complicates with abscess and fistula formation, surgical management and drainage is warranted, especially in life threatening presentations.

7. Conclusion

Actinomycosis is a chronic bacterial infection able to cause pathological invasion and destruction of multiple tissue mimicking other conditions as systemic and inflammatory diseases, other infections and cancer. Diagnosis is difficult and often take prolonged time, with the need of invasive procedures, and the requirement of tissue samples for diagnosis. From the microscopic to the macroscopic findings, this bacteria is capable of simulating being other that what really is, and has a unique physical and pathogenic characteristics that allow it to survive and resist in inhospitable environments, making its identification a real challenge when invading the human host.

Author details

Onix J. Cantres-Fonseca*, Vanessa Vando-Rivera, Vanessa Fonseca-Ferrer,
Christian Castillo Latorre and Francisco J. Del Olmo-Arroyo
Pulmonary and Critical Care Medicine Department, VA Caribbean Health Care
System, San Juan, Puerto Rico

*Address all correspondence to: onixcantres@gmail.com

IntechOpen

References

[1] Mabeza GF, Macfarlane J. European Respiratory Journal. 2003;**21**(3):545-551. DOI: 10.1183/090311936.03.00089103

[2] Chatterjee RP, Shah N, Kundu S, Mahmud SA, Bhandari S. Cervicofacial Actinomycosis mimicking osseous neoplasm: A rare case. Journal of Clinical and Diagnostic Research. 2015;**9**(7):ZD29-ZD31. DOI: 10.7860/JCDR/2015/12825.6249

[3] Han JY, Lee KN, Lee JK, et al. An overview of thoracic actinomycosis: CT features. Insights Into Imaging. 2013;**4**(2):245-252. DOI: 10.1007/s13244-012-0205-9

[4] Wong V, Turmezei T, Weston V. Actinomycosis. BMJ. 2011;**343**:d6099. DOI: 10.1136/bmj.d6099

[5] Ariel I, Breuer R, Kamal NS, Ben-Dov I, Mogel P, Rosenmann E. Endobronchial actinomycosis simulating bronchogenic carcinoma. Diagnosis by bronchial biopsy. Chest. 1991;**99**(2):493-495

[6] Valour F, Sénéchal A, Dupieux C, et al. Actinomycosis: Etiology, clinical features, diagnosis, treatment, and management. Infection and Drug Resistance. 2014;**7**:183-197. Published 2014 Jul 5. DOI: 10.2147/IDR.S39601

[7] Urban E, Gajdacs M. Microbiological and clinical aspects of Actinomyces infections: What have we learned? Antibiotics. 2021;**10**:151

[8] Mandell GL, Bennett JE, Dolin R, editors. Mandell, Douglas, and Bennett's Principles and Practice of Infectious Diseases. 7th ed. Philadelphia, PA: Churchill Livingstone Elsevier; 2010

[9] Roth J, Ram Z. Intracranial infections caused by Actinomyces species. World Neurosurgery. 2010;**74**(2-3):261-262. DOI: 10.1016/j.wneu.2010.06.011

[10] Smego RA Jr. Actinomycosis of the central nervous system. Reviews of Infectious Diseases. 1987;**9**(5):855-865. DOI: 10.1093/clinids/9.5.855

[11] Schaal KP, Lee HJ. Actinomycete infections in humans – A review. Gene. 1992;**115**(1-2):201-211

[12] Oostman O, Smego RA. Cervicofacial Actinomycosis: Diagnosis and management. Current Infectious Disease Reports. 2005;**7**(3):170-174

[13] Cheon JE, Im JG, Kim MY, Lee JS, Choi GM, Yeon KM. Thoracic actinomycosis: CT findings. Radiology. 1998;**209**(1):229-233

[14] Hirschfield LS, Graver LM, Isenberg HD. Broncholithiasis due to Histoplasma capsulatum subsequently infected by actinomycetes. Chest. 1989;**96**(1):218-219

[15] Piper MH, Schaberg DR, Ross JM, Shartsis JM, Orzechowski RW. Endoscopic detection and therapy of colonic actinomycosis. The American Journal of Gastroenterology. 1992;**87**(8):1040-1042

[16] Yegüez JF, Martinez SA, Sands LR, Hellinger MD. Pelvic actinomycosis presenting as malignant large bowel obstruction: A case report and a review of the literature. The American Surgeon. 2000;**66**(1):85-90

[17] Dayan K, Neufeld D, Zissin R, et al. Actinomycosis of the large bowel: Unusual presentations and their surgical treatment. The European Journal of Surgery. 1996;**162**:657

[18] Agarwal K, Sharma U, Acharya V. Microbial and cytopathological study of intrauterine contraceptive device users. Indian Journal of Medical Sciences. 2004;**58**(9):394-399

[19] Fiorino AS. Intrauterine contraceptive device-associated actinomycotic abscess and Actinomyces detection on cervical smear. Obstetrics and Gynecology. 1996;**87**(1):142-149. DOI: 10.1016/0029-7844(95)00350-9

[20] Mahadevappa M, Kulkarni P, Poornima KS. Disseminated Actinomycosis presenting as chronic right-heart failure due to right ventricular and pericardial infiltration. JACC Case Reports. 2020;**2**(12):1992-1998. Published 2020 Sep 2. DOI: 10.1016/j.jaccas.2020.07.015

[21] Weese WC, Smith IM. A study of 57 cases of actinomycosis over a 36-year period. A diagnostic 'failure' with good prognosis after treatment. Archives of Internal Medicine. 1975;**135**(12): 1562-1568

[22] Smego RA Jr, Foglia G. Actinomycosis. Clinical Infectious Diseases. 1998;**26**:1255-1261

[23] Martin MV. The use of oral amoxycillin for the treatment of actinomycosis. A clinical and in vitro study. British Dental Journal. 1984;**156**:252

[24] Skoutelis A, Petrochilos J, Bassaris H. Successful treatment of thoracic actinomycosis with ceftriaxone. Clinical Infectious Diseases. 1994;**19**:161

[25] Padmanabhan A, Thomas AV. Recurrent endobronchial actinomycosis following an interventional procedure. Lung India. 2017;**34**:189

Chapter 12

Diagnosis of Actinomycosis in the Physician's Clinic

Frans Maruma and Rhulani Edward Ngwenya

Abstract

The actinobacteria have an important role to play in humans. On one hand the actinobacteria, especially the *streptomyces spp.* are used medicinaly for synthesis of some novel anti-microbial medicines whilst on the other hand, *Actinomyces species* also exist as normal commensals as well as pathogens that cause disease in humans. The bacteria play a role in the maintenance of healthy human organs where they are considered normal inhabitants. The genus, *Actinomyces* is normally found in the mouth, gut and the female genital tract of human beings. This chapter elaborates on different clinical manifestations of actinomycosis in humans, making use of visual illustrations depicting the vast array of pathology therein. Finally, the diagnosis and treatment of these infections in humans is also fully discussed.

Keywords: actinomycosis, clinical phenotypes, diagnosis, and treatment

1. Introduction

Human actinomycosis is an unusual but insidiously chronic granulomatous bacterial infection caused by Actinomyces genera. These pathogens are filamentous gram-positive anaerobic bacteria from the Actinomycetaceae family [1]. Due to its insidious chronicity and rarity, human actinomycosis is often misdiagnosed for other more common conditions such as tuberculosis and even malignancy [2]. Once astutely identified in a timely manner, it is however treatable with good prognostic outcomes.

Actinomycosis in humans was first described in 1878 by Israel and Wolfe [3, 4]. In early reports of infections caused by filamentous gram-positive organisms, no distinction was made between diseases caused by actinomyces and nocardia's. As more taxonomy was developing, Kruse in 1896 described these organisms as streptothrix Israeli [5]. It was not until 1943 that the genera were clearly differentiated by Waksman and Henrici, enabling separation of the diseases they caused [6]. Today, *Actinomycosis Israeli* is a well-known pathogen that can cause human disease.

It must be noted that the human beings remain the natural reservoirs of these organisms. Given that the *Actinomyces Israeli* is found in abundance within the oral cavity, gastro-intestinal tract and female genital tract, as part of normal flora, it is of no surprise that this organism is commonly aetiologic to the infections affecting these anatomical regions. In rare cases, other species of the actinomyces family such as *Actinomyces odontolyticus*, *A. viscosus*, *Actinomyces naeslundii*, *Actinomyces radingae*, *Actinomyces turicensis*, and *A. meyeri* have also been aetiologically implicated. For actinomyces to cause disease, these organisms appear to require a synergistic relation with other accompanying bacteria, both from within and outside the genera [7–10].

In clinical practice, the diagnosis of actinomycosis requires a physician to be highly suspicious of the disease. Other confounders include the irrational use of penicillin containing antimicrobials leading to delayed diagnosis of the actinomycosis. The penicillin's are generally used for any suspected upper and lower aerodigestive tract infections particularly in low income countries. Actinomycosis commonly affect the structures of the head and neck region though it can affect virtually any organ of the body. The central nervous system, pelvic, abdominal, and thoracic involvements are generally very rare, with cervicofacial actinomycosis representing the most common (more than 50% cases) clinical form of actinomycosis disease manifestation [11, 12].

Cervicofacial actinomycosis is a chronic suppurative disease that results in typical formation of fistulae, sinus tracts, and fibrosis of the affected tissue. The characteristic spread of the disease tends to evade fascial planes thereby forming the typical sinuses, fistulae and tracts. In the physicians clinic, one of the most challenging aspects in the diagnosis of cervicofacial actinomycosis is the variation in phenotypical presentation of the disease. It is in accordance with the authors that the diagnosis often rests in the physicians highest index of suspicion. Once such a suspicion is raised, then necessary investigations should be undertaken. A tissue sample is always necessary for the diagnosis that almost always depends on a combination of clinical and characteristic histopathological features. If a physician becomes mistaken to think that the patient has odontogenic abscess, the empiric treatment is by means of broad spectrum penicillin based antibiotics. In this case, there exists a serious dilemma in the diagnosis because the genus actinomyces generally respond to broad spectrum antimicrobial (albeit prolonged courses) that are often used in treatment for odontogenic abscesses. Also, to isolate the genus actinomyces, prolonged anaerobic cultures are required. However, the recurrence and progression of the disease in this case should sensitise the treating clinician into thinking of the possibility of cervicofacial actinomycosis. Actinomycosis treatment typically require prolonged (6-12 months) penicillin containing antimicrobial agents as compared to short course (7–14 days) therapies typically used in odontogenic abscesses [8].

Herein we describe the 4 main clinical forms recognised, namely, cervicofacial (typically known as a lumpy jaw), primary cutaneous, thoracic, and abdominopelvic actinomycosis. In this chapter, we will give a detailed account on various clinical forms of forms of presentation, their investigations and management principle from a clinical point of view. There shall also be an account on synopsis regarding the classes of antibiotic that may be used to successfully treat actinomycosis in general from a physician's point of view guided by the scientific literature. Whilst other anatomical classifications include musculoskeletal and disseminated clinical forms exist, we have elected to focus our brief discussion on the aforementioned four clinical forms and elaborate on extended manifestations therein. The other forms of actinomycosis is pretty rare.

2. Clinical phenotypes

2.1 Cervicofacial actinomycosis

This clinical variant of human actinomycosis is the commonest form encountered owing to its typical clinical presentation, it is often referred to as a 'lumpy jaw" [7, 10]. See **Figures 1–3**. Generally, the disease can present acutely or chronically to the physician's clinic.

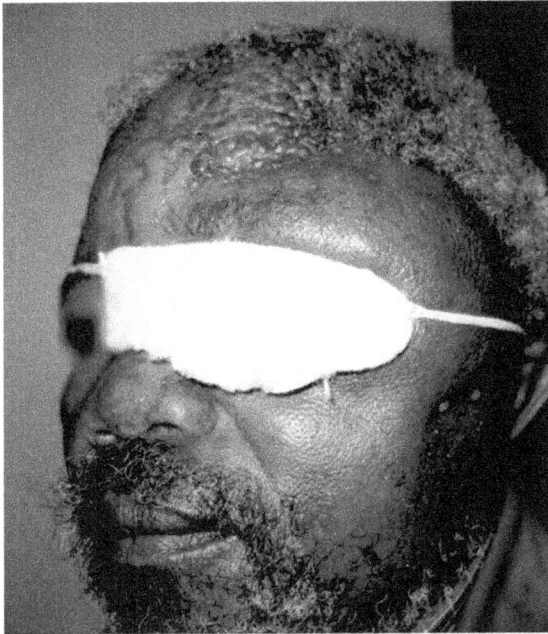

Figure 1.
Massive induration on the left side of the face extending into the scalp. Small nodules visible in the preauricular region.

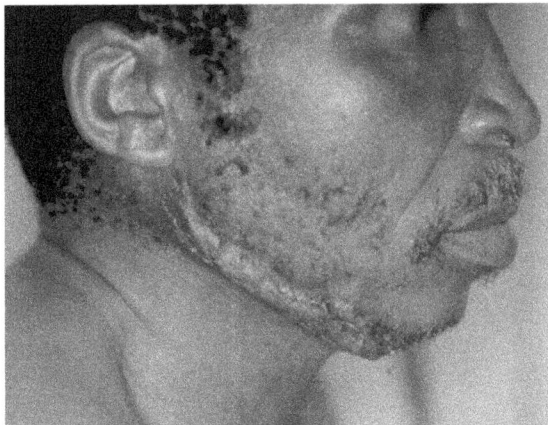

Figure 2.
Actinomycosis: Fibrotic, indurated linear plaque overlying right jawline with surface erosions.

It typically presents with a slow growing non-tender firm lump located around the neck, cheek and/or jaw, depending on primary site of inoculation. Clinical signs within the jaw region may be trismus as well as facial asymmetry, as a result of muscles of mastication undergoing spasms and fibrosis. With further chronicity, the infection may extend to adjacent anatomical structures such as lymph nodes, tongue, bones, oropharynx and rarely meninges of brain and spinal cord. The acute disease process is rare and typically can present with pain, trismus and rapid progression towards abscess formation on the affected site.

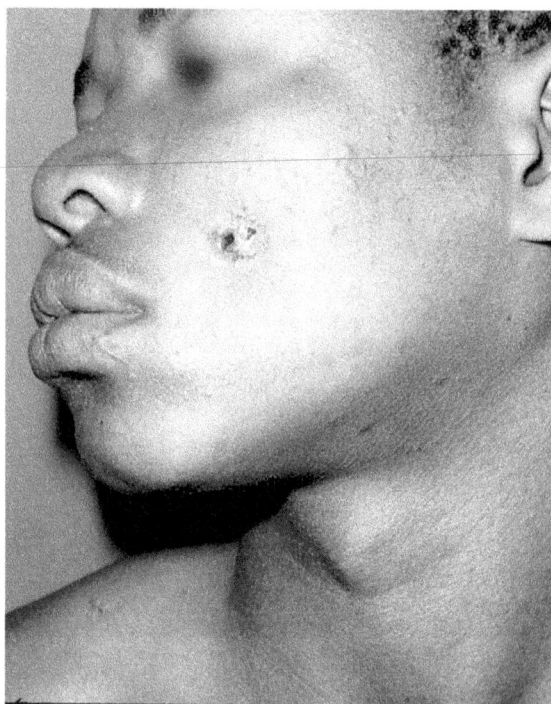

Figure 3.
Early lesion of cervicofacial actinomycosis represented by an ulcerated plaque on the left cheek.

Risk factors for cervicofacial actinomycosis

- Poor oral hygiene

- Farm workers, as well as residence in tropical countries

- Post tooth extraction or other invasive dental procedures

- Trauma of any form to the oral cavity

- Immunosuppressed host, secondary to HIV/AIDS, diabetes mellitus, chemotherapy and other immunosuppressive therapy [13–19].

2.2 Primary cutaneous actinomycosis

Primary cutaneous disease may result from direct inoculation of skin by
actinomyces organisms. Although very rare, this type of clinical manifestation
can be seen following trauma to the site of inoculation (**Figures 4** and **5**). The
anatomical sites that can be involved are variable and include the hands, feet, and
head and neck region (ears and nose). Clinical examination may reveal, localised
non-healing ulcers, suppurative discharging abscesses, and sinuses. It cannot be
overemphasised that disseminated cutaneous manifestation of this disease is of
extreme rarity especially when the patient does not have any concomitant systemic
infections [8, 9].

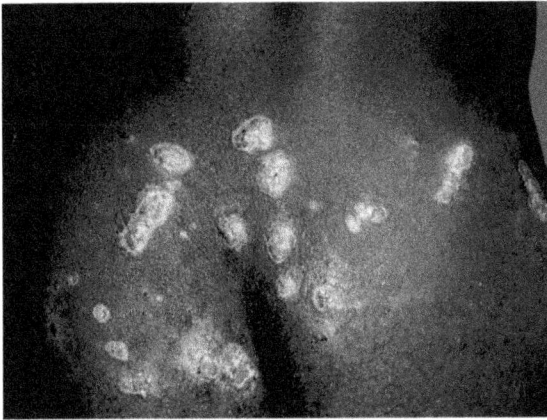

Figure 4.
Cutaneous actinomycosis presenting as crusted and indurated multiple plaques on lumbo-sacral region.

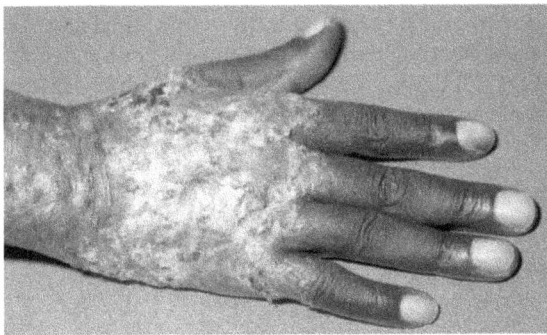

Figure 5.
Primary cutaneous actinomycosis of the hand. This represents an area of primary inoculation.

2.3 Thoracic actinomycosis

Accounting for 15–20% of anatomically classified actinomycosis clinical patterns [20], thoracic actinomycosis requires a particularly high index of suspicion to successfully diagnose and manage.

2.3.1 Pathogenesis

The anatomical location and construct of the thoracic cavity makes it inherently vulnerable to being infected. Located between the Head-and-Neck region and abdominal cavity, it may succumb to direct regional extension of cervicofacial infection (above), Oesophageal breach (within), and abdominopelvic infection (below). Other routes of infection may involve, *inter alia*, inhalation of contaminated aerosol particles, aspiration of gastric content or oropharyngeal secretions, and hematogenous dissemination [15]. Risk factors for thoracic actinomycosis would thus include, but limited to, dental caries, swallowing problems for aspiration, immunosuppression, and previous surgery with resultant clean-contaminated wound.

2.3.2 Clinical manifestation

There are no clear demographic predilections for thoracic actinomycosis. Though several studies have noted a 2–4 times greater prevalence in males aged 30 to 50 years [21, 22], still others have reported cases in children as young as 14 years old and as the elderly over 70 years [23]. This follows also for both sex and gender.

Patients may present symptomatic or asymptomatic. Naturally, structures within the thoracic cavity that may be affected include the lungs, pleura, mediastinum, or chest wall [15]. Depending on the anatomical structure affected, so will the clinical manifestations results.

The most affected structures are the lungs and pleura (either primarily or secondary to a chronic lung foci contiguously manifesting as empyema). Therefore, it is no surprise why the most common symptoms are haemoptysis, cough, sputum production, chest pain, fever, and weight loss [23, 24]. This makes the diagnosis of particular difficulty as these symptoms are equally common in most thoracic pathologies, tuberculosis and malignancy being the more prevalent. Suffice to say, Actinomycosis can also co-exist with the aforementioned diseases, making diagnosis and further management even more challenging.

Patients may also present asymptomatically, with vicinal destruction of ribs or sternum, sinus tracts extending to the skin (**Figure 6**), or incidental finding of an endobronchial mass [15]. Endo-, myo-, and pericardial disease are not uncommon, often resulting from haematogenous spread or direct extension of mediastinal disease – These patients commonly present with pericarditis related signs and

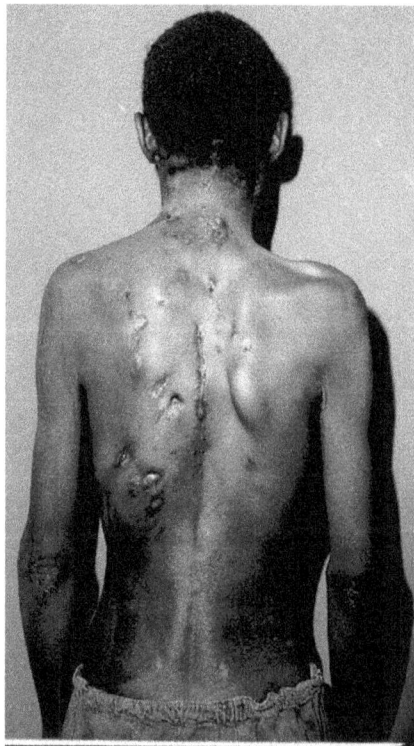

Figure 6.
Pulmonary actinomycosis with infiltration into the chest wall. Multiple fibrotic sinus tracts and ulcerated, indurated plaques and nodules are evident mostly on left upper posterior chest.

symptoms [25]. Investigations such as chest radiographs, CT-scan, cardiac echo-grams, bronchoscopy with biopsies, may be equivocal even with findings of sulphur granules in sputum or sinus tract discharge. It would thus follow that Actinomycosis remains a diagnosis of exclusion, after eliminating the more common and even aggressive diagnosis of malignancy.

2.3.3 Abdominopelvic actinomycosis

Abdominopelvic disease accounts for 20% of human actinomycosis [26], 65% of which is as a result of acute appendicitis [25]. Abdominopelvic actinomycosis may be the most difficult to diagnose due to its indolence and latency, often presenting months to years after the initial insult [15].

2.3.4 Pathogenesis

Already established as being mainly an endogenous infection, in that the pathogenic *Actinomyces* species are commensals and normal inhabitants of the oropharynx, gastrointestinal tract, and female genital tracts in humans. It thus follows that pathogenicity in the abdominopelvic region would involve a breach in intestinal or internal genital mucosal lining [15], thus affecting the local structures and organs. Causes would therefore include penetrating trauma, gastrointestinal or gynaecological surgery, neoplasia, and foreign bodies in the gastrointestinal tract or genitourinary tract, with or without erosion through the mucosal barrier [25].

In the abdomen the disease may be localised, often with a strong predilection for the ileocecal region as a result of acute appendicitis. However, the vast array of more common pathologies within this region such as tuberculosis, carcinoid, typhlitis, ameboma, regional enteritis, chronic appendicitis, or malignancy, makes diagnosis extremely challenging. The disease may also spread extensively with very little respect for fascial and connective tissue planes. This represents one of the ways in which pelvic actinomycosis may develop- direct extension from the abdomen.

Primary involvement of pelvic structures may also occur and is predominantly associated with intrauterine contraceptive devices (IUDs) as a result of colonisa-tion and/or infection [27]. Other primary causes may include septic abortions and retained sutures from previous surgery [15].

2.3.5 Clinical manifestation

As with thoracic actinomycosis, there is no clear demographic predilections for abdominopelvic actinomycosis. Risk factors generally remain the same, that being males (except for pelvic actinomycosis, where woman are mainly affected), age 20–60 years, immunosuppression (with diabetes mellitus ranking highest) [14, 26–28]. However, caution must be exercised, as many patients seldom present with any of the mentioned risk factors.

With a predilection for the ileocecal region, patients may present with localised symptoms of right iliac fossa pain. Other symptoms may be nonspecific such as fever, weight loss, and generalised abdominal pain. Clinical examination may only elicit tenderness and or an abdominal mass (of which less than 10% are diagnosed preoperatively [14], or a completely normal abdominal thus making the diagnosis difficult of abdominal actinomycosis extremely challenging. It is no surprise that less than 10% are diagnosed preoperatively.

Anorectal disease is not uncommon, with a clinical picture as vast as rectal strictures, draining sinuses and fistulas, and perirectal or ischiorectal abscesses [15].

The diagnosis of Pelvic Actinomycosis is of equal perplexity, with patients often presenting with nonspecific symptoms (lower abdominal or suprapubic pain, weight loss, vaginal discharge, and low-grade fever if at all). These symptoms may persist from months to years [28]. The biggest clue in these patients is a prolonged use of an intrauterine contraceptive devices (IUDs), as it is the predominant cause.

3. Overarching mechanism of disease and diagnosis

Actinomyces spp. reside on their respective mucosal surfaces of the anatomical sites that they inhabit. A gain into deeper adjacent tissues require a disruption of the mucosal surface usually following some mechanical factor such as trauma or surgery [29, 30].

The gold standard diagnostic method is through a tissue sample plus clinical correlation in some cases. Tissue biopsy must be performed usually with two separate specimens. One specimen should be sent for microbiological microscopy, culture and sensitivity for confirmation, or to exclude other possible infections that could mimic the actinomycosis. The second specimen is then sent for histopathological sectioning and assessment.

Microbiological assessment of the tissue specimens is often braised with challenges. It is generally difficult to obtain bacteriological identification of Actinomyces as a result of various factors. The heterogenicity of the actinomyces, frequent presence of other micro-organisms, short incubation period, as well as recent antibiotic therapy of the patient, all contribute to the negative microbiological tests in majority of cases. The sulphur granules, although strongly suggestive, are also not diagnostic as these may also be found in other bacteria, mainly nocardia spp. [29–31].

Histopathological sections of affected tissue may depict abscesses and sinuses in the midst of mixed inflammatory cell infiltrate and fibrotic changes. The bacterial colonies and their sulphur granules may also be seen in the centre of the lesions reminiscent of the concept of Splendore-hoeppli phenomenon. Radiological examination is important for extra-cutaneous disease. X-rays and computed tomography scans are useful in delineating lesions affecting the soft tissue and bony structures [7, 8].

4. The differential diagnosis

In clinical practice, it is always crucial to rule out most mimickers of actinomycosis. These disorders range from chronic inflammatory, infectious, to malignant diseases. The authors find it useful to categorise the differential diagnosis (which is by no means exhaustive) according to the site of involvement. Therefore, for abdominal and pelvic disease, the following differential diagnosis must be considered:

- Abdominal abscess

- Appendicitis

- Liver abscess

- Colon cancer

- Chron's disease

- Diverticulitis

- Pelvic inflammatory disease

- Uterine cancers

- Abdominal tuberculosis;

And for the thoracic disease, the clinician ought to rule out the following:

- Bacterial pneumonia including pulmonary tuberculosis

- Lung abscess

- Fungal pneumonia

- Lung cancer (non- small cell type)

- Small cell carcinoma of the lung

- Non Hodgkin lymphoma

- Aspiration pneumonia

- Blastomycosis

For cutaneous disease, a clinician ought to exclude the following differential diagnosis by means of tissue sample for culture, microscopy, and where necessary polymerase chain reaction tests:

- Botriomycosis

- Syphilitic gummata

- Fungal infections (e.g., Sporotrichosis, Blastomycosis, Eumycetoma, Aspergillosis, etc)

- Nocardiosis

- Osteomyelitis

- Squamous cell carcinoma

- Cutaneous Tuberculosis

- Mycobacteria other than tuberculosis (MOTT)

- Secondary metastasis

Finally, the cervical, facial and central nervous system involvement can also mimic a wide variety of diseases that are of interest in terms of differential diagnosis. These are:

- Tuberculosis (for example tuberculoma's in the brain)

- Brain abscess

- Lymphoma's

- Fungal infections of the brain, for example, cryptococcoma's

- Non Hodgkins Lymphoma

- Nocardiosis

- Odontogenic abscess

5. Management

The management of human actinomycosis is comprised of antimicrobial agents as well as surgery as indicated. The choice to use combination treatment is necessitated by the severity and clinical condition of the patient. Fortunately, antimicrobial resistance are generally not considered an issue when it comes to actinomyces spp. The organisms are favourably sensitive to penicillin based antibiotic therapy as they do not produce beta lactamase. Amoxycillin or penicillin G are considered as the drugs of choice for treating actinomycosis [7, 10]. Macrolides and clindamycin have all been used with good efficacy. Cloxacillin, oxacillin, doxycycline ciprofloxacin, metronidazole, cephalexin are not considered effective in treating human actinomycosis.

Although Piperacillin plus tazobactam, carbapenems such as meropenem and imipenem, are also effective, they should not be used to avoid acquisition of resistant normal flora to these agents. Notably, *Actinomyces graevenitzii* and *A. europaeus* have been particularly reported to be resistant to third generation cephalosporins suggesting that there is limitations regarding the use of this group of antimicrobials [14–16, 32].

6. Conclusion

Human infection with actinomyces spp. remains rare however an important clinical entity for physicians to recognise and treat promptly. The cervicofacial disease has a predilection towards tropical countries, farm workers, and those with poor dental hygiene. Other risk factors include post-surgery of the anatomic site where actinomyces spp. are a part of normal flora as well as the prolonged use of intrauterine contraceptive device in female patients.

Early disease can be treated successfully, albeit for a long time (6-12 months) with antimicrobial agents such as penicillin's, tetracyclines and macrolides.

Advanced disease often requires a combination of antimicrobial agents as well as surgical intervention.

Primary prevention seems to be of great value in management of the human actinomycosis. Measures such as maintaining good oral health, reduction of alcohol abuse, may limit occurrence of extra-cutaneous actinomycosis.

Women should also consider frequent changing of the intrauterine contraceptive devices according to the manufacturers' stipulation in order to circumvent the occurrence of pelvic actinomycosis.

It is the authors view that actinomycosis remains a diagnosis of exclusion, therefore this calls for a high index of clinical suspicion, and an astute physician for the successful diagnosis and management therein.

Acknowledgements

The authors would like to acknowledge the archives of photographs used in this chapter as obtained from the department of dermatology, Faculty of health sciences, University of Free State, Bloemfontein, South Africa.

Conflict of interest

The authors declare no conflict of interest.

Notes/thanks/other declarations

The two authors would like to thank their spouses, Thato and Molline respectively, including their children, for their understanding and moral support during the production of this work.

Author details

Frans Maruma[1]* and Rhulani Edward Ngwenya[2]

1 Faculty of Health Sciences, Department of Dermatology, University of the Free State, Bloemfontein, South Africa

2 Faculty of Health Sciences, Department of Plastic and Reconstructive Surgery, University of Pretoria, Pretoria, South Africa

*Address all correspondence to: marumaf@ufs.ac.za; fransmaruma2@icloud.com

IntechOpen

References

[1] Russo TA. Agents of actinomycosis. In: Mandell GL, Bennett JE, Dolin R, editors. Principles and Practice of Infectious Diseases . 7th ed. Philadelphia PA: Churchill Livingstone Elsevier; 2010. pp. 3209-3219

[2] Acevedo F, Baudrand R, Letelier LM, Gaete P. Actinomycosis: A great pretender. Case reports of unusual presentations and a review of the literature. International Journal of Infectious Diseases. 2008;**12**:358-362

[3] Israel J. Neue Beobactungen auf dem Bebiete der Mykosen des Menshen. Virchows Archiv fur Pathologische Anatomie. 1878;**74**:15-53

[4] Wolfe M, Israel J. Ueber Reincultur des Actinomyces and seine Ueber-tragbarkeit auf Thiere. Virchows Archiv fur Pathologische Anatomie. 1891;**126**:11-59

[5] Kruse W. Systematik der Streptothricheen und Bakterien. In: Flugge C, editor. Die Mikroorganismen. 3rd ed. Vol. 2. Leipzig, Germany: Vogel; 1896. pp. 48-96

[6] Waksman SA, Henrici AT. The nomenclature and classification of the actinomycetes. Journal of Bacteriology. 1943;**46**:337

[7] Valour F, Sénéchal A, Dupieux C, Karsenty J, Lustig S, Breton P, et al. Actinomycosis: Etiology, clinical features, diagnosis, treatment, and management. Infect Drug Resist. 2014;**5**(7):183-197. DOI: 10.2147/IDR.S39601

[8] Eduardo CJ, Brenn T, Lazar AJ, Billings SD. Mckee's Pathology of the Skin. 5th ed. London: Elsevier; 2019

[9] Bolognia JL, Schaffer JV, Cerroni L. Dermatology. 4th ed. London: Elsevier; 2017

[10] Burns T, Breathnach SM, Cox NH, Griffiths C. Rook's Text of Dermatology. Chicester: Wiley; 2010

[11] Kwartler JA, Limaye A. Pathologic quiz case 1. Cervicofacial actinomycosis. Archives of Otolaryngology – Head & Neck Surgery. 1989;**115**:524

[12] Könönen E, Wade WG. Actinomyces and related organisms in human infections. Clinical Microbiology Reviews. 2015;**28**:419

[13] Abdalla J, Myers J, Moorman J. Actinomycotic infection of the oesophagus. The Journal of Infection. 2005;**51**(2):E39-E43

[14] Wong VK, Turmezei TD, Weston VC. Actinomycosis. BMJ. 2011;**343**:d6099

[15] Smego RA Jr, Foglia G. Actinomycosis. Clinical Infectious Diseases. 1998;**26**(6):1255-1261

[16] Mandell GL, Bennett JE, Dolin R, editors. Mandell, Douglas, and Bennett's Principles and Practice of Infectious Diseases. 7th ed. Philadelphia, PA: Churchill Livingstone Elsevier; 2010

[17] Schaal KP, Lee HJ. Actinomycete infections in humans – A review. Gene. 1992;**115**(1-2):201-211

[18] Pulverer G, Schütt-Gerowitt H, Schaal KP. Human cervicofacial actinomycoses: microbiological data for 1997 cases. Clinical Infectious Diseases. 2003;**37**(4):490-497

[19] Shikino K, Ikusaka M, Takada T. Cervicofacial actinomycosis. Journal of General Internal Medicine. 2015;**30**(2):263. DOI: 10.1007/s11606-014-3001-z

[20] Mabeza GF, Macfarlane J. Pulmonary actinomycosis.

The European Respiratory Journal. 2003;**21**:545-551

[21] Bennhoff DF. Actinomycosis: Diagnostic and therapeutic considerations and a review of 32 cases. Laryngoscope. 1984;**94**:1198-1217

[22] Brown JR, Human actinomycosis. A study of 181 subjects. Human Pathology. 1978;**4**:319-330

[23] Baik JJ, Lee LL, Yoo CG, Han SK, Shim YS, Kim YW. Pulmonary Actinomycosis in Korea. Respirology. 1999;**4**:31-35

[24] Lu MS, Liu HP, Yeh CH, Wu YC, Liu YH, Hsieh MJ, et al. The role of surgery in hemoptysis caused by thoracic actinomycosis; a forgotten disease. European Journal of Cardio-Thoracic Surgery. 2003;**24**:694-698

[25] Fowler RC, Simpkins KC. Abdominal actinomycosis: A report of three cases. Clinical Radiology. 1983;**34**:301-307

[26] Weese WC, Smith IM. A study of 57 cases of actinomycosis over a 36-year period. A diagnostic "failure" with good prognosis after treatment. Archives of Internal Medicine. 1975;**135**:1562-1568

[27] Fiorino AS. Intrauterine contraceptive device-associated actinomycotic abscess and Actinomyces detection on cervical smear. Obstetrics and Gynecology. 1996;**87**:142-149

[28] Brook I. Actinomycosis: Diagnosis and management. Southern Medical Journal. 2008;**101**:1019-1023

[29] Vandeplas C, Politis C, Van Eldere J, Hauben E. Cervicofacial actinomycosis following third molar removal: Case-series and review. Oral and Maxillofacial Surgery. 2021;**25**(1):119-125. DOI: 10.1007/s10006-020-00896-x

[30] Karanfilian KM, Valentin MN, Kapila R, Bhate C, Fatahzadeh M, Micali G, et al. Cervicofacial actinomycosis. International Journal of Dermatology. 2020;**59**(10):1185-1190. DOI: 10.1111/ijd.14833

[31] Boyanova L, Kolarov R, Mateva L, Markovska R, Mitov I. Actinomycosis: A frequently forgotten disease. Future Microbiology. 2015;**10**(4):613-628. DOI: 10.2217/fmb.14.130

[32] Smith AJ, Hall V, Thakker B, Gemmell CG. Antimicrobial susceptibility testing of Actinomyces species with 12 antimicrobial agents. The Journal of Antimicrobial Chemotherapy. 2005;**56**(2):407-409

Immune Response during *Saccharopolyspora rectivirgula* Induced Farmer's Lung Disease

Jessica Elmore and Avery August

Abstract

Repeated exposures to Saccharopolyspora rectivirgula in some individuals can lead to a hypersensitivity reaction where a pro-inflammatory feedback loop can occur in the interstitial space in the alveoli of the lungs that can ultimately lead to granuloma formation and fibrosis, referred to as Hypersensitivity pneumonitis or Farmer's Lung Disease. The pathogenesis of FLD is complex and incompletely understood. *S. rectivirgula* induces an immune response, triggering neutrophil influx into the lung followed by lymphocyte influx of CD8+ and CD4+ T cells. The cytokine IL17A has been shown to be critical for the development of *S. rectivirgula* induced Hypersensitivity pneumonitis. This chapter will review the immune response leading to the development of *S. rectivirgula* induced Hypersensitivity pneumonitis.

Keywords: *Saccharopolyspora rectivirgula*, hypersensitivity pneumonitis, farmer's lung disease, lung inflammation, immune response

1. Introduction

Bacteria were one of the very first organisms to inhabit the Earth three billion years ago [1, 2]. Using today's current technological approaches to determine the total mass of all the organisms on Earth, would reveal that bacteria are second most dominant organism on the planet [3]. Bacteria are important for human health, providing nutrients, protecting against developing diseases, aiding in food production, energy storage, helping to train the immune system and have the potential to capture carbon dioxide to reduce greenhouse gas emissions [4]. One of the most important discoveries of bacteria is that they can make antibiotics to treat bacterial diseases. The discovery of novel bacteria was prominent throughout the 1920's until the 1970's where the antibiotic resistant diseases were starting to take a foothold in the world [5, 6]. The discovery of new bacteria that have the potential to alleviate antimicrobial resistance is still an issue today. The phylum *Actinobacteria* are among the largest groups of bacteria and are one the major sources of antibiotics [7]. *Actinobacteria* are found in a wide variety of environments such as soils, deep oceans, plants, caves, and stones, among others [7, 8]. Rare actinobacteria such as *Saccharopolyspora* have the potential to be tapped for novel therapeutics [9]. However, some *Saccharopolyspora* species are also the cause of disease in humans [10].

1.1 History of *Saccharopolyspora* and its isolation

The genus *Saccharopolyspora* was first described by Lacey and Goodfellow in 1975 and later revised by Korn-Wendisch in 1989 [11, 12]. In 1971, Lacey was researching bacteria that may be associated with causing Bagassosis, a respiratory disease similar to farmer's lung induced by the inhalation of dust from bagasse or sugar cane that had been crushed, liquid removed, dried, and baled for further processing [13–15]. The storage of bagasse at high humidity and high temperature encouraged the growth of mold and bacteria [15]. They suspected that thermophilic *Actinomycetes* may be the culprit due other thermophilic *Actinomycetes* involvement in farmer's lung or mushroom worker's lung [13, 16]. Furthermore, previous work had shown that serum from patients that were diagnosed with Bagassosis reacted against extracts of thermophilic *Actinomycetes* [15]. Lacey was able to isolate *Thermoactinomyces vulgaris* from moldy baggase, along with an unknown bacterium similar to other thermophilic *Actinomycetes*, but that was different enough from *T. vulgaris* to be considered a new species. Lacey named it *Thermoactinomyces sacchari* [13]. Three years later Lacey and Goodfellow reported on their observation of isolates of *T. sacchari* when grown at 40°C also had growth of an unidentified actinomycete, which was different from other taxa suggesting this was a new species. Lacey and Goodfellow called the unknown isolate *Saccharopolyspora hirsute* gen. et. sp. nov [11]. What followed Lacey and Goodfellow's discovery was a reclassification of strains to other genera by the scientific community. In 1989 Korn-Wendisch completed a detailed taxonomic study comparing the genera *Faenia* and *Saccharopolyspora* finding that *Faenia rectivirgula* should be transferred to the genus *Saccharopolyspora* with its nomenclature now being *S. rectivirgula* [12].

S. rectivirgula is an aerobic, non-motile aerobic gram-positive bacteria part of the class *Actinomycetes* [9]. It grows up to 5 mm in diameter at temperature ranges of 50–55°C [17], in a variety of environments such as soil, plants and moldy hay [9, 18]. *S. rectivirgula* is one of the causative agents of a type of extrinsic allergic alveolitis, farmer's lung disease. This review will focus on the immune response to *S. rectivirgula* induced hypersensitivity pneumonitis (HP) better or farmer's lung disease [19, 20].

1.2 Hypersensitivity pneumonitis (HP) or farmer's lung disease (FLD)

HP was first described by Bernardino Ramazzini da Capri in the early 1700s when he encountered workers that sifted grain developed a dry cough coupled with edema and weight loss [21]. He suggested the workers developed the cough due to the humid environments inside the wheat dust [21]. In the early 1930s farmer's lung disease was first described by John Campbell after seeing five patients whom had pulmonary difficulties due to their working conditions [22]. All five patients were farmer laborers working with hay that was improperly stored to keep out moisture. Dr. Campbell obtained samples of "dust" from the hay and discovered fungi present. However, he was unable to discover the origin of disease, and found no correlation with the presence of fungi and his patients' sputum samples. Today, we know HP as a collection of lung diseases that can occur in an occupational, recreational or home settings. HP is caused by the repeated exposure to organic or inorganic agents such as avian proteins, chemicals, molds, and bacteria, among others where the subset of HP is categorized by the offending agent [19, 20, 23]. Individuals are asymptomatic or symptomatic leading to a hypersensitive reaction. Historically, HP has been characterized into three stages, acute, sub-acute and chronic but the field seems to be moving towards categorizing HP in a sensitization phase and challenge phase [24]. The onset of symptoms are fever, cough, wheezing, chills and if left untreated patients can experience dyspnea, fatigue, weight loss and fibrosis [23]. Some individuals develop poorly formed granulomas that advance to pulmonary fibrosis [25–28], which can lead to irreversible dysfunction with an increase in mortality [29], and some patients may even develop emphysema [30, 31]. Early diagnosis and treatment of HP and farmer's lung disease may prevent long term effects. However, HP is difficult to diagnose due to patients' symptoms that have similar pathologies to other respiratory diseases such as asthma or idiopathic pulmonary fibrosis coupled with a lack of a standardized diagnostic testing [23, 26, 32].

Diagnosis of patients with HP including farmer's lung disease is a multi-factorial process. A clinical history is taken to determine the patients' symptoms, descriptions of the environment they interact with such as work and home [33]. Samples from the patients' environment may be undertaken to undergo microbiological identification. When an offending agent is unknown, in some cases diagnosis would include provoking a physiological response from the patient by conducting an inhalation challenge with various potential triggers to try to determine the nature of the offending agent [19, 33–36]. Serum antibody tests are used to determine if the patient has been sensitized to the offending agent. IgG has been shown to be associated with farmer's lung patients that develop symptoms or those that asymptomatic [19, 33]. A standardized enzyme-linked immunosorbent assay (ELISA) test has yet to be developed and adapted in the clinical setting or able to differentiate between patients who have been exposed from those that have actual manifestation of the disease [23, 37]. Sputum or serum samples may be collected to determine lymphocytosis or alveolitis in bronchial lavage fluid (BAL) [33]. High resolution CT of the chest or bronchoscopy may be used to differentiate FLD from other diseases that have similar clinical features [19, 33]. Lung biopsy is reserved for patients where there has been no identification of the specific antigen exposure [37].

Treatment options are limited for all subsets of HP. The best treatment for patients is to avoid the offending antigen or utilize personal protective equipment. Determining the offending agent can be challenging and avoiding it may not be feasible for resolution for all [23, 38]. Personal protective equipment is successful in reducing inhalation of antigens but might be financially challenging or have low compliance rates [38]. Even in the cases of FLD, improvement of hay packing techniques meant to reduce bacterial or fungi growth do not always work [39, 40]. Furthermore, in some cases when avoidance of the offending agent is successful some patients may still have a decline in lung function [23, 35]. Corticosteroids may be prescribed for certain patient demographics as they reduce the immune response but is not beneficial for long term usage [41].

Farmer's lung affects approximately up to 20 percent of individuals that typically work in agricultural settings (i.e. dairy farms) where environmental and genetic factors influence the susceptibility of workers, and even in some cases that of their families [23, 27, 35, 42–44]. However, not all exposed individuals develop FLD suggesting a genetic component that lead to developing symptoms [42]. The environmental factors and genetic factors that contribute to the development of FLD remains incompletely understood.

1.3 Environmental impact of FLD

Secondary infections, pesticides, air quality, among others have all been attributed to increased susceptibility to developing HP including FLD [45–47]. Interestingly, smokers have a decreased risk of developing HP but non-smokers have a lower risk of developing emphysema [30, 48]. The reduction of risk with smokers developing HP has been attributed to nicotine's ability to suppress immune responses [48]. There are a variety of antigens that can lead to the development of

HP. In some cases, patients are exposed to a mixture of antigens [49–51]. However, few studies evaluate the immune response when there is a mixture of antigens involved in disease pathogenesis, but there have been some studies looking at co-infection with viruses [46, 51–53]. For example, in mouse models, Cormier et al., found that single round of infection with Sendai virus following *S. rectivirgula* exposure enhances the immune response to *S. rectivirgula*, with increased BAL cell number, TNF-α and IL-1α compared control and *S. rectivirgula* only exposed mice [47]. Histological analysis of lungs also showed clear formation of granulomas in *S. rectivirgula* mice inoculated with Sendai virus. Thus, viral or secondary infections may contribute to the triggering of an immune response to *S. rectivirgula*.

Our environmental exposure during upbringing impacts our risk of developing allergic diseases whereby increased microbial exposure in early life can reduce the risk of developing allergies or allergic disease in adulthood [54–56]. The microbiome can modify host-immune responses and dysbiosis of the microbiome can lead to disease [2, 57]. Previous research has shown children that grow up in an agricultural setting have a reduction in developing allergic diseases. Interestingly children have been diagnosed with HP albeit there a very few studies that examine children with HP [58]. Thus, environmental upbringing may not fully explain risk factors to developing FLD. A genetic component may play a role in those with healthy microbiomes that leads to their susceptibility of disease. The microbiome has also been suggested to be able to affect the development of HP including farmer's lung disease. Russell et al. investigated this effect on the development of FLD using antibiotic-mediated microbial shifts [59]. In experiments where mice were treated prenatally or perinatally with vancomycin or streptomycin and exposed to *S. rectivirgula* antigen for 3 days per week for 3 weeks, they found that streptomycin treated mice had increased IFN-γ and IL-17A production, increased lymphocytes in the BAL and more severe lung pathology compared to untreated or vancomycin treated mice. However, the vancomycin treated group had the biggest change microbial shift in the gut at the family level, with reduction of *Bacteroidetes*, but streptomycin treated mice had an increase in *Bacteroidetes*. Taken together antibiotic treatments that shifts the microbiome can result in increased severity of FLD. Increasing the diversity or specific group of bacteria may result in less severity of FLD but further testing needs to be conducted.

1.4 Genetic contribution to FLD

HP including FLD is such a rare disease the studies that genetic analysis of susceptibility to disease are few and far in between. In addition, the majority of studies that have evaluated genetic susceptibility have small sample size in their respective patient cohorts, and do not delineate the causative agent of HP. These reasons, among others, partly explains why there is no consensus on the genetic polymorphisms that may cause individuals to be sensitive to *S. rectivirgula* induced HP, yet they still may offer clues to a better understanding of specific intricacies FLD. Studies that were conducted hypothesize that polymorphisms in tumor necrosis factor alpha (TNF-α), major histocompatibility complex (MHC)/human leukocyte antigen (HLA) and transporter for antigen presentation (TAP), or pulmonary surfactant may play a role in sensitivities to *S. rectivirgula* [60, 61].

Polymorphisms in the TNF-α gene have been linked to elevated inflammation, and single nucleotide polymorphisms (SNPs) at −308 position in the promoter region of the TNF gene has been associated with an increased risk to inflammatory diseases. Similarly to TNF-308A, TNF-238G has also been shown to be associated with increased production of plasma TNF-α in systemic lupus erythematous patients [62]. Furthermore, TNF-308A has been shown to be associated with high levels of TNF-α *in vitro* investigated in connection with susceptibility to developing FLD [61]. Patients with farmer's lung disease that had been exposed to moldy hay and developed elevated responses had a higher frequency of both TNFA2 homozygous or heterozygous alleles compared to healthy controls. There were no significant differences in the allele frequencies of TNF-β intron 1 gene polymorphisms between the healthy controls compared to patients with HP. These studies suggest patients that develop HP including farmer's lung may have a genetic predisposition to elevated TNF-α production.

MHC class I (HLA-DR) and II (HLA-DQ) play a critical role in the adaptive immune response by presenting epitopes from foreign or self-antigens to $CD8^+$ or $CD4^+$ T cells respectively [63]. MHC class I utilizes ATP-binding cassette proteins, transporters associated with antigen processing-1 and 2 (TAP1 and TAP2) to aide in the process of antigen presentation [57]. The genetic regions that encode for MHC are polygenic and incredibly polymorphic, allowing for the recognition of many different epitopes; polymorphism that occur in these gene regions also serve as risk factors in susceptibility to disease [36, 57]. Both TAP1 and TAP2 genes also map within the MHC class II region [64]. Polymorphisms in HLA-DR and DQ have been associated with susceptibility to developing HP, however it is unclear if this association extends to those who develop farmer's lung [36]. Patients that have HP have been reported to have a significant increase in the frequency of the TAP1 genotypes Asp-637/Gly-637 and Pro-661/Pro-661 among patient cohorts that had bird fancier's lung compared to healthy controls [64]. Familial studies of HP investigated polymorphisms in MHC class II genes HLA -DRB1/2/3/4/5, -DQA1, -DQB1, -DPA1, -DPB1, -DMA, and -DMB in healthy and HP patients with one patient having been diagnosed with FLD, have found that DRB1*04 alleles and haplotypes could be contributing the susceptibility to HP [65]. This suggests MHC and TAP might have a role in patient susceptibility to disease. However, we do not know if these polymorphisms are only prevalent for those whom may be afflicted with bird fancier's lung compared to FLD.

Phospholipids and glycoproteins play a large role in maintaining healthy lung function. Dysfunction of these material can increase risk of developing pulmonary diseases. Surfactant is a complex material composed of phospholipids and proteins that reduce the surface tension in the alveoli and bronchiole of fully developed lungs secreted by type II alveolar cells. Pulmonary surfactants, SP-A, SP-B, SP-C and SP-D are critical for normal lung function by maintaining the integrity of alveoli [32, 66]. Surfactants can regulate phagocytosis in alveolar macrophages or bind to pathogens and allergens [67, 68]. The decrease or absence of surfactant that may occur during inflammatory conditions can lead to respiratory failure [32]. There is variability in the levels of surfactant reported in HP [69–72]. One study found elevated levels of SP-A in patients with farmer's lung [71]. Surfactants from HP patients have a reduce capacity to inhibit the proliferation of PBMCs [73]. The surfactant protein C gene has been associated with familial interstitial lung diseases.

2. Immune response to *S. rectivirgula* during development of FLD

The immune response to *S. rectivirgula* induced FLD is complex and remains incompletely understood. It is still unclear which innate immune cells trigger the inflammatory process in FLD. It is generally accepted that upon inhalation of an offending agent leads to its deposition in the interstitial spaces of the lung in the alveoli where alveolar macrophage most likely recognizes *S. rectivirgula*, leading to cytokine production to recruit neutrophils or induce the differentiation of T_H1, T_H2, or T_H17 cells (**Figure 1**) [20, 57]. It has been reported that the cytokines: IFN-γ,

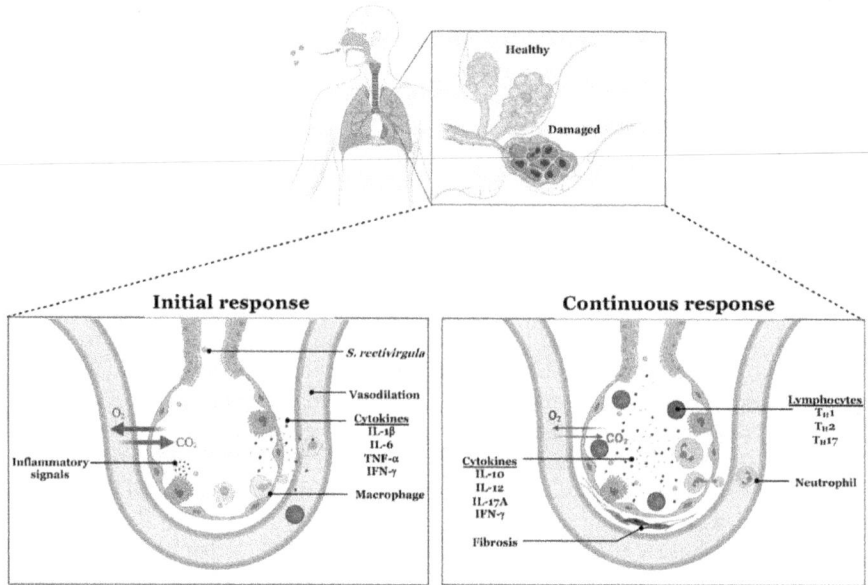

Figure 1.
Overview of FLD pathogenesis. Saccharopolyspora rectivirgula is inhaled in large quantities and deposited in the interstitial spaces of the lung. Alveolar macrophages produce inflammatory mediators upon recognition of S. rectivirgula by TLRs and potentially other pattern recognition receptors, triggering an immune response by producing pro-inflammatory mediators. In response to these inflammatory signals, neutrophils are recruited and lymphocytes, such as T_H17, Tregs, or T_H1 cells differentiate and produce their respective cytokines. As a susceptible individual continually comes into contact with S. rectivirgula, a pro-inflammatory feedback loop commences which can lead to fibrosis or granuloma formation, resulting in reduced lung function. (adaptive from "SARS-CoV-2: What We Know About Its Effects on Respiration", by BioRender.com (2021) retrieved from https://app.biorender.com/biorender-templates).

TNF-α, IL-1β, IL-6, IL-8, IL-10, IL-12, IL-13, IL-17A are all implicated in the disease [23, 28, 74, 75]. However, the cellular sources of these cytokines are still ambiguous [76], and there are conflicting reports as to which cells and cytokines are important for disease pathogenesis. Neutrophils, T_H2, T_H1, T_H17 and Tregs along with the cytokines IFN-γ, TNF-α, IL-10, or IL-17A have all been considered critical for disease progression [74, 77]. In addition, B cells develop into plasma cells that can secrete IgG to form antigen-antibody complexes with antigens from *S. rectivirgula* [77]. The varying experimental approaches for inducing FLD, such as mouse strains, the amount and route of *S. rectivirgula* exposure may account for discrepancies. Understanding the immune response to FLD will reveal mechanisms that enable better treatments for this disease as well as other pulmonary allergic diseases.

Mouse models used have been utilized to study the immune response and pathology of FLD caused by *S. rectivirgula*. These models exposed mice to varying doses of *S. rectivirgula* and for varying lengths of time [78]. There are variations in the way *S. rectivirgula* is prepared for subsequent exposure to mice. In some cases, *S. rectivirgula* is collected and lyophilized or sonicated. This is often done so as to break down the bacterial cell wall and release internal contents for exposure to the immune system in PBS [47, 53, 76, 79–85]. Intranasal, oropharyngeal aspiration or intratracheal inoculations are common techniques used to exposing mice to *S. rectivirgula* [59, 74, 79]. The timeline of exposure is not uniform. Exposure times vary anywhere from 3 consecutive days per week for 3 weeks up to 12 weeks [59, 74, 76, 85, 86]. Regardless of the vagaries of the model, studies have shown that inhalation of *S. rectivirgula* results in the recognition of pathogen associated molecular patterns found on *S. rectivirgula* proteins or lipoproteins toll like receptors (TLR). TLRs are transmembrane glycoproteins that can recognize pathogen associated molecular patterns important for microbial functions, and can trigger host immune responses [87]. There are 12 different TLRs in mice and 10 in humans [88]. *In vitro* studies suggest *S. rectivirgula* can interact with TLR2 [84]. An unknown TLR ligand of *S. rectivirgula's* can activate TLR2 or TLR9; TLR2 is can heterodimerize with TLR1 or TLR6 and detect lipoproteins from bacteria, whereas TLR9 recognizes unmethylated CpG motifs [87, 89, 90]. Through the immune signaling adaptor, MyD88, TLR2 or TLR9 activate different downstream signaling pathways that lead to the production of a variety of cytokines and chemokines such as TNF-α, IL-6, IL-1β, or CXCL2 by macrophages or dendritic cells; the cytokine and chemokine production leads to the recruitment of neutrophils, CD4+ or CD8+ T cells that further produce inflammatory mediators that may result in inflammation in the lung [27, 57, 91, 92].

TLR2 can modulate neutrophil responses [93]. The absence of TLR2 in mice did not affect neutrophil influx in response to a single exposure to *S. rectivirgula* [84]. However, the absence of TLR2 led to a significant reduction in MIP-2 production suggesting another pathway may be more important for the recruitment of neutrophils. Signaling through TLR2 is dependent on MyD88, and MyD88 may be important for neutrophil recruitment into the lung during *S. rectivirgula* exposure. MyD88$^{-/-}$ mice exhibited a reduction in neutrophils in the lungs, along with a decrease in MIP-2/CXCL2 and TNFα in the bronchoalveolar lavage fluid, that may partly explain the reduction of neutrophils. Further studies found that a single exposure of *S. rectivirgula* results in a significant decrease in CXCL2 mice lacking TLR2, TLR9, or both found, and upon repeated exposure to *S. rectivirgula*, TLR2$^{-/-}$ and TLR2/9$^{-/-}$ mutant mice exhibited reduced levels of lung CXCL2 and neutrophils. However, TLR2/9$^{-/-}$ mice still developed granuloma as a result of exposure to *S. rectivirgula*. These data suggest TLR2 or TLR9 are not critical for the pathogenesis of FLD [91, 94]. The serine/threonine protein kinase (PK)D, can be activated downstream of MyD88, and has been shown to be activated by *S. rectivirgula* [94]. PKD1 activation leads to activation of MAPKs and NF-κB and other downstream signals that leading to the production of pro-inflammatory mediators in macrophages or dendritic cells that can recruit leukocytes [94]. When Young-In Kim et al. used a PKD inhibitor in a mouse model of *S. rectivirgula* induced famer's lung disease, they found a significant reduction of neutrophils, MPO activity and that PKD inhibition also suppressed the development of alveolitis. This suggest PKD1 may be a potential target to reduce inflammation in FLD. Taken together, these data suggest that while TLR2 or TLR9 may not be critical, MyD88 dependent TLRs may play a role in the pathogenesis of FLD. Understanding which TLRs and the pathways they use that lead to disease progression could potentially lead to a therapeutic target.

TLRs are not the only way pathogenesis of FLD may occur. Antibodies such as IgG are able to tag *S. rectivirgula* as foreign also initiating an immune response. Patients with FLD have IgG antibodies and have been reported to have circulating immune complexes, suggesting antigen-antibody aggregates forms upon inhalation of SR [95]. In allergic diseases that provoke an IgG response, immune complexes may form when there is a large dose of inhalation of an allergen. When re-exposure of another large dose of allergen occurs, immune complexes, composed of antigen-antibody formation, may form in the walls of the alveoli, resulting in a reduction of O2 exchange due to possible fluid, protein or cells accumulation [57]. Upon exposure to *S. rectivirgula*, glycoproteins on *S. rectivirgula's* surface are recognized by IgG antibodies, forming antigen-antibody complexes [96]. Antigen-antibody complexes mark *S. rectivirgula* as foreign to the host and can trigger alveolar macrophage or dendritic cells to produce pro-inflammatory mediators that can recruit neutrophils, CD8$^+$, CD4$^+$ cells. *S. rectivirgula* activation of dendritic cells or macrophages through TLR and the formation of antigen-antibody complexes induces a pro-inflammatory environment via the production of cytokines from CD4$^+$ T helper cells.

CD4$^+$ T helper cells can fine tune an immune response with their unique abilities to produce a wide variety of cytokines. Naïve CD4$^+$ T cells encounter antigen and can differentiate into T helper (T$_H$) subsets such as T$_H$1, T$_H$2, T$_H$9, T$_H$17, T$_H$22, T follicular cells (T$_{FH}$), or T regulatory cells (Tregs) and Type 1 regulatory cells (Tr1), depending on signals from the microenvironment. Previous studies were unable to reach a consensus on which T$_H$ cell(s) type are important for disease pathogenesis in FLD [28, 74, 76, 85, 86, 97, 98], however, subsequent studies have found a prominent role for T$_H$17 cells. The presence of lung lymphocytes and granuloma formation are classic hallmarks of FLD, and CD4$^+$ and CD8$^+$ T cells are found in the BAL of patients with FLD, suggesting they play a role in pathogenesis [79]. Granuloma formation has been historically associated with macrophages, T$_H$1 and T$_H$2 cells [99, 100]. T$_H$1 cells produce IFN-γ and promote macrophage responses against extracellular or intracellular bacteria [57]. IFN-γ may play an important role since IFN-$\gamma^{-/-}$ mice (on a Balb/c background) that were exposed to *S. rectivirgula* did not develop granulomas, and when given exogenous IFN-γ, developed granulomas in the airways [81]. However, the absence of IFN-γ did not affect the *S. rectivirgula* induced increase in the number of cells that infiltrated the airways. The cytokine IL-12 can regulate IFN-γ [101], and mice (C57Bl/6) exposed to *S. rectivirgula* have elevated IL-12 and IFN-γ when compared to DBA/2 mice that are resistant against developing FLD. However, when DBA/2 mice are given recombinant IL-12 intranasally, they become sensitive to developing FLD [83]. While BAL and blood samples taken from patients with FLD or summer-type HP had elevated IFN-γ levels, it is unclear whether these were patients who were sensitive to *S. rectivirgula* [98]. The cellular sources of IFN-γ were likened to neutrophils and a small proportion of NK cells, and in the absence of neutrophils there is a reduction of the severity of alveolitis, although the recruitment of immune cells into the lung following *S. rectivirgula* is not completely eliminated [28], although more recent work suggest that neutrophils may not be important for the development of inflammation in the lung [102]. Experiments with Rag-1$^{-/-}$ mice (lacking B and T cells) reconstituted with spleen cells from IFN$\gamma^{-/-}$ or WT mice and exposed to *S. rectivirgula*, suggest that IFN-γ plays an important role in disease pathogenesis.

2.1 Th17 immune response to *S. rectivirgula* induced FLD

In 2005, a new T$_H$ cell type was discovered and determined to be an independent subset from T$_H$1 and T$_H$2 cells [103]. This cell type, the T$_H$17 cell, develops in the presence of TGF-β, IL-1, IL-6, IL-21, or IL-23, and has been shown to protect against extracellular bacteria by producing the pro-inflammatory cytokines IL-17A, IL-17F, and IL-22 [57, 104]. Analysis of gene expression profiles of lung biopsies of patients with HP found an upregulation of the IL-17 receptor IL-17RC, suggesting that T$_H$17 cells may play a role in HP pathogenesis [105]. In a mouse model of *S. rectivirgula* induced FLD, Joshi et al. found an increase in IFN-γ and IL-12p35 mRNA in *S. rectivirgula* exposed mice, along with significant increase in IFN-γ, IL-4, IL-13, IL-6 and IL-17A in lung homogenates. However, when they exposed IL-17$^{-/-}$ mice they found a decrease in alveolitis and lower production of cytokines and chemokines compared to WT mice [85]. Simonian et al. investigated the role of T$_H$17 cells in *S. rectivirgula* induced FLD over the course of 4 weeks [74], comparing WT, TCR$\beta^{-/-}$ (that lack CD4$^+$ or CD8$^+$ T cells), or IL-17ra$^{-/-}$ mice for collagen deposition and cytokine expression in lung homogenates. They found that T cells are critical for disease since TCRβ$^{-/-}$ mice do not develop pulmonary fibrosis, but still had collagen deposits albeit significantly less than WT mice. Reconstitution of mice lacking both αβ and γδ T cells (TCRβ$^{-/-}$δ$^{-/-}$ mice) with CD4$^+$ or CD8$^+$ T cells followed by exposure to *S. rectivirgula* indicated that αβ CD4$^+$ T cells contribute significantly to collagen deposit, influx of cells into the lung, increase in IL-17 production, along with pulmonary fibrosis compared to CD8$^+$ T cells. Analysis of T$_H$1, T$_H$2, or T$_H$17 cytokines and TGF-β revealed that IL-17A was predominant in the airways of *S. rectivirgula* exposed mice. Furthermore, IL17ra$^{-/-}$ mice exposed to *S. rectivirgula* exhibited a significant decrease in lymphocyte influx and lung fibrosis, and little production of IFN-γ by T cells after 2 weeks of *S. rectivirgula* exposure. Interestingly, *S. rectivirgula* exposure of T-bet$^{-/-}$ mice led to more severe disease, with an enhanced T$_H$17 response and increases in collagen production compared to WT mice [97]. However, it should be noted that IFN-γ production is not entirely dependent on T-bet [97]. Taken together these data suggest T$_H$17 cells and their production of IL-17A are important for the pathogenesis of FLD.

T$_H$17 cells are not the only cell type that can produce IL-17A. Neutrophils, γδ T cells, invariant natural killer cells (iNKT), and group 3 innate lymphoid cells (ILC3s) have all been shown to produce IL-17A [106–109]. Studies with γδ T cells have shown that they produce IL-17A in *S. rectivirgula* induced FLD but are not important for disease pathogenesis [29]. However, neutrophils and their production of IL-17A has been implicated in *S. rectivirgula* FLD as well [110]. When mice are exposed to *S. rectivirgula* over the course of 6 weeks and during the final two weeks of exposure neutrophils were depleted using anti-GR-1 antibody, it was found that neutrophil depletion reduces neutrophils, lymphocytes, and collagen levels in the airways when compared to isotype controls (although more recent work suggest that neutrophils may not be important for the development of inflammation in the lung [102]). This group also utilized an IL-17A enrichment kit to determine which cells produced IL-17A secreting in lung homogenates of mice exposed *S. rectivirgula* for 3 weeks. They found that GR1^{int} and GR1^{high} populations were the most significant populations in the IL-17A positive fraction compared to B cells, CD4$^+$ or CD8$^+$ T cells. Stimulating these cells in vitro followed by flow cytometric analysis also revealed the expression of intracellular IL-17A in lung neutrophils and not lymphocytes, along with expression of Il-17A mRNA. While these data suggest neutrophils, monocytes, and macrophages are the major source of IL-17A, more recent work using IL-17A cytokine reporter mice suggest instead that CD4+ T cells are the major producers of this cytokine, and that neutrophils do not produce IL17A during development of inflammation in the lung [102]).

2.2 Anti-inflammatory response during *S. rectivirgula* induced FLD

The anti-inflammatory responses of *S. rectivirgula* induced FLD remains largely unexplored. The anti-inflammatory cytokine IL-10 can reduce inflammatory responses [57]. IL-10$^{-/-}$ mice have been shown to have increased inflammatory cells in response to *S. rectivirgula* exposure, and histological evidence suggested an increase in granuloma formation [82]. This suggest that IL-10 may be important for reducing the inflammatory responses in FLD, but are unable to completely neutralize the inflammation under usual circumstances. Immune cells that produce IL-10 include T regulatory cells, such as forkhead box 3 (Foxp3$^+$ Tregs) and Foxp3$^-$ type 1 T regulatory cells (Tr1). These T regulatory cells are critical for controlling inflammatory responses including allergic responses by producing the anti-inflammatory cytokines IL-10 and TGF-β. In humans, Tregs can be identified as CD4$^+$CD25$^+$CD127$^-$Foxp3$^+$ [111]. Subjects that are asymptomatic but have farmer's lung might have functional T regulatory cells [36]. Analysis of isolated Tregs from blood or BAL from healthy, asymptomatic patients or diagnosed with FLD (n = 6) found no differences in the proportion of Tregs among the patient populations [112]. However, when Treg suppression assays were performed, they found that Tregs from patients that had FLD were unable to suppress the proliferation of activated T cells. These data suggest patients that have FLD may have impaired ability to suppress effector T cell responses, resulting in unresolved inflammation with the potential to develop irreversible lung damage. More studies need to be conducted to fully elucidate the role of T regulatory cells in this disease.

3. Conclusion

Actinobacteria are a source of natural metabolites that can be important for human health. However, some species cause disease such as HP, a collection of rare interstitial lung disease caused by the inhalation of (in)organic agents including *S. rectivirgula* leading to farmer's lung, one of the most well studied forms of HP. It is unclear why some individuals develop farmer's lung upon exposure to *S. rectivirgula*, and others do not. Environmental factors and genetics may play a role in the development of disease, and polymorphisms in the immune genes including TNF-α and MHC II have been found in patients diagnosed with HP [60]. However, genetic studies have not been able to fully identify specific mutations in patients predisposed to develop FLD. The pathogenesis of HP including FLD is challenging to fully elucidate. *S. rectivirgula* exposure results in the development of an immune response that includes macrophage TLRs recognition of *S. rectivirgula*, leading to the production of chemokines and cytokines that recruit neutrophils or induce the differentiation of T_H1, T_H2, or T_H17 cells (**Figure 1**) [20, 57]. In addition, B cell recognition of *S. rectivirgula* antigens lead to the differentiation of plasma cells that secrete IgG that form antigen-antibody complexes with *S. rectivirgula* antigens [77]. Thus neutrophils, T_H2, T_H1, T_H17 and Tregs along with the cytokines IFN-γ, TNF-α, IL-10, or IL-17A have all been considered critical for disease progression [74, 77]. While there are conflicting reports on the main cellular source of IFN-γ or IL-17A, neutrophils and T_H17 cells have been implicated as sources of IL-17A production which has been shown to be most critical for the development of farmer's lung in response to exposure to *S. rectivirgula* [74, 110]. Furthermore, understanding the role of IL-10, and T regulatory cells that produce this cytokine to suppress inflammation may provide an avenue to treat disease. Newer technologies and techniques will also help gain a better understanding of disease pathogenesis in *S. rectivirgula* induced FLD. For instance, genome-wide association studies can help determine what genetic components may be associated with developing symptoms. Next generation sequencing such as RNA sequencing, or immune cell profiling could help identify which cells are present and important for disease pathogenesis. Advances in techniques such as multi-color flow cytometry allow for identification of many different immune cells than ever before [113, 114].

It is important to gain a better understanding of rare diseases such as HP. Global climate change caused by human activities are anticipated to affect respiratory diseases with increases in rainy, humid, hot weather which can lead to humid environments which may make environments suitable for the growth of actinobacteria such as *S. rectivirgula* [115–118]. Understanding how *S. rectivirgula* causes HP such as FLD will lend to better therapeutics for patients.

Author details

Jessica Elmore[1] and Avery August[1,2*]

1 Department of Microbiology and Immunology, Cornell University, Ithaca, New York, United States of America

2 Cornell Center for Immunology, Cornell Institute for Host-Microbe Interactions and Disease, and Cornell Center for Health Equity, Cornell University, Ithaca, New York, United States of America

*Address all correspondence to: averyaugust@cornell.edu

IntechOpen

References

[1] Panno J. The Cell: Evolution of the First Organism. New York, NY: Infobase Publishing; 2014

[2] Wilson BA, Winkler M, Ho BT. Bacterial Pathogenesis: A Molecular Approach. Washington, D.C.: John Wiley & Sons; 2020

[3] Bar-On YM, Phillips R, Milo R. The biomass distribution on earth. Proceedings of the National Academy of Sciences. 2018;**115**(25):6506-6511

[4] Gleizer S, Ben-Nissan R, Bar-On YM, Antonovsky N, Noor E, Zohar Y, et al. Conversion of Escherichia coli to generate all biomass carbon from CO_2. Cell. 2019;**179**(6):1255-1263.e12

[5] Demain AL, Sanchez S. Microbial drug discovery: 80 years of progress. The Journal of Antibiotics. 2009;**62**(1): 5-16

[6] Aminov RI. A brief history of the antibiotic era: Lessons learned and challenges for the future. Frontiers in Microbiology. 2010;**1**:134

[7] Miao V, Davies J. Actinobacteria: The good, the bad, and the ugly. Antonie Van Leeuwenhoek. 2010;**98**(2):143-150

[8] Azman A-S, Othman I, Velu S, Chan K-G, Lee L-H. Mangrove rare actinobacteria: Taxonomy, natural compound, and discovery of bioactivity. Frontiers in Microbiology. 2015;**6**:856

[9] Sayed AM, Abdel-Wahab NM, Hassan HM, Abdelmohsen UR. Saccharopolyspora: An underexplored source for bioactive natural products. Journal of Applied Microbiology. 2020;**128**(2):314-329

[10] Regal JF, Selgrade M. Hyper-sensitivity Reactions in the Respiratory Tract. Immune System Toxicology: Elsevier Inc.; 2010. pp. 375-395

[11] Lacey J, Goodfellow M. A novel actinomycete from sugar-cane bagasse: Saccharopolyspora hirsuta gen. Et sp. nov. Microbiology. 1975;**88**(1):75-85

[12] Korn-Wendisch F, Kempf A, Grund E, Kroppenstedt R, Kutzner H. Transfer of Faenia rectivirgula Kurup and agre 1983 to the genus Saccharopolyspora Lacey and Goodfellow 1975, elevation of Saccharopolyspora hirsuta subsp. taberi Labeda 1987 to species level, and emended description of the genus Saccharopolyspora. International Journal of Systematic and Evolutionary Microbiology. 1989;**39**(4):430-441

[13] Lacey J. Thermoactinomyces sacchari sp. nov., a thermophilic actinomycete causing bagassosis. Microbiology. 1971;**66**(3):327-338

[14] Lemone DV, Scott WG, Moore S, Koven AL. Bagasse disease of the lungs. Radiology. 1947;**49**(5):556-567

[15] Seabury J, Salvaggio J, Buechner H, Kundur V, Bagassois III. Isolation of thermophilic and mesophilic Actinomycetes and fungi from moldy bagasse. Proceedings of the Society for Experimental Biology and Medicine. 1968;**129**(2):351-360

[16] Hargreave F, Pepys J, Holford-Strevens V. Bagassosis. The Lancet. 1968;**291**(7543):619-620

[17] Goodfellow M, Kempfer P, Busse HJ, Trujillo ME, Suzuki K-I, Ludwig W, et al. Bergey's Manual of Systematic Bacteriology: Volume Five The Actinobacteria, Part A. New York, NY: Springer New York; 2012

[18] Schäfer J, Kämpfer P, Jäckel U. Detection of Saccharopolyspora rectivirgula by quantitative real-time PCR. Annals of Occupational Hygiene. 2011;**55**(6):612-619

[19] Cano-Jiménez E, Acuña A, Botana MI, Hermida T, González MG, Leiro V, et al. Farmer's lung disease. A review. Archivos de Bronconeumología (English Edition). 2016;**52**(6):321-328

[20] Bourke S, Dalphin J, Boyd G, McSharry C, Baldwin C, Calvert J. Hypersensitivity pneumonitis: Current concepts. European Respiratory Journal. 2001;**18**(Suppl. 32):81s-92s

[21] Ramazzini B. Diseases of Workers. New York: Hafner Pub. Co.; 1964

[22] Campbell JM. Acute symptoms following work with Hay. The British Medical Journal. 1932;**2**(3755):1143-1144

[23] Jose J, Craig TJ. Hypersensitivity pneumonitis. Allergy and Asthma: Springer; 2016. pp. 311-331

[24] Churg A. Hypersensitivity pneumonitis: New concepts and classifications. Modern Pathology. 2022;**35**:15-27

[25] Dickie HA, Rankin J. Farmer's lung: An acute granulomatous interstitial pneumonitis occurring in agricultural workers. Journal of the American Medical Association. 1958;**167**(9): 1069-1076

[26] Barnes H, Jones K, Blanc P. The hidden history of hypersensitivity pneumonitis. European Respiratory Journal. 20 Jan 2022;**59**(1):2100252

[27] Barnes H, Troy L, Lee CT, Sperling A, Strek M, Glaspole I. Hypersensitivity pneumonitis: Current concepts in pathogenesis, diagnosis, and treatment. Allergy. 2022;**77**(2):442-453

[28] Nance S, Cross R, Yi AK, Fitzpatrick EA. IFN-γ production by innate immune cells is sufficient for development of hypersensitivity pneumonitis. European Journal of Immunology. 2005;**35**(6):1928-1938

[29] Simonian PL, Roark CL, Born WK, O'Brien RL, Fontenot AP. γδ T cells and Th17 cytokines in hypersensitivity pneumonitis and lung fibrosis. Translational Research. 2009;**154**(5): 222-227

[30] Soumagne T, Chardon M-L, Dournes G, Laurent L, Degano B, Laurent F, et al. Emphysema in active farmer's lung disease. PLoS One. 2017;**12**(6):e0178263

[31] Erkinjuntti-Pekkanen R, Rytkonen H, Kokkarinen JI, Tukiainen HO, Partanen K, Terho EO. Long-term risk of emphysema in patients with farmer's lung and matched control farmers. American journal of respiratory and critical care medicine. 1998;**158**(2):662-665

[32] Shah PL, Herth FJ, Lee YG, Criner GJ. Essentials of Clinical Pulmonology. Boca Raton: CRC Press; 2018

[33] Morell F, Ojanguren I, Cruz M-J. Diagnosis of occupational hypersensitivity pneumonitis. Current Opinion in Allergy and Clinical Immunology. 2019;**19**(2):105-110

[34] Kokkarinen J, Tukiainen H, Terho EO. Severe farmer's lung following a workplace challenge. Scandinavian Journal of Work, Environment & Health. 1992;**18**(5): 327-328

[35] Koster MA, Thomson CC, Collins BF, Jenkins AR, Ruminjo JK, Raghu G. Diagnosis of hypersensitivity pneumonitis in adults, 2020 clinical practice guideline: Summary for clinicians. Annals of the American Thoracic Society. 2021;**18**(4):559-566

[36] Selman M, Pardo A, King TE Jr. Hypersensitivity pneumonitis: Insights in diagnosis and pathobiology. American Journal of Respiratory and Critical Care Medicine. 2012;**186**(4):314-324

[37] Myers JL. Hypersensitivity pneumonia: The role of lung biopsy in diagnosis and management. Modern Pathology. 2012;**25**(1):S58-S67

[38] Aronson KI, O'Beirne R, Martinez FJ, Safford MM. Barriers to antigen detection and avoidance in chronic hypersensitivity pneumonitis in the United States. Respiratory Research. 2021;**22**(1):1-10

[39] Ranalli G, Grazia L, Roggeri A. The influence of hay-packing techniques on the presence of Saccharopolyspora rectivirgula. Journal of Applied Microbiology. 1999;**87**(3):359-365

[40] Roussel S, Reboux G, Dalphin JC, Laplante JJ, Piarroux R. Evaluation of salting as a hay preservative against farmer's lung disease agents. Annals Of Agricultural And Environmental Medicine. 2005;**12**(2):217-221

[41] Kokkarinen JI, Tukiainen HO, Terho EO. Effect of corticosteroid treatment on the recovery of pulmonary function in farmer's lung 1-3. The American Review of Respiratory Disease. 1992;**145**:3-5

[42] Cano-Jiménez E, Rubal D, de Llano LAP, Mengual N, Castro-Añón O, Méndez L, et al. Farmer's lung disease: Analysis of 75 cases. Medicina Clínica (English Edition). 2017;**149**(10):429-435

[43] Liu S, Chen D, Fu S, Ren Y, Wang L, Zhang Y, et al. Prevalence and risk factors for farmer's lung in greenhouse farmers: An epidemiological study of 5,880 farmers from Northeast China. Cell Biochemistry and Biophysics. 2015;**71**(2):1051-1057

[44] Hoppin JA, Umbach DM, Long S, Rinsky JL, Henneberger PK, Salo PM, et al. Respiratory disease in United States farmers. Occupational and Environmental Medicine. 2014;**71**(7): 484-491

[45] Hoppin JA, Umbach DM, Kullman GJ, Henneberger PK, London SJ, Alavanja MC, et al. Pesticides and other agricultural factors associated with self-reported farmer's lung among farm residents in the agricultural health study. Occupational and Environmental Medicine. 2007;**64**(5):334-341

[46] Cormier Y, Israel-Assayag E, Fournier M, Tremblay GM. Modulation of experimental hypersensitivity pneumonitis by Sendai virus. The Journal of Laboratory and Clinical Medicine. 1993;**121**(5):683-688

[47] Cormier Y, Tremblay G, Fournier M, Israël-Assayag E. Long-term viral enhancement of lung response to Saccharopolyspora rectivirgula. American Journal of Respiratory and Critical Care Medicine. 1994;**149**(2): 490-494

[48] Costabel U, Miyazaki Y, Pardo A, Koschel D, Bonella F, Spagnolo P, et al. Hypersensitivity pneumonitis. Nature Reviews Disease Primers. 2020;**6**(1):65

[49] Millerick-May M, Mulks M, Gerlach J, Flaherty K, Schmidt S, Martinez F, et al. Hypersensitivity pneumonitis and antigen identification– an alternate approach. Respiratory Medicine. 2016;**112**:97-105

[50] Kotimaa MH, Husman KH, Terho EO, Mustonen MH. Airborne molds and actinomycetes in the work environment of farmer's lung patients in Finland. Scandinavian Journal of Work, Environment & Health. 1984;**10**(2):115-119

[51] Marx JJ Jr, Kettrick-Marx MA, Mitchell PD, Flaherty DK. Correlation of exposure to various respiratory pathogens with farmer's lung disease. Journal of Allergy and Clinical Immunology. 1977;**60**(3):169-173

[52] Dakhama A, Hegele RG, Laflamme G, Israel-Assayag E,

Cormier Y. Common respiratory viruses in lower airways of patients with acute hypersensitivity pneumonitis. American Journal of Respiratory and Critical Care Medicine. 1999;**159**(4):1316-1322

[53] Cormier Y, Samson N, Isräel-Assayag E. Viral infection enhances the response to Saccharopolyspora rectivirgula in mice prechallenged with this farmer's lung antigen. Lung. 1996;**174**(6):399-407

[54] Sordillo JE, Hoffman EB, Celedón JC, Litonjua AA, Milton DK, Gold DR. Multiple microbial exposures in the home may protect against asthma or allergy in childhood. Clinical & Experimental Allergy. 2010;**40**(6):902-910

[55] Perkin MR, Strachan DP. Which aspects of the farming lifestyle explain the inverse association with childhood allergy? Journal of Allergy and Clinical Immunology. 2006;**117**(6):1374-1381

[56] Michel S, Busato F, Genuneit J, Pekkanen J, Dalphin JC, Riedler J, et al. Farm exposure and time trends in early childhood may influence DNA methylation in genes related to asthma and allergy. Allergy. 2013;**68**(3):355-364

[57] Murphy K. In: Weaver C, editor. Janeway's Immunobiology. 9th ed. New York, NY: Garland Science/Taylor & Francis Group, LLC; 2017

[58] Venkatesh P, Wild L. Hypersensitivity pneumonitis in children. Pediatric Drugs. 2005;7(4):235-244

[59] Russell SL, Gold MJ, Reynolds LA, Willing BP, Dimitriu P, Thorson L, et al. Perinatal antibiotic-induced shifts in gut microbiota have differential effects on inflammatory lung diseases. Journal of Allergy and Clinical Immunology. 2015;**135**(1):100-9. e5

[60] Vasakova M, Selman M, Morell F, Sterclova M, Molina-Molina M,

Raghu G. Hypersensitivity pneumonitis: Current concepts of pathogenesis and potential targets for treatment. American Journal of Respiratory and Critical Care Medicine. 2019;**200**(3):301-308

[61] Schaaf BM, Seitzer U, Pravica V, Aries SP, Zabel P. Tumor necrosis factor-α− 308 promoter gene polymorphism and increased tumor necrosis factor serum bioactivity in farmer's lung patients. American Journal of Respiratory and Critical Care Medicine. 2001;**163**(2):379-382

[62] Mahto H, Tripathy R, Meher BR, Prusty BK, Sharma M, Deogharia D, et al. TNF-α promoter polymorphisms (G-238A and G-308A) are associated with susceptibility to systemic lupus erythematosus (SLE) and P. falciparum malaria: A study in malaria endemic area. Scientific Reports. 2019;**9**(1):1-11

[63] Wieczorek M, Abualrous ET, Sticht J, Álvaro-Benito M, Stolzenberg S, Noé F, et al. Major histocompatibility complex (MHC) class I and MHC class II proteins: Conformational plasticity in antigen presentation. Frontiers in Immunology. 2017;**8**:292

[64] Aquino-Galvez A, Camarena Á, Montaño M, Juarez A, Zamora AC, González-Avila G, et al. Transporter associated with antigen processing (TAP) 1 gene polymorphisms in patients with hypersensitivity pneumonitis. Experimental and Molecular Pathology. 2008;**84**(2):173-177

[65] Falfán-Valencia R, Camarena Á, Pineda CL, Montaño M, Juárez A, Buendía-Roldán I, et al. Genetic susceptibility to multicase hypersensitivity pneumonitis is associated with the TNF-238 GG genotype of the promoter region and HLA-DRB1* 04 bearing HLA haplotypes. Respiratory Medicine. 2014;**108**(1): 211-217

[66] Bernhard W. Lung surfactant: Function and composition in the context of development and respiratory physiology. Annals of Anatomy-Anatomischer Anzeiger. 2016;**208**: 146-150

[67] Chroneos Z, Sever-Chroneos Z, Shepherd V. Pulmonary surfactant: An immunological perspective. Cellular Physiology and Biochemistry. 2010;**25**(1):13-26

[68] Nayak A, Dodagatta-Marri E, Tsolaki AG, Kishore U. An insight into the diverse roles of surfactant proteins, SP-A and SP-D in innate and adaptive immunity. Frontiers in Immunology. 2012;**3**:131

[69] Pantelidis P, Veeraraghavan S, Du Bois RM. Surfactant gene polymorphisms and interstitial lung diseases. Respiratory Research. 2001;**3**(1):1-7

[70] Cormier Y, Israel-Assayag E, Desmeules M, Lesur O. Effect of contact avoidance or treatment with oral prednisolone on bronchoalveolar lavage surfactant protein a levels in subjects with farmer's lung. Thorax. 1996;**51**(12): 1210-1215

[71] Hamm H, Lührs J, y Rotaeche JG, Costabel U, Fabel H, Bartsch W. Elevated surfactant protein a in bronchoalveolar lavage fluids from sarcoidosis and hypersensitivity pneumonitis patients. Chest. 1994; **106**(6):1766-1770

[72] Okamoto T, Fujii M, Furusawa H, Tsuchiya K, Miyazaki Y, Inase N. The usefulness of KL-6 and SP-D for the diagnosis and management of chronic hypersensitivity pneumonitis. Respiratory Medicine. 2015;**109**(12): 1576-1581

[73] Israël-Assayag E, Cormier Y. Surfactant modifies the lympho-proliferative activity of macrophages in hypersensitivity pneumonitis. American

Journal of Physiology-Lung Cellular and Molecular Physiology. 1997;**273**(6): L1258-L1L64

[74] Simonian PL, Roark CL, Wehrmann F, Lanham AK, del Valle FD, Born WK, et al. Th17-polarized immune response in a murine model of hypersensitivity pneumonitis and lung fibrosis. The Journal of Immunology. 2009;**182**(1):657-665

[75] Greenberger PA. Hypersensitivity pneumonitis: A fibrosing alveolitis produced by inhalation of diverse antigens. Journal of Allergy and Clinical Immunology. 2019;**143**(4):1295-1301

[76] Andrews K, Ghosh MC, Schwingshackl A, Rapalo G, Luellen C, Waters CM, et al. Chronic hypersensitivity pneumonitis caused by Saccharopolyspora rectivirgula is not associated with a switch to a Th2 response. American Journal of Physiology-Lung Cellular and Molecular Physiology. 2016;**310**(5):L393-L402

[77] Costabel U. The alveolitis of hypersensitivity pneumonitis. European Respiratory Journal. 1988;**1**(1):5-9

[78] Cottin V. Interstitial lung disease. European Respiratory Review. 2013;**22**(127):26-32

[79] Schuyler M, Gott K, Shopp G, Crooks L. CD3+ and CD4+ cells adoptively transfer experimental hypersensitivity pneumonitis. The American Review of Respiratory Disease. 1992;**146**(6):1582-1588

[80] Schuyler M, Crooks L. Experimental hypersensitivity pneumonitis in Guinea pigs. The American Review of Respiratory Disease. 1989;**139**:996-1002

[81] Gudmundsson G, Hunninghake GW. Interferon-gamma is necessary for the expression of hypersensitivity pneumonitis. The Journal of Clinical Investigation. 1997;**99**(10):2386-2390

[82] Gudmundsson G, Bosch A, Davidson BL, Berg DJ, Hunninghake GW. Interleukin-10 modulates the severity of hypersensitivity pneumonitis in mice. American Journal of Respiratory Cell and Molecular Biology. 1998;**19**(5):812-818

[83] Gudmundsson G, Monick MM, Hunninghake GW. IL-12 modulates expression of hypersensitivity pneumonitis. The Journal of Immunology. 1998;**161**(2):991-999

[84] Nance SC, Yi AK, Re FC, Fitzpatrick EA. MyD88 is necessary for neutrophil recruitment in hypersensitivity pneumonitis. Journal of Leukocyte Biology. 2008;**83**(5):1207-1217

[85] Joshi AD, Fong DJ, Oak SR, Trujillo G, Flaherty KR, Martinez FJ, et al. Interleukin-17–mediated immunopathogenesis in experimental hypersensitivity pneumonitis. American Journal of Respiratory and Critical Care Medicine. 2009;**179**(8):705-716

[86] Matsuno Y, Ishii Y, Yoh K, Morishima Y, Haraguchi N, Kikuchi N, et al. Overexpression of GATA-3 protects against the development of hypersensitivity pneumonitis. American Journal of Respiratory and Critical Care Medicine. 2007;**176**(10):1015-1025

[87] de Oliviera NL, Massari P, Wetzler LM. The role of TLR2 in infection and immunity. Frontiers in Immunology. 2012;**3**:79

[88] Christmas P. Toll-like receptors: Sensors that detect infection. Nature Education. 2010;**3**(9):85

[89] Ishii KJ, Akira S. 3 - innate immunity. In: Rich RR, Fleisher TA, Shearer WT, Schroeder HW, Frew AJ, Weyand CM, editors. Clinical Immunology. Third ed. Edinburgh: Mosby; 2008. pp. 39-51

[90] Lamphier MS, Sirois CM, Verma A, Golenbock DT, Latz E. TLR9 and the recognition of self and non-self nucleic acids. Annals of the New York Academy of Sciences. 2006;**1082**(1):31-43

[91] Andrews K, Abdelsamed H, Yi A-K, Miller MA, Fitzpatrick EA. TLR2 regulates neutrophil recruitment and cytokine production with minor contributions from TLR9 during hypersensitivity pneumonitis. PLoS One. 2013;**8**(8):e73143

[92] Lentini G, Famà A, Biondo C, Mohammadi N, Galbo R, Mancuso G, et al. Neutrophils enhance their own influx to sites of bacterial infection via endosomal TLR-dependent Cxcl2 production. The Journal of Immunology. 2020;**204**(3):660-670

[93] Sabroe I, Prince LR, Jones EC, Horsburgh MJ, Foster SJ, Vogel SN, et al. Selective roles for toll-like receptor (TLR) 2 and TLR4 in the regulation of neutrophil activation and life span. The Journal of Immunology. 2003;**170**(10):5268-5275

[94] Kim Y-I, Park J-E, Brand DD, Fitzpatrick EA, Yi A-K. Protein kinase D1 is essential for the proinflammatory response induced by hypersensitivity pneumonitis-causing thermophilic actinomycetes Saccharopolyspora rectivirgula. The Journal of Immunology. 2010;**184**(6):3145-3156

[95] Marcer G, Simioni L, Saia B, Saladino G, Gemignani C, Mastrangelo G. Study of immunological parameters in farmer's lung. Clinical Allergy. 1983;**13**(5):443-449

[96] Mundt C, Becker W-M, Schlaak M. Farmer's lung: Patients' IgG2 antibodies specifically recognize Saccharopolyspora rectivirgula proteins and carbohydrate structures. Journal of Allergy and Clinical Immunology. 1996;**98**(2):441-450

[97] Abdelsamed HA, Desai M, Nance SC, Fitzpatrick EA. T-bet controls severity of hypersensitivity pneumonitis. Journal of Inflammation. 2011;**8**(1):1-11

[98] Yamasaki H, Ando M, Brazer W, Center DM, Cruikshank WW. Polarized type 1 cytokine profile in bronchoalveolar lavage T cells of patients with hypersensitivity pneumonitis. The Journal of Immunology. 1999;**163**(6): 3516-3523

[99] Flynn JL, Chan J, Lin P. Macrophages and control of granulomatous inflammation in tuberculosis. Mucosal Immunology. 2011;**4**(3):271-278

[100] Pagán AJ, Ramakrishnan L. The formation and function of granulomas. Annual Review of Immunology. 2018;**36**:639-665

[101] Trinchieri G. Interleukin-12 and the regulation of innate resistance and adaptive immunity. Nature Reviews Immunology. 2003;**3**(2):133-146

[102] Elmore JP, Carter C, Redko A, Koylass N, Bennett A, Mead M et al. Itk independent development of Th17 responses during Hypersensitivity pneumonitis". Communications Biology. 2022;**5**:162. DOI: 10.1038/s42003-022-03109-1

[103] Marwaha A, Leung N, McMurchy AN, Levings M. TH17 cells in autoimmunity and immunodeficiency: Protective or pathogenic? Frontiers in Immunology. 2012;**3**:129

[104] Wu X, Tian J, Wang S. Insight into non-pathogenic Th17 cells in autoimmune diseases. Frontiers in Immunology. 2018;**9**:1112

[105] Selman M, Pardo A, Barrera L, Estrada A, Watson SR, Wilson K, et al. Gene expression profiles distinguish idiopathic pulmonary fibrosis from hypersensitivity pneumonitis. American Journal of Respiratory and Critical Care Medicine. 2006;**173**(2):188-198

[106] Montaldo E, Juelke K, Romagnani C. Group 3 innate lymphoid cells (ILC3s): Origin, differentiation, and plasticity in humans and mice. European Journal of Immunology. 2015;**45**(8):2171-2182

[107] Papotto PH, Ribot JC, Silva-Santos B. IL-17+ γδ T cells as kick-starters of inflammation. Nature Immunology. 2017;**18**(6):604-611

[108] Yu J-S, Hamada M, Ohtsuka S, Yoh K, Takahashi S, Miaw S-C. Differentiation of IL-17-producing invariant natural killer T cells requires expression of the transcription factor c-Maf. Frontiers in Immunology. 2017;**8**:1399

[109] Hu S, He W, Du X, Yang J, Wen Q, Zhong X-P, et al. IL-17 production of neutrophils enhances antibacteria ability but promotes arthritis development during mycobacterium tuberculosis infection. eBioMedicine. 2017;**23**:88-99

[110] Hasan SA, Eksteen B, Reid D, Paine HV, Alansary A, Johannson K, et al. Role of IL-17A and neutrophils in fibrosis in experimental hypersensitivity pneumonitis. Journal of Allergy and Clinical Immunology. 2013;**131**(6): 1663-73.e5

[111] Żabińska M, Krajewska M, Kościelska-Kasprzak K, Jakuszko K, Bartoszek D, Myszka M, et al. CD4+ CD25+ CD127− and CD4+ CD25+ Foxp3+ regulatory T cell subsets in mediating autoimmune reactivity in systemic lupus erythematosus patients. Archivum Immunologiae et Therapiae Experimentalis. 2016;**64**(5):399-407

[112] Girard M, Israel-Assayag E, Cormier Y. Impaired function of regulatory T-cells in hypersensitivity pneumonitis. European Respiratory Journal. 2011;**37**(3):632-639

[113] Behbehani GK. Immuno-phenotyping by Mass Cytometry. Immunophenotyping: Springer; 2019. pp. 31-51

[114] Holmberg-Thyden S, Grønbæk K, Gang AO, El Fassi D, Hadrup SR. A user's guide to multicolor flow cytometry panels for comprehensive immune profiling. Analytical Biochemistry. 2021;**15**;627:114210

[115] Barnes CS, Alexis NE, Bernstein JA, Cohn JR, Demain JG, Horner E, et al. Climate change and our environment: The effect on respiratory and allergic disease. The Journal of Allergy and Clinical Immunology: In Practice. 2013;**1**(2):137-141

[116] Barnes CS. Impact of climate change on pollen and respiratory disease. Current allergy and asthma reports. 2018;**18**(11):1-11

[117] D'amato G, Pawankar R, Vitale C, Lanza M, Molino A, Stanziola A, et al. Climate change and air pollution: Effects on respiratory allergy. Allergy, Asthma & Immunology Research. 2016;**8**(5):391-395

[118] Casadevall A. Climate change brings the specter of new infectious diseases. The Journal of Clinical Investigation. 2020;**130**(2):553-555

* 9 7 8 1 8 0 3 5 5 0 9 6 1 *